HUMAN IMPACT ON ENVIRONMENT AND SUSTAINABLE DEVELOPMENT IN AFRICA

Contemporary Perspectives on Developing Societies

JOHN MUKUM MBAKU, Series General Editor
Weber State University
MWANGI S. KIMENYI, Series Associate Editor,
The University of Connecticut & KIPPRA, Kenya

Between 1989 and 1991, there were several changes in the global political economy that have had significant impact on policy reform in developing societies. The most important of these were the collapse of socialism in Eastern Europe, the subsequent disintegration of the Soviet Union, the cessation of superpower rivalry, and the demise of apartheid in South Africa. These events have provided scholars a new and challenging research agenda: To help the peoples of the Third World participate more effectively in the new global economy. Given existing conditions in these societies, the first line of business for researchers would be to help these countries establish and maintain transparent, accountable and participatory governance structures and, at the same time, provide themselves with more viable economic infrastructures. The *Contemporary Perspectives on Developing Societies* series was founded to serve as an outlet for such policy relevant research. It is expected that books published in this series will provide rigorous analyses of issues relevant to the peoples of the Third World and their efforts to improve their participation in the global economy.

Also in this Series

Ike Udogu, E. (ed.) (2001) *The Issue of Political Ethnicity in Africa.*

Mainuddin. Rolin, G. (ed.) (2002) *Religion and Politics in the Developing World: Explosive Interactions.*

Saitoti, G. (2002) *The Challenges of Economic and Institutional Reforms in Africa.*

Human Impact on Environment and Sustainable Development in Africa

Edited by
Michael Darkoh
and
Apollo Rwomire
University of Botswana

ASHGATE

Published by
Ashgate Publishing Limited
Gower House
Croft Road
Aldershot
Hampshire GU11 3HR
England

Ashgate Publishing Company
Suite 420
101 Cherry Street
Burlington, VT 05401-4405
USA

Ashgate website: http://www.ashgate.com

British Library Cataloguing in Publication Data
Human impact on environment and sustainable development in
Africa. - (Contemporary perspectives on developing
societies)
1. Human ecology - Africa 2. Sustainable development - Africa
3. Africa - Environmental conditions
I. Darkoh, Michael B. K. II. Rwomire, Apollo, 1945-
333.7'096

Library of Congress Cataloging-in-Publication Data
Human impact on environment and sustainable development in Africa / edited by Michael B.K. Darkoh and Apollo Rwomire.
p.cm.
Includes bibliographical references and index.
ISBN 0-7546-3154-0 (alk. paper)
1. Environmental degradation--Africa. 2. Nature--Effect of human beings on--Africa. 3. Human ecology--Africa. 4. Sustainable development--Africa. I. Darkoh, M.B. Kwesi. II. Rwomire, Apollo, 1945-

GE160.A35 H86 2002
338.96--dc21

2002074719

ISBN 0 7546 3154 0

Printed and bound in Great Britain by MPG Books Ltd, Bodmin, Cornwall

Contents

List of Tables

List of Figures

List of Contributors

Avwunudiogba, A.
Lecturer, Department of Geography & Regional Planning, University of Benin, Nigeria.

Cardy, W. F. G.
Former Deputy Assistant Executive Director, Environment Management Programme & Director, Desertification Control Programme, United Nations Environment Programme, Nairobi, Kenya and currently, Water Resources Management Specialist, Water and Urban Africa Region, World Bank, Washington D. C.

Darkoh, M. B. K.
Professor of Human Geography and Head, Department of Environmental Science, University of Botswana, Gaborone, Botswana.

Kgabung, B. T.
Lecturer, Department of Environmental Science, University of Botswana, Gaborone, Botswana.

Koala, S.
Global Coordinator, Desert Margins Progamme (DMP), and Principal Scientist, International Crops Research Institute for the Semi-arid Tropics (ICRISAT), Niamey, Niger.

Letamo, G.
Senior Lecturer and Head, Department of Demography and Population Development, University of Botswana, Gaborone, Botswana.

Mbaiwa, J. E.
Research Fellow, Harry Oppenheimer Okavango Research Centre (Maun), University of Botswana, Gaborone, Botswana.

Mohamed Salih, M. A.
Professor of Politics of Alternative Development Strategies, Institute of Social Studies, The Hague, The Netherlands.

Mushala, H. M.
Professor of Geography, Department of Geography, Planning and Environmental Science, University of Swaziland, Kwaluseni Campus, Swaziland.

Nakizito, J.
Teaching Assistant, Department of Environmental Science, University of Botswana, Gaborone, Botswana.

Nuwagaba, A.
Senior Lecturer, Department of Social Work and Social Administration, Makerere University, Kampala, Uganda.

Obi, C.
Senior Research Fellow, Nigerian Institute of International Affairs, Lagos, Nigeria.

Ogbudinkpa, R. N.
Senior Lecturer, Department of Economics, University of Nigeria, Nsukka, Nigeria.

Okafor, S. O.
Professor of Public Administration and Dean, Faculty of Social and Management Sciences, University of Maidugiri, Nigeria.

Opschoor, J. B.
Professor of Development Studies and Rector, Institute of Social Studies, The Hague; Professor of Environmental Economics, Vrije Universiteit, Amsterdam.

Rwomire, A.
Associate Professor, Department of Social Work, Faculty of Social Sciences, University of Botswana, Gaborone, Botswana.

Smith, O. B.
Senior Programme Specialist, International Development Research Centre (IDRC), Ottawa, Canada.

Totolo, O.
Senior Lecturer, Department of Environmental Science, University of Botswana, Gaborone, Botswana.

Van Wyk, J. K.
Lecturer, International Politics, Department of Political Sciences, University of South Africa, Pretoria, Republic of South Africa.

Preface

The African environment is under stress, with land degradation problems becoming increasingly apparent. Since the twentieth century, human activities in the African environment have intensified, causing localised damage and degradation. In some parts of Africa like the Sahel, human impacts appear to be more sustained and far-reaching.

This book focuses on the economic, social and political processes that generate environmental change and processes in Africa. An attempt has been made to analyse some of the major social, economic and political determinants and consequences of environmental problems in the continent. Blending knowledge and perspectives from a variety of disciplines and based on the principle that theory, research and practice are inextricably intertwined, the book contains substantial contributions on several of the manifold environmental problems facing Africa. It is anticipated that the book will be of interest and use to students who need to acquire some understanding of environmental matters in the African context. We also hope that the book will be of interest and help to supporters of conservation and environmental organisations who wish to understand more of the background to the issues which concern them and academic and professional researchers as well as policy makers and practitioners concerned with environmental problems in Africa.

The Editors would like to express their sincere thanks to Dr M Khayesi formerly of Kenyatta University, Nairobi, Kenya and now with WHO, Geneva and Mr R. J. Sebego of the University of Botswana, Gaborone for valuable assistance with the preparation of the initial draft manuscript.

M B K Darkoh
A Rwomire

Chapter 1

Introduction: Ideas on Human Impact on Environment and Prospects for a Sustainable Future in Africa

M. B. K. Darkoh and A. Rwomire

Development of Ideas on Human-Environmental Relationships

The quality of the environment continues to decline in many parts of the world. The last great forests in Africa and the Amazon are rapidly disappearing. Rangelands in the tropics are rapidly being desertified. There seems to be few, if any, parts of the developing world that are immune to this general trend of decline or environmental despoliation. The problem is not restricted to the developing countries either. European lakes and forests continue to be destroyed by the phenomenon of acid rain and nitrate contamination of water is a major problem in the USA. Oil spills and chemical and toxic production and disposal are posing intractable environmental problems in developed economies. Important activities of humans such as industrialisation, mining, urbanisation and agriculture are increasingly posing a threat not only to the quality of the physical environment but also human health and socio-economic well-being. Human diseases related to pollution resulting from mining, industrialisation and transport are emerging in different parts of the world, including the developing world. This state of affairs leads to some critical reflections:

- Why is this universally apparent decline in the quality of the environment occurring?
- In what manner have humans in their long tenure of the earth related to their environment?
- What have been the detrimental changes and consequences of human activities on environment?
- What sustainable solutions can be offered to mitigate these impacts?

The issue of human relationship with and impact on the environment is an old one, spanning centuries of human civilisation and knowledge. Though this issue is old, each generation has contended and continues to contend with it in terms of diagnosis and trying to find answers to problems emerging from human-environment interaction. The cultural geographer Clarence J. Glacken in his monumental and path-breaking treatise on *Traces on the Rhodian Shore* (1967) has comprehensively examined the history of Western thought on the habitable earth (*ecumene*) and human relationships from the Greek era to modern times. Clarence J. Glacken has addressed three important questions related to: a) whether the earth, which is obviously a fit environment for humans and other organic life, is a purposefully made creation, b) whether the earth's climates, relief, and configuration of its continents have influenced the moral and social nature of individuals and moulded the character and nature of human culture, and c) in what way humans in their long tenure of the earth have changed it from its hypothetical pristine condition.

Earlier work by the social anthropologist George Perkins Marsh in the mid-nineteenth century had eloquently considered the third issue raised by Glacken by surveying the extent of human influence on the environment in *Man and Nature* published in 1864. This book dealt with the impact of human activities and their consequences on the woods, soils and water of Marsh's own native Vermont in the USA. The works of both Glacken and Marsh were more concerned with Western civilisation and did not touch upon human influence in the non-Western societies. Although the theme of human impact on the environment was central in subsequent studies by Western scholars, the emphasis continued to be on the Western world, as epitomised by the works of Western historical geographers such as Darby (1956) and McKnight (1959) who respectively studied the clearing of woodlands in Europe and introduction of alien plants and animals in Anglo-America. Other scholars explored the impacts that European settlement and development had had on the landscape of North America.

On the lower latitudes or the tropical regions, however, much of Western scholarship on human-land relationship before the twentieth century has been speculative and largely based on the doctrine of environmental determinism, which viewed human behaviour and activity as conditioned by the environment. European mindset tried to explain the poverty and backwardness of the tropical lands in terms of the constraining and debilitating effects of the environment. Protagonists of this doctrine of environmental determinism, typified by the works of Ellsworth Huntington (1914) and Ellen Semple (1911), propounded theories of 'natural explanation' of the broad persistent backwardness of this region of the world. From such pseudo-theories of human-environment relationship were spin-offs of other concepts such as the 'vicious circle' of poverty which were said to explain the origin and persistence of primitive forms of land use and low level of social and economic development. Perhaps, the work that best reflects this negative mindset and how environment influenced humans rather than vice versa is Pierre Gourou's *The Tropical World* (1953), which went into about eight editions and which only recently went out of print. Although environmental determinism was replaced by possibilism and other later orthodoxies that emphasised the significance of human choices of activity rather than terrestrial limitations to it and saw humans as geomorphological agents, the prejudice and myth surrounding environmentalism continued to have a powerful influence in shaping the way Westerners perceived other alien societies and their relationship to the land.

Before the 1960s, one of the most outstanding scholars whose works on the human impact on the environment transcended the orthodox Western stereotypes was Carl Ortwin Sauer (1938, 1952). Sauer was an American cultural and historical geographer whose studies examined such central themes as domestication, the effects of resource exploitation, conservation and the evolution of cultural landscapes. He was one of the few scholars eager for an alternative to naïve environmental determinism who recognised the ecological virtues of some so-called primitive peoples. Nevertheless, much of Sauer's work was limited to Latin America. But perhaps, the publication which overshadowed existing orthodoxies or stereotypes and made considerable impact on Western thought in the late 1950s (and probably heralded the ecological movement of the 1960s) was a major symposium volume, entitled *Man's Role in Changing the Face of the Earth* (Thomas, 1956). This book, drawing upon a much broader store of knowledge than its predecessors, identified the human forces that had changed the state of the biosphere. Several components of the biosphere as well as aspects of society responsible for changes in the environment were discussed directly or indirectly with emphasis on technology, urbanisation, industry and trade.

It was from about the 1970s that the human impact on the natural environment became a central concern for many disciplines and many scientists (Goudie, 1997). It became increasingly clear in this period, as Goudie (1997) puts it, that humans not only caused change at the local level (e.g. by accelerating rates of erosion in an area that had been stripped of forest) but also promoted change at the global level. Themes like biological magnification of pesticides became of common concern following Rachel Carson's *Silent Spring* (1963). More recent works by geographers such as Goudie (1981, 1997), Barrow (1991, 1995), Mannion (1997, 1995), Mannion and Bowlby (1992) and Ahmed and Doeleman (1995) have tended to focus on global environmental change and to advocate better approaches to development, especially sustainable resource management. Economists have also contributed to the analysis of the environmental issues. This is evident in the edited work by Swanson *The Economics of Environmental Degradation: Tragedy for the Commons?* (1996). This book utilizes an institutional economics approach to analyse the underlying causes of continuing environmental degradation: poverty, population, poor policies and trade. The overall position reached in this book is that environmental degradation will persist because it is the static result of the ongoing dynamic contest between opportunistic individuals and institutional adaptation.

Other recent works taking interdisciplinary approaches have focused specifically on the tropics. They have examined the complex relationships between humans and environment, trying to understand the causes of their environmental problems. For example, Randall Baker in his *Environmental Management in the Tropics* (1993) provides an elucidating historical expose of the exploitation of natural resources which has brought destruction and degradation to peoples and societies in the tropics. Similarly, Michael J. Eden and John T. Parry have insightfully addressed the issue of environmental change in their reader on *Land Degradation in the*

Tropics (1996). The environments, which are analysed in the latter two studies range from rain forests through tropical, dry forests and wetlands to cities, and examples are drawn from all the principal cultural regions, which make up the developing world, including Africa. Friday and Laskey (1989) have also examined the impact of such factors as deforestation, acid rain and pollution, desertification, over-exploitation of natural resources, and the greenhouse effect on the environment in their multidisciplinary reader on *The Fragile Environment.*

In the case of Africa, for the first time, a number of significant publications on environment, range ecology, politics, development and policy have begun to emerge. These include: *Ecology and Politics* (Hjort-af-Ornäs and Salih, 1989) which demonstrates the complexity of the interrelationships between ecology and politics, *Security in African Drylands* (Hjort-af-Ornäs, 1992) which deals with diverse issues related to environment and secure livelihoods including research, development and policy issues, *Social Aspects of Sustainable Dryland Management* (Stiles, 1995) which addresses the socio-economic aspects of land degradation and dryland management, *African River Basins and Dryland Crises* (Darkoh, 1992) which is a collection of case studies highlighting the social, economic, political and environmental dimensions of African river basins and drylands, *Approaching Nature from Local Communities* (Hjort-af-Ornäs, 1996) which, like the preceding one, presents case studies of security perceived and achieved by local communities and land users and their relationships to nature, and *Environmental Politics and Liberation in Contemporary Africa* (Salih, 1999) which explores the intricate relationship between liberation movements and struggles on the one hand, and environmental degradation on the other. Another publication with contributions from scholars who are all of African origin is *Environmental Planning, Policies and Politics in Eastern and Southern Africa* (edited by Salih and Tedla, 1999). This book provides nine case studies illustrating the politics, policies and planning in environment and natural resources in Eastern and Southern Africa.

The period since the 1970s has also seen studies challenging mainstream views of range science. The best example of this is *Range Ecology at Disequilibrium* (Behnke *et al.*, 1993). This book reviews recent biological research on African rangeland and highlights its management implications for donor and national government policy. The contributors to this book argue that the mainstream view of range science is fundamentally flawed in its application to certain rangeland ecologies and forms of pastoral production. The relationship between livestock numbers and range degradation is debated and more appropriate techniques for the assessment of rangeland carrying capacity and degradation have been proposed. Kgabung in Chapter 14 has outlined these new concepts in range ecology. Most recently, however, the new and so-called 'disequilibrium theories of range ecology' (DTRE) have been challenged by Lykke (2000) as Darkoh has pointed out in Chapter 12 in this book.

A central theme that has received much concern in the literature since the 1970s is desertification and land degradation, as attested by some of the papers in this volume and the innumerable publications by multilateral agencies and international organisations such as UNEP, ECA, FAO, IFAD, IUCN, UNDP, UNESCO, UNSO, SADC/ELMS and the World Bank on the subject. Desertification and land degradation have become recognized as key issues for world conservation in the twenty-first century, thanks largely to the efforts of UNEP (Darkoh, 1999, 2000) and the United Nations Conference on Environment (UNCED). Other themes that have become increasingly important in environmental research and programmes are green development (Adams, 1990), indigenous knowledge (Brokensha, *et al.*, 1986; Richards, 1985; Warren, 1991; Darkoh, 1996), gender (Chingono, 2001; Fairhurst, *et al.*, 1999), governance (Maathai, 1995; Weaver, Rock and Kusterer, 1997) and refugees (El-Hinnawi, 1985).

From much of the interdisciplinary studies emerging since the 1970s on human-environment relationships in the developing world, it would appear that the root cause of the environmental problems today lies in the way resources are used or managed under an expanding global economy and especially the issue of power, control and access to resources and institutional failures. As the global economy is expanding, local ecosystems are collapsing at an accelerated pace (Brown, 2000: 4). The challenge of human survival and development is leading to the emergence of dominant groups and an ever increasing struggle over control of and access to resources. The dilemma of tackling environmental problems is that the very process of development that is dependent on the environment creates situations that are destroying the basis of human survival. Thus, the unequal access to resources and unsustainable utilisation of natural resources is not only a threat to the ecological balance but also the very existence of human livelihood. It goes without saying that humankind is greatly dependent on resources in the environment.

In the case of Africa, because of globalisation, pressures are mounting on people and environment with each passing year. Not only are local ecosystems collapsing but the continent as a whole has been experiencing a breakdown in the relationship between people and their natural support systems (Brown and Wolf, 1985). As a result, living standards are deteriorating and many Africans are no longer able to feed themselves (Darkoh, 1989). The typical list of the ever growing environmental problems includes deforestation, soil erosion, drought, desertification, loss of biodiversity, water and air pollution, improper disposal of commercial and domestic wastes, poor sanitation and spread of diseases. Famines and the threat of famines are among the manifestations of these ever growing problems. Many of these problems may appear physical in nature but they are underlain by a number of significant socio-economic processes such as growing human population, stagnant and declining economic growth, poverty, migration, foreign debt and political instability.

Environmental problems facing Africa have to be placed in the historical and developmental experience of the continent. Walter Rodney (1972), in his book *How Europe Underdeveloped Africa*, has provided insights into how Africa has been separated from the indigenous momentum of its own history and thrown into Europe's backyard and how its incorporation into the global economy has worked to hold the continent back in technological and socio-economic development. The grafting or imposition of Western technology and socio-economic systems on African indigenous systems was not well thought out. Africa now has to cope with the unreconciled clash of European and indigenous systems, as well as the legacy of its grossly unequal incorporation into an economic system created by and for the Western world (Baker, 1993: 204). Whereas the Western world itself is slowly coming to grips with reconciling technology, development and environment in its realm, this is far from the case in Africa. Unfortunately for the African continent, the driving force for rapid development is powered by policies of exploitation of its forests, soils, minerals and other natural resources for quick gain, and, in all too many instances, the culprits are the capitalists of the Western world (Eden and Parry, 1996). African countries are now having to deal with the legacies of exploitation and environmental mismanagement, and at the same time cope with the processes of global change, such as climatic warming, sea-level rise and desertification (*ibid.*). The challenge before scholars and policy makers in Africa is to analyse the problems that emanate from such a development process and search for meaningful solutions. This book is an effort towards this direction.

There are books on environment in Africa, as indicated in the preceding review. The question that immediately arises is: What does the current book add to the existing literature? While the themes presented are related to those discussed in other works, this book is enriched by being based on a blend of knowledge and perspectives from a variety of disciplines. The contributors come from different academic and professional areas, and they have brought these backgrounds to bear on the discussion and analysis of the political, social, economic and technological determinants and consequences of human impact on environment in Africa. Thus, unlike some other books which have utilised a single disciplinary approach (e.g. economics), the perspective of the current book is interdisciplinary and multidisciplinary, drawing knowledge and ideas from not only academics (sociologists, economists, environmental scientists, geographers, political economists and international relation specialists) but also policy makers and implementers. Looking at issues on human impact on environment in Africa from this perspective contributes significantly towards unravelling different layers of meaning and experience. Furthermore, this book has benefited from current knowledge and research findings, which are based on revisits and reinterpretation of earlier views. Currently, there are controversies on and challenges to such concepts and issues as environment-human relationship, ecological resilience, desertification, sustainable development, globalisation and North-South dialogue. This book has drawn on past and present research findings to discuss these issues. The volume has a perspective, which not only seeks to answer some of the old questions but also opens up new ones for further research and publication. The findings presented therefore provide a rich source of information for researchers, policy makers and implementers in environment and development matters in Africa.

Summary of Contributions

The chapters in this book cover a wide spectrum of cross-cutting environmental themes. Below is a summary of the chapters and the issues discussed.

One of the major problems facing Africa today is the degradation of its biophysical environment. Some see this as a social problem (e.g. Blaikie and Brookfield, 1987; Stiles, 1995) and in Chapter 2, Ogbudinkpa discusses the broad socio-economic dimensions of environmental degradation in the continent. This author presents examples of how population growth, urbanisation, industrialisation, use of chemical fertilizers and affluence lead to environmental degradation. He also outlines the socio-economic effects of environmental degradation.

In Chapter 3, Obi illustrates the socio-economic and political dimension of environmental crisis of Nigeria. The issues highlighted include population explosion, rapid urbanisation, rural decay, rising crime rates, environmental stress and conflict, improper waste management, industrial pollution, inefficient land use and economic problems. Obi points out that at the heart of the environmental crisis in Nigeria is the issue of distributive inequity which denies the majority access to a fair share of resources and the authoritarian mould of state power and power relations in society. Other issues pointed out include the persistence of unsustainable forms of production and consumption. Obi observes that although there have been efforts by the government of Nigeria to put in place institutional structures to contain environmental crises, there has been limited success because of a complex set of political and social forces.

Environmental security is one of the serious issues in the drylands of Africa and it is the next theme highlighted by two chapters in the text. According to Darkoh, the clearest manifestation of the unfolding environmental insecurity in the drylands of Africa is the famines and food shortages that severely hit people in the 1970s and 1980s resulting in mass starvation, death, disease and the exodus of millions of environmental refugees. Darkoh perceives environmental security from two perspectives: first in terms of those environmental conditions that enable individuals and groups to satisfy their basic needs and second, as a precondition for peace. In this author's view, the depletion of natural resources or their appropriation by a few wealthy individuals or groups jeopardises peace and undermines security. The major environmental security issues examined relate to climate, water, population growth, human activities, land degradation and urbanisation. The author sees science and technology as ways forward in resolving environmental problems.

In his view, sound technological transfer and acquisition have an increasing role to play in African development and he points out the need to enhance in particular the human, institutional and legal capacity in technology transfer. The chapter also identifies some specific areas posing the greatest challenge for scientific and technical co-operation in arid lands development in Africa. It is Darkoh's view that solutions to our environmental problems have much to do with our systems of values and issues of social justice, and he says that, in finding solutions to environmental problems, it is important to understand what our values are and which potential solutions are socially just.

Van Wyk takes the environmental security issue further by examining the South African situation in Chapter 5. This author discusses the impact of scarcity of environmental resources on security within the Southern African region. Van Wyk observes that environmental degradation and competition for natural resources are leading to massive migration of people and putting new pressures on state boundaries of the Republic of South Africa, as elsewhere in Africa. The author singles out and discusses the following environmental security issues: climate change, water, boundaries and soil erosion. Van Wyk looks at the environmental security of not only the Republic of South Africa but also the entire Southern African region. The author notes that there has been a traditional reluctance to include environment on the security agenda in Southern Africa. In other words, environment has been given a low profile or relegated to 'low politics'. The author examines a number of options to enhance environmental security which each individual state or more than one state can pursue in the region.

Letamo and Totolo provide an overview of the interrelationships between population and environment in Botswana. They observe in Chapter 6 that the country has experienced rapid economic growth and this has led to enormous pressure on the use of natural resources. These two authors note that with an annual population growth of 3.5%, the fragile semi-arid environment of Botswana has been placed under increasing stress from rangeland degradation, loss of woodlands, veld products depletion and reduction of wildlife resources. As Botswana's population continues to increase, more demands are placed on the use of natural resources and more non-biodegradable waste is generated which makes it difficult to improve the quality of the environment. Drilling of more boreholes for water in the rural areas has also led to rangeland degradation. In an effort to address the major environmental constraints, the Government of Botswana has introduced a number of policies and strategies such as the National Conservation Strategy, the Accelerated Development Programme, the Financial Assistance Policy, the Water Resources Policy and the Industrial/Urban Pollution Policy. The authors are optimistic that these policies will go a long way in solving environmental problems in Botswana if well implemented.

One of the causes and effects of environmental degradation is migration. Two chapters have taken up this theme. In Chapter 7, Cardy provides a review of environmental causes and effects of forced migration. The author notes that the conventional definition of the UN of refugees is very restrictive and does not include environmental migrants and especially the internally displaced. Cardy describes three broad categories of environmental refugees proposed by El-Hinnawi (1985). The first group is made of those who have been temporarily displaced because of an environmental stress. Once the environmental disruption is over and the environment rehabilitated, they return to their original habitat. The second group consists of those who have to be permanently displaced and re-settled in a new area. The third category comprises individuals or groups of people who migrate from their original habitats temporarily or permanently in search of better quality life. It is evident that these groups of refugees initiate their migration due to constraints in the environment. The environmental impacts of migration discussed by Cardy are overexploitation of natural vegetation, pollution of water and soils, depletion of water resources, changes in land use, loss of 'reserve lands' and spread of diseases and decreased resistance to diseases. The overall conclusion reached by Cardy is that the environmental causes and effects of migration need to be considered and addressed within the broader context of development and management of sustainable environments. Consideration needs to be given to access to resources and resource use as these are important keys to addressing the environmental causes and consequences of migration. Ensuring sustainable management is essential in the long run. To achieve this, planning and socially rational management are required.

The other chapter on socio-economic and environmental dimensions of migration is Chapter 8 by Rwomire who looks at the theory and practice of international migration with special reference to Southern Africa, especially during the apartheid era. Rwomire points out that international migration is one of the most enduring and controversial issues which countries in Southern Africa have to grapple with. The chapter reviews this migration process from economic, political and social perspectives, investigating the linkages between immigration and development and underdevelopment in the region. It describes the characteristics, determinants, dynamics and consequences of migration. The author observes that to a great extent, the political economy of the region revolves around the pervasive system of migrant labour operated by the Republic of South Africa (RSA) as the centre of attraction and the Southern Africa Development Community (SADC) countries as the peripheral providers of relatively cheap labour. He concludes that due to uneven development and South Africa's dominant position in the region, migration could adversely affect the development process in the supplier countries. The author is of the view that despite recent political changes in the region, SADC countries are unlikely, at least in the short run, to extricate themselves from South African based economic dependence and migratory labour entanglements.

As already highlighted, desertification and land degradation have received significant attention in research on man-environment relationship in recent years. In Chapter 9, Smith and Koala discuss the myths and realities of desertification. These two authors observe that an intense debate over the phenomenon of desertification has been raging for well over two decades. The debate has generated some new conceptual and empirical insights, raised interest in, and stimulated commitment to controlling the phenomenon, but has also generated some myths around it, to the extent that some have begun to question the existence of desertification. The issues around which certain myths have been built and which tend to detract attention from the pressing realities of combating and alleviating its effects include:

- the historical perspective (is it a recent or an old and time-tested phenomenon);
- the definition and its extent (what is desertification, and how common or serious is it);
- the causes and process (what are the responsible factors and how does it develop);
- the consequences (what are its visible and not so visible effects); and
- the remedies and control (how can we combat it).

The chapter by Smith and Koala presents an overview of the above issues, with the goal of identifying and contrasting the myths and realities, because some of the former have the potential to confuse issues and negatively affect policy decisions, actions and approaches required to effectively tackle the problem.

Chapter 10 by Darkoh examines the nature, causes and consequences of desertification in the drylands of Africa. The chapter provides a detailed conceptual analysis of the term desertification, and its causes and linkages. It concentrates on the human causes and consequences of desertification, unraveling the changing views on these causes. The status of desertification and efforts to combat are also highlighted. Darkoh observes that a significant drawback in combating desertification and drought is the failure of African governments to devolve power to the people who are affected, and to link environmental degradation to economic policy. Consequently, many programmes lack local support or are undermined by conflicting trade and agricultural policies pursued by governments. He contends that for sustainable development strategies to work, policies should put the welfare of the people in these drylands at the centre of the development agenda and give them the rights and power to determine their future. The chapter points out that presently many interventions from outside are weakening instead of strengthening local adaptive strategies, livelihood and production systems. For example, the adoption of Structural Adjustment Programmes (SAPs) and the subsequent policies of trade liberalisation to achieve equilibrium and economic growth in many of the African countries have worsened the standards of living and undermined efforts at achieving food security. The ability of the poor to sustain adaptive strategies towards sustainable livelihoods is constrained by their vulnerability to loss of access and assurances over resources, as well as the lack of relevant and timely information to transform access and assurances into capacities to evolve strategies to deal with change. The need to facilitate some measures of empowerment to give the people of the drylands opportunities to shoulder the risk of change towards sustainable livelihoods is critical. It is also argued that tackling desertification requires action at both the national and international levels and that greater access to developed country markets and reduced subsidisation of developed country exports could do more for development and environment in Africa than a tripling of development assistance.

In Chapter 11, Smith and Koala provide a review of available wind erosion control processes and strategies in the Sahel and highlight a number of successful interventions to combat desertification. According to these authors, because of its peculiar climate and agricultural systems, land degradation by wind erosion is perhaps the most important type of desertification in the Sahel. The processes and effects of wind erosion, as well as its direct and indirect consequences on people such as the loss of agricultural land, food insecurity, forced migration and the accompanying poverty and human misery are presented. The authors believe that access to land, credit, other inputs (e.g. appropriate tree species), information and opportunities to generate alternative sources of income could provide incentives for communities to invest in wind erosion control measures. Collective action from all stakeholders including government, communities and farmers is, however, necessary. Lessons learned from successful interventions exemplify the value of these incentives and may well serve development planners and policy-makers in developing sustainable schemes to combat the menace of desertification.

In another chapter (Chapter 12), Darkoh examines the closely related theme of biodiversity in the drylands of Africa. He provides a regional overview of the impact of agriculture on biodiversity with examples drawn from Northern Africa, the West African Sahel, East Africa and the Horn, and Southern Africa. It is evident from the discussion in this chapter that agriculture accelerates loss of biodiversity in all these regions because of attempts by farmers to increase crop and animal production to feed the increasing population and to contribute to the growth of the national economies. As in the case of desertification, it is noted that in all these regions, harmful agricultural practices such as overcultivation, overgrazing, selective grazing, bush fires, cultivation of marginal and easily eroded land, widespread use of chemicals and pesticides, destumping to introduce mechanisation on fields that are often too large, lopping and leaf-thinning of trees for fodder and human use have intensified the degradation of the soil and vegetation and led to the alarming decline of species types and their numbers. The drylands of Africa have the potential to be more productive but they are fragile and must be wisely managed. In order to avoid a situation where agricultural activity leads to loss of biological resources, a broad landscape-ecosystem approach is required to ensure biodiversity sustainability. Successful biodiversity conservation strategies must respect and incorporate African values, knowledge systems and priorities. It is contended that the single most important activity that will enhance biodiversity conservation in the medium to long-term is to raise well-informed future generations with a strong commitment to sustain management of natural resources and biodiversity.

Land evaluation is a theme taken up by Mushala. Land evaluation seeks to assess the suitability of land for sustainable production taking into account inherent characteristics of land and the socio-economic milieu of production. In Chapter 13, Mushala has attempted to characterise modern and indigenous systems of land evaluation as a basis for proposing an integrated approach. He has shown that modern systems of land evaluation have their inherent weaknesses that make them ineffective in solving problems of the small-scale farmers. Some problems also exist within indigenous systems of land evaluation, but these systems are more realistic to the farmer because they are based on gained experiences, which, unfortunately are not found in most literature. However, there are attributes from both evaluation systems, which complement each other. An integrated approach is proposed taking into account the simplicity of the indigenous systems and effective methods of collecting data. It is emphasised that the experiences of the local farmers should be utilised to make scientific knowledge more relevant to affected farming communities.

Kgabung discusses natural resource use in Africa's rangelands with particular reference to the livestock industry in Botswana in Chapter 14. The chapter first analyses at both the regional and national levels, past and present management strategies or policies in order to determine how the relationship between livestock rearing and hunting and gathering has evolved amidst resource use changes. Second, Kgabung outlines ecological change theories of semi-arid environments in order to provide an understanding of their functions and dynamics, which is an inevitable prerequisite for ecologically sound management and optimal use of resources in semi-arid rangelands. Finally, the impact of the livestock industry on the subsistence use of wildlife and veld products in the semi-arid rangelands of Botswana is examined as an example to illustrate the effect that more intensive forms of agriculture are having on ecologically marginal environments in Africa.

Mbaiwa, Darkoh and Nakizito examine globalisation and wildlife resource sustainability in Botswana. The basic argument presented in their chapter (16) is that although globalisation, viewed broadly as integration into the world market economy, is important in the economic development of a developing country such as Botswana, it has not always had a positive impact on sustainable resource use. As Botswana's wildlife resources are currently in a state of decline, there is a considerable concern about the prospects of the globalisation process. This chapter, therefore, traces the historical development of the wildlife industry in Botswana. It highlights the impact of commercialisation and globalisation of the wildlife industry. The authors examine *inter alia* the issues of globalisation and Botswana's elephant problem and land use conflicts and wildlife tourism. They propose alternative policies that need to be considered if the sustainability of wildlife resource utilisation is to be achieved in the country. Overall, the chapter reveals that despite the fact that the globalisation of the wildlife sector in Botswana has wrought serious environmental problems such as the decline of the country's wildlife resource base, it has had a considerable positive effect on the economy of Botswana. This is illustrated by the fact that the country's tourist industry which is mostly wildlife-based contributes significantly to the Gross Domestic Product (about 16 per cent). The wildlife industry also remains important to the socio-economic livelihoods of the rural communities living in wildlife areas. To stem the constant decline in Botswana's wildlife resources, the authors recommend that wildlife utilisation should be governed by the principles and ideals of wildlife conservation and environmental sustainability.

As already indicated in Chapter 15 by Mbaiwa, Darkoh and Nakizito, there is a constant conflict between wildlife-based tourism and the adjacent communities over the use of land and other resources in Botswana. One of the ways of dealing with this problem is to share benefits accruing from tourism with the adjacent communities. Incorporation of adjacent communities in the planning and management of tourism becomes key in securing their involvement towards sustainable utilization and conservation of wildlife resources. Mbaiwa takes up this theme and examines the development of community-based tourism in Ngamiland district, Botswana in Chapter 16. He traces the evolution of community-based resource management in Southern Africa and Botswana in general and Ngamiland district specifically. The chapter assesses environmental and socio-economic impacts of community-based tourism and reveals that involvement of local communities in natural resource utilization and management has been possible in

Ngamiland through the establishment of community-based organisations (CBOs), of which there are nine such trusts in the district. Mbaiwa notes that these CBOs generate revenue and provide employment opportunities to the local people through consumptive and non-consumptive tourism activities. These socio-economic benefits have partly contributed to the development of positive attitudes and perceptions of the local communities towards natural resource conservation and tourism in Ngamiland district. These CBOs are, however, faced with problems related to lack of empowerment, marketing and entrepreneurial skills in the tourism business. In addition, community-based tourism is facing competition from the predominantly foreign owned elite tourism industry that has already developed in Ngamiland district.

Urbanisation has led to important environmental issues with regard to the quality of both the human and physical environments. Some of the environmental issues of urbanisation were touched upon in Darkoh's earlier chapter (4) on environmental security in the drylands of Africa. Two authors examine environmental concerns in urban areas in Africa further. Nuwagaba examines in Chapter 17 the relationship between urban poverty and environmental health in the city of Kampala, Uganda. Urban poverty is conceptualized in this chapter as a culmination of the skewed spatial distribution of socio-economic opportunities, resulting in rural - urban drift *en masse* of the rural poor, seeking better means of eking out a living in the urban economy. According to Nuwagaba, the problems of environmental sanitation seem to be the major urban environmental problems in Uganda. These have resulted from rapid population growth, overcrowding and high density of human settlements coupled with congestion amidst poor social infrastructure services. The peri-urban areas are particularly in a poor state, saddled with overcrowded dwelling units lacking private toilet facilities and resulting in the disposing of human excreta in open drains in Kampala city. Poor governance regimes that have resulted in macro-economic disequilibria with dwindling capacity of the municipal authorities which are unable to provide social infrastructure services have exacerbated the problems.

Okafor examines the impact of government policy on the management of environmental waste in Maiduguri urban area of Nigeria in Chapter 18. The chapter reviews some of the major federal government policy directives on management of environmental waste in Nigeria. The aspects of these directives, which have been implemented in Borno State, in which Maiduguri is located, have led to the establishment of certain institutions required for the management of waste. An important finding of the chapter is that though officially the management of environmental waste is entrusted to the Sanitation Board, the Board does not carry out all the functions stated on its own as there are some non-governmental groups which make their voluntary contributions to waste disposal. There are definitely some functional gaps, which should be filled by the Board. The chapter recommends the need for decentralisation of waste management in Maiduguri and the active participation of communities at the grassroots, not only in the control but also in the management of solid waste.

In Chapter 19, Avwunudiogba examines adaptive strategies of small fishermen in view of environmental degradation induced by oil pollution in the Niger delta. The increasing rate of crude oil exploration, exploitation and the associated pollution of some areas of the Niger Delta (Nigeria) have led to the disruption of the delicate aquatic ecosystem and the traditional fishery resource base of the local communities. As a consequence, the local people have adopted several survival strategies, the main ones being multiple occupational practices (e.g. a combination of a secondary occupation to fishing, farming, etc.), out-migration, and induced income from compensations due to oil spillage.

In Chapter 20, Salih examines the global and national context of conservation strategies and environmental action plans, highlighting the conditions necessary for their success and constraints faced. The chapter traces actors and major developments in conservation strategies and environmental action plans at the global level. The major global actors described in the chapter are the United Nations Environment Programme, International Union for the Conservation of Nature, United Nations Educational, Scientific and Cultural Organisation, Food and Agriculture Organisation and World Wildlife Fund. The major developments outlined are: the publication of the World Conservation Strategy in 1981 by the IUCN (which stressed the importance of sustainable development of species and their ecosystems), publication of the World Commission on Environment and Conservation (WCED) in 1987 and Environmental Perspective to the Year 2000 (which popularised the concept of sustainable development), and hosting of the Earth Summit Meeting of 1992 (which adopted three major documents: Rio Declaration on Environment and Development, Agenda 21 and a non-legally binding authoritative statement of principles for a global consensus on the management, conservation and sustainable development for all types of forests). The chapter notes that in most countries, environmental institutions have concerned themselves with what Salih calls 'dart-board theory'. This means that these institutions are basically involved in overlapping circles on interrelated national, provincial, district and local (or village) institutions but strictly speaking are not effective, highly responsive and committed to implementing environmental policies. Salih calls for a commitment on the part of institutions and researchers to take up environmental concerns. He argues for a committed natural and social science fraternity to transform itself into a liberation science that empowers people to conserve their environment.

An important aspect, which contributes to the deterioration of the environment and the socio-economic condition of Africa, is the unequal partnership between the North and South. Though there are programmes and concepts that can be used to narrow the gap between the North and South, a critical assessment reveals that this is not always the case. Opschoor in the last chapter of this book examines one such concept, which is at the centre of many environmental and socio-economic activities. This is the concept of sustainable development and the North-South dialogue. This chapter traces the origin and changes in the conceptualisation of sustainable development. It is revealed that the definition of sustainable development that was put forth by the Brundtland Commission in 1987 was a compromised one and represents 'weak sustainability'. As a basis for North-South dialogue, the chapter notes that:

- There are marked differences in positions vis-à-vis what constitutes a relevant issue. Ecological constraints or the notion of a finite environmental space are thus regarded with suspicion.
- There are marked differences with respect to the relative urgency even within a set of accepted environmental problems.
- There are strong ethical cases made with respect to the right to development and hence the right to use more environmental space: the North has always done it.
- Similar arguments can be put forward to underpin positions about cost sharing ratios etc. After all: the North is largely to blame AND the North has the means (the South NOT).
- There is divergence of opinion on environmental matters and their

relevance in the North and between countries or regions in the North, as much as there is in the South.

Opschoor concludes that given the fact that many in the North are concerned about these global environmental problems and that co-operation with developing countries is essential, countries in the South are actually sitting on a diamond mine. They can cash in on this Northern concern, whether that concern is justified or not. They thus have a strong set of cards to play the game with. Some would conclude that they are already using these cards very smartly.

Towards a Sustainable Future

Looking ahead with respect to the prospects for a sustainable future, one is tempted to agree with Opschoor that African countries are sitting on a diamond mine in view of the fact that many in the North are concerned about global environmental problems and can cash in on this Northern concern for better or worse, justifiable or not justifiable. Unfortunately, the opportunities appear foreclosed for many players, and no one has a free ticket!

The hard and harsh reality is that we still live today in a divided and unequal world. If indeed, there is to be a real change in international economic relations with better prospects for a sustainable future for African countries, our attention must be targeted upon removing the hurdles that perpetuate this world order. As Baker (1992: 204) has pointed out, the keys to real sustainability are access to resources and markets, fair recompense for labour and a stake in a lasting future. The capital needed to close the gap between the North and South and address environmental problems is flowing in net terms in the wrong way - namely out of the South, particularly Africa. This capital outflow, together with the albatross debt as Baker has pointed out is not going to allow developing African countries to do anything other than take a short-term, survivalist view, for which nature will go on paying the price. Baker puts the argument more succinctly for tropical countries in general as follows:

> Having been frozen out of much of the industrial revolution by the colonial and neocolonial systems, the developing countries have exported most of the added value of their raw materials to the industrialized nations.

> Thus, for example, only around 10c/1b of coffee, or around 1.5 to 2.5 per cent of the final selling price is retained in the exporting country. The same picture holds true for timber and most other primary products. The real wealth accumulates downstream. Until the tropical countries can truly participate in the value-adding process, which means being free to export processed and manufactured goods without giving away the store through duty-free export-processing zones, they will remain the poor relations. From the perspective of the tropics, it is alarming to see the reconfiguration of the world into (protectionist?) trading blocs of 'rich peoples' clubs' such as the European Community or the North American Free Trade Area (Baker 1992: 205).

A number of scholars (e.g. Williams, 1989; Baker, 1993; Darkoh, 1993 a and b, 1996) have observed that if there is to be real change in the management of Africa's environment and the general development of the continent, then several things need to happen a long way from Africa and in Africa itself; that is, there is need for change or reforms at two levels: external and internal.

At the external plane, there is need for collective action to resolve the debt problems, strengthening international financial stability, redirect resources away from wasteful armaments and generally establish conditions for more equitable terms of trade (Williams, 1989: 9; Darkoh, 1993a: 19). There is need to facilitate real capital formation and the creation of meaningful opportunities away from the land. This requires serious industrialisation which should follow a different pattern from that which has blighted the environment and imposed heavy social costs in many developed countries. It also requires a fair deal with the developed countries. As Baker points out, this means global expansion of the free trade concept, a considerable amount of debt elimination, and capital availability on preferential terms- in other words, reversing the present flow, opening the trade doors, and increasing concessionary lending. Without these measures, there will be no easy way out of the short-term debt, balance-of-payments crisis, and nature will go on paying the bill.

At the internal or domestic front, there is an urgent need for African countries to put their houses in order. The fairly well documented African crisis has an internal dimension and there is need for a new domestic order individually and collectively to overcome the dependency-syndrome, the endemic sloth and government bureaucracies inherent in these countries and build up efficient, democratic and self-reliant economies (Darkoh, 1993a: 13;

1996: 61-63). Further changes needed are improvement of access to resources (especially reforming land tenure and addressing the perennial problem of land redistribution), availing credit to farmers at fair interest rates, genuine involvement of people in development planning, redefining national sovereignty, promotion of regional co-operation and improvement of agriculture (*ibid.*). As pointed out by Obi in Chapter 3, there is need to meaningfully address the political and social factors that lead to environmental crisis in Africa, especially democratisation of the state and breaking the hold of the power elite over it. This reform is necessary as a basis to open up the political space for popular empowerment, debates and participation in decision-making in matters which affect the lives and livelihoods of people.

It ought to be pointed out that if laissez-faire philosophy and economic growth are pursued at the expense of equity and human rights, the environment suffers, human life is endangered, and becomes neither sustainable nor desirable (Weaver, Rock and Kusterer, 1997: 2). Meaningful environmental protection requires broad-based sustainable development (BBSD) strategies, which include the following components (*ibid.*):

- economic growth that enhances standards of living;
- equitable distribution of national wealth, and ensuring that the needs of women, minorities, immigrants, the poor, and people with disabilities are met;
- promotion of human rights and democratic governance; and
- promotion of economic growth that is environmentally sensitive and sustainable.

It should further be noted that environmental damage resulting from resource exploitation imposes costs, often on large segments of the community. Efficient resource use requires that the prices paid by producers for inputs, and by consumers for final goods and services, accurately reflect the full costs (IISD, 1994: 16). An often neglected responsibility of governments in Africa is to see that the full costs of environmental damage are imposed upon the perpetrators (Darkoh, 1997). Polluting firms, and consumers of polluting products, should bear the cost of pollution prevention and cleanup in accordance with the 'polluter pays' principle. This will provide incentives for firms to alter their production methods and for consumers to switch to alternative products, thereby safeguarding the environment and increasing efficiency. It will also avoid trade and investment distortions, which occur when goods and services are sold at less than their full costs. Internalising environmental costs is an important step towards the establishment of real comparative advantages on an international scale. Governments should ensure that environmental resources within their national boundaries are not misused because of failure to internalise costs. As much as possible the price of goods should be made to include the costs of producing and disposing of them without damage to the environment (*ibid.*). There is therefore the need for project appraisal and cost-benefit analysis to internalize the environmental components of conservation sustainability and the cost of resource loss into conventional appraisal methodologies. Formulating sustainability economics would be the true expression of changed reality (Baker, 1993).

As Wayne has observed, some of the environmental problems that humankind has to grapple with are traceable to corporate sins. He points out that environmental liabilities, which are the legacy of past actions, are very costly and 'What society has gained from the industrial revolution in comforts and efficiency, it has generally sacrificed in the health of the ecosystem on which it depends' (2000: 80). Environmental liabilities refer to the obligations and commitments to incur future expenditure to prevent, reduce, or repair environmental damage. The costs incurred constitute environmental liabilities such as legislative liability, rehabilitation and clean-up costs, community and public liability, and macro-economic instruments (*ibid.*). Visser contends that professionals in environmental and financial fields need to understand the characteristics and magnitude of environmental liabilities and to establish an integrated environmental accounting system, which can ensure effective environmental management decisions.

Research has an important role to play in the reform being called for at the international and local levels. There are a number of issues related to the human agency in environment that social science research should unravel and bring to the attention of policy makers, intervention agents and communities. Three areas are particularly in need of social science research: a) how human activities may alter the environment; b) how societies and individuals are affected by environmental change, and c) how human societies may adapt to or intervene to modify its rate of direction. These three areas are clearly the domains of social science inquiry because they are closely related to human behaviour, political activity, economic structure, social organisation and institutions. As argued by Salih (Chapter 20), natural and social sciences have to transform themselves into a liberation science, not just an instrument for the justification of the global nature of the environmental crisis but rather as an instrument for empowering people to conserve their environment.

In concluding, we observe that this book has an important message, which is a call to action at the research, policy and programme fronts at all spatial scales to reverse the deterioration of the African environment, which has been brought about by human activity from without and within the continent. As the world matches into the twenty-first century, riding on galvanizing concepts of sustainability, globalisation and empowerment, it is important to remember that these global trends may bypass and could remain a dream to many in Africa, who are hindered from meaningful participation because of the divided and unequal world we live in today. Hence, the need to fully address the besieged African environment and socio-economic systems to give hope to many.

References

Adams, W. M. 1990. *Green Development*, Routledge, London.

Ahmed, I and Doeleman, J. A. (Eds), 1995. *Beyond Rio: The Environmental Crisis and Sustainable Livelihoods in the Third World*, Macmillan, London.

Baker, R. 1993. *Environmental Management in the Tropics: An Historical Perspective*, CRC Press, Boca Raton, Florida.

Barrow, C. J. 1991. *Land Degradation: Development and Breakdown of Terrestrial Environments*, Cambridge University Press, Cambridge.

Barrow, C. J. 1995. *Developing the Environment: Problems & Management*, Longman, Harlow Essex.

Behnka, R.J., Scoones, I. and Kerven, C. (Eds), 1993. *Range Ecology at Disequilibrium*, Overseas Development Institute, London.

Blaikie, P. and Brookfield, H. 1987. *Land Degradation and Society*, Methuen, London.

Brokensha, *et al.* 1980. *Indigenous Knowledge Systems and Development*, University Press of America, Lanham, Maryland.

Brown, L. R. 2000. 'Challenges of the new century', in *State of the World 2000,* The Worldwatch Institute, New York, pp. 3-21.

Brown, L. R. and Wolf. E. C. 1985. *Reversing Africa's Decline,* Worldwatch Paper 65, Worldwatch Institute New York.

Carson, R. 1963. *Silent Spring,* Hamish Hamilton, London.

Chingono, M. 2001. 'Women, knowledge, and power in environmental and social change', Rwomire, A. (Ed), *African Women and Children: Crisis and Response*. Preager Publishers, Westport, USA, pp. 225-251.

Darby, H. C. 1956. 'The clearing of the woodland in Europe' in Thomas, W. L. *Man's Role in Changing the Face of the Earth,* University of Chicago Press, Chicago, pp. 183-216.

Darkoh, M. B. K. 1989. 'The underlying causes of the food crisis in Africa', *Trans-African Journal of History,* 8, 54-79.

Darkoh, M. B. K. (Ed), 1992. *African River Basins and Dryland Crises,* Organization for Social Science Research in Eastern Africa and the Research Programme on Environment and International Security, Departments of Human and Physical Geography, Uppsala University.

Darkoh, M. B. K. 1993a. 'Towards sustainable development and environmental conservation in African drylands', *Journal of Eastern African Research and Development,* 23, 1-23.

Darkoh, M. B. K. 1993b. 'Land degradation and soil conservation in eastern and Southern Africa: A research agenda', *Desertification Control Bulletin*. No. 22, pp. 60-68.

Darkoh, M. B. K. 1996. 'Sub-Saharan Africa in crisis and the need for a new domestic order', In Yeung, Y. (Ed), *Global Change and the Commonwealth,* Don Bosco Printing Company Limited, Hong Kong, pp. 36-52.

Darkoh, M. B. K. 1997. 'Trade, environment and sustainable development', *Malaysian Journal of Tropical Geography,* 28, 2, 67-82.

Darkoh, M. B. K. 1999. *Evaluation of the Caring for Land Resources Subprogramme of UNEP,* UNEP, Nairobi.

Darkoh, M. B. K. 2000. 'UNEP and caring for land resources', Desertification Control Bulletin, 36, 107-117.

Eden, M. J. and Parry, J. T (Eds), 1996. *Land Degradation in the Tropics: Environmental and Policy Issues*. Pinter and The Commonwealth Foundation, London.

El-Hinnawi, E. 1985. *Environmental Refugees*, UNEP, Nairobi.

Fairhurst, U. J. *et al.* (Eds), 1999. *Integrating Environment and Society: The Thohoyandou Environs Northern Province, South Africa*. Departments of Geography, Universities of Pretoria and Venda, South Africa.

Friday, L. and Laskey, R. (Eds), 1989. *The Fragile Environment: The Darwin College Lectures*, Cambridge University Press, Cambridge.

Glacken, C. J. 1967. *Traces on the Rhodian Shore: Nature and Culture in Western Thought from Ancient Times to the End of the Eighteenth Century*, University of California Press, Berkeley.

Goudie, A. 1981. *The Human Impact on the Natural Environment*, Blackwell Publishers, Oxford.

Goudie, A. 1997. *The Human Impact Reader: Readings and Case Studies*, Blackwell Publishers, Oxford.

Gourou, P. 1953. *The Tropical World: Its Social and Economic Conditions and Its Future Status*, John Wiley and Sons, New York.

Hjort-af-Ornäs, A. (Ed), 1992. *Security in African Drylands: Research, Development and Policy*, Research Programme on Environment and International Security, Departments of Human and Physical Geography, Uppsala University.

Hjort-af-Ornäs, A. (Ed), 1996. *Approaching Nature from Local Communities: Security Perceived and Security Achieved*. EPOS, Research Programme on Environmental Policy and Programme, Institute of Tema Research, Linkoping University, Sweden.

Hjort-af-Ornäs, A. and Salih, M. A. M. (Eds), 1989. *Ecology and Politics: Environmental Stress and Security in Africa*. Scandinavian Institute of African Studies, Uppsala.

Huntington, E. 1914. *The Climatic Factor as illustrated in Arid America*, Carnegie Institution of Washington Publication, 192.

IISD. 1994. *Trade and Sustainable Development Principles*, International Institute for Sustainable Development, Winnipeg.

Lykke, A. M. 2000. 'Refining the ecological aspects of disequilibrium theories for Africa's pastoral drylands', *Desertification Control Bulletin* 36, 23-33.

Maathai, W. 1995. 'Bottle-Necks of Development in Africa', Paper presented at the 4th UN World Women's Conference in Beijing, China, 30 August – 15 September 1995.

Mannion, A. M. 1995. *Agriculture and Environmental Change: Temporal and Spatial Dimensions,* John Wiley and Sons Ltd, Chichester.

Mannion, A. M. 1997. *Global Environmental Change,* Longman, Harlow, Essex.

Mannion, A. M. and Bowlby, S. R. (Eds), 1992. *Environmental Issues in the 1990s.* Wiley and Sons, Chichester.

Marsh, G. P. 1864. *Man and Nature,* Scribner, New York.

McKnight, T. L. 1959. 'The feral horse in Anglo-America', *Geographical Review,* 49: 506-25.

Richards, P. 1985. *Indigenous Agricultural Revolution,* Hutchinson and Company, London.

Salih, M. A. M. 1999. *Environmental Politics and Liberation in Contemporary Africa,* Kluwer Academc Publishers, Dordrecht.

Salih, M. A. and Tedla, S. (Eds), 1999. *Environmental Planning, Policies and Politics in Eastern and Southern Africa.* Macmillan Press Ltd, London.

Sauer, C. O. 1938. 'Destructive exploitation in modern colonial expansion', International Geographical Congress, Amsterdam, vol. III, sect. IIIC, 494-499.

Sauer, C. O. 1952. *Agricultural origins and dispersals,* Isaiah Bowman lecture series 2, American Geographical Society, New York.

Semple, E. C. *Influences of Geographical Environment,* Henry Holt, New York.

Stiles, D. (Ed) 1995. *Social Aspects of Sustainable Dryland Management,* John Wiley and Sons, Chichester.

Swanson, T. M. (Ed) 1996. *The Economics of Environmental Degradation: Tragedy for the Commons?,* Edward Elgar Publishing Limited, Cheltenham.

Thomas, W. L. (Ed) 1956. *Man's Role in Changing the Face of the Earth,* University of Chicago Press, Chicago.

Visser, W. 2000. 'Facing environmental legacies of past corporate sins: understanding environmental liabilities', *Msafiri,* 31, 80–81.

Warren, D. M. 1991. *Using Indigenous Knowledge in Agricultural Development,* World Bank Discussion Paper No. 127, The World Bank, Washington, DC.

Weaver, J. H., Rock, M. T. and Kusterer, K. 1997. *Achieving Broad-based Development: Governance, Environment, and Growth with Equity,* Kumarian Press, West Hartford, Connecticut.

Williams, M. 1989. 'Sustainable development: A SID perspective', *Development*, 2, 3, 7-9.

World Commission on Environment and Development, 1987. *Our Common Future*, Oxford University Press, New York.

Chapter 2

A Socio-Economic Analysis of Environmetal Degradation in Africa

R. N. Ogbudinkpa

Introduction

Throughout history, human beings have continuously loaded the earth, waterways and atmosphere with wastes and by so doing, have placed such stress on the ecosystem that the latter has lost its capability for self-adjustment. The stress on the environment causes it to be degraded, with the inevitable threat to the lives of its inhabitants. Van Nostrand (1989: 846) and Liniger (1995: 47) see environmental degradation as reduction in the quality of the components of the environment, originating from both biotic and abiotic influences. Such degradation can be seen as the result of human activities in the exploitation and use of natural resources. Certainly, human exploitation of natural resources generates wastes, which degrade the environment. Such wastes reduce the quality of the air, land and water, wherever they are dumped. It should be pointed out that environmental degradation occurs whenever wastes of all kinds, from life and industrial activities, cannot be absorbed by the ecosystem as fast as they are generated. Heller (1975: 147) notes that the level of environmental degradation varies from place to place according to the way we grow and make use of our growth.

Environmental degradation is basically the negative impacts on the quality of the environment. It is bad and should, therefore, be controlled. The ideal way to eliminate it entirely would be to have a world of neither life nor industry. But Africa is a world with human beings and therefore must have industrial activities. As such, we cannot eliminate environmental degradation entirely. The implication of this state of affairs is that we cannot restore the continent to its original purity. The goal of African governments and communities, therefore, should be to minimise the level of environmental degradation that results from human pursuit of economic and social life. Controlling and, therefore, minimising all types of wastes dumped on the continent's land, waters and air can do this. The African governments should also focus their attention on activities which threaten the quality of their environments.

Evidence of Environmental Degradation in Africa

African countries, like many other countries of the world, suffer from environmental degradation, arising largely from their population densities, levels of urbanisation and industrialisation. Markandya and Munoz (1993: 20) have pointed out that African countries' environmental degradation may be among the worst in the world because their average rates of population growth fell between 2.5 per cent and 3.1 per cent in the period 1981-1990. The World Bank (1994) shows that there is increasing urbanisation going on in many countries. Furthermore, more and more manufacturing and tertiary industries are taking place as African countries try to jump hurdles so as to catch up with the more advanced industrial countries. Air pollution, an aspect of environmental degradation, is most often associated with the highly industrialised countries but has become a significant and increasing problem in many of the less developed countries. This has arisen from economic growth, which leads to the consumption of fossil fuels in industry and transport. Air pollution levels in some of the cities of developing countries, with poorly planned industries and laxity and/or no control regulations, are already worse than in some of those of the highly industrialised countries. Ngugi and Mbaria (1995: 12) examined the degrading effects of chemical fertilisers and biocides.

These scholars argue that the application of massive quantities of increasingly expensive fertilisers and potent biocides to coffee plantations in Kiambaa, Central Kenya, not only poisons the environment but also involves extreme exploitation of the low paid employees, who out of sheer need for survival, live and work within this type of environment. Inhabitants of the less developed countries are more likely to suffer the detrimental health effects resulting from breathing polluted air, in part because they spend more time outdoors and in part because they are more likely to travel in open vehicles.

Omoregie and Okpanachi (1992: 283) studied the growth of *Tilapia zilli* exposed to sublethal concentration of crude extracts of *Azadirachta indica*, a degrader in Nigeria. They came to the conclusion that such exposure not only retards the fish growth but also leads to their abnormal behavioural responses. From this, they concluded that the use of *Azadirachta indica* for fishing is a health hazard. Onianwa (1995: 345) studied the effect of petrol pollution, a hydrocarbon degrader, on the African environment. He used the city of Ibadan in Nigeria as a case study. His study led him to the conclusion that hydrocarbon contaminates very seriously the topsoil around petrol stations and mechanical workshops. The factors of hydrocarbon accumulation in both cases are 10.1 and 4.7, respectively. He went on to show that the degree of hydrocarbon pollution decreases in intensity from the petrol station to the mechanical workshop to the refuse dump to high area railroad tracks and to the control zones of the residential areas. This study implicitly advises that constant monitoring be made of hydrocarbon in the urban environment, however remote from petroleum exploration activities.

Sal *et al.* (1995: 318) studied hazards resulting from setting up a cement manufacturing plant at Kaymore in India. The two scholars found that people living up to 12 km downwind from the plant were affected by respiratory diseases of maximum incidence. Livestock was affected with gastrointestinal disorders of types which were almost nonexistent in non-cement manufacturing rural areas. It appears reasonable to expect the above two types of ailments in man and livestock around Yandev cement manufacturing plant in Nigeria's savannah. Coal is used to fire power plants in Zambia, Botswana and Zimbabwe. Singh *et al.* (1994: 89) observe that emissions from such coal-fired plants considerably reduce chlorophyll and ascorbic contents of foliage area at the site receiving high pollution load.

It seems reasonable to narrow down the contributors to environmental degradation to a few broad factors. The first is population growth. Goldman (1970: 375) notes that high population growth and its often inevitable urbanisation lead to human clusters in more concentrated areas like Lagos in Nigeria, Nairobi in Kenya, Arusha in Tanzania and Dakar in Senegal. More people means more wastes, both organic and inorganic, discharged on to the land, in the air and in water, beyond what humans and nature can recycle within a short time. The inevitable outcome is environmental degradation.

Since the early 1980s, African countries have experienced rapid population increases. With few exceptions, the growth rate of population in African countries over the period 1975-1990 is equal to or greater than 2.5 per cent. The average growth rate of the African countries over the period is inordinately 2.8 per cent (United Nations Demographic Yearbook, 1990). Urbanisation rates in Africa lie between 10 per cent and 71 per cent of the total population of respective countries. With more urbanisation, the direct and indirect degradation of the environment is increased by people's consumption habits. The resulting wastes degrade because they reduce the quality of life.

The second contributor to environmental degradation can be traced to industrial activities. Though the role of population and urbanisation in causing environmental degradation has been recognised, Yoskihara and Tanaka (1995: 769) argue that industrial activities, meant to satisfy the needs of the rising population, have a part to play in causing environmental degradation. Industrial activities, which involve the use of machines to produce, have in many instances led to reduced air quality. In particular, diesel engine exhausts are known to discharge nitric oxide into the air. Liaskos (1994:28) observes that the rubber tyre industry, a typical African industry, produces rubber tyres whose disposal constitutes a growing problem. Agriculture and logging cause deforestation. Environmental crisis from agriculture and deforestation is responsible for the *harmattan* dust storm, which blows off a considerable portion of the top soil in the Sudan. The dust pollutes rivers, cities, lakes, and the atmosphere, thereby reducing the quality of life. Furthermore, strip-mining and quarrying of tin and columbite in Nigeria, bauxite in Ghana, coal in Botswana, Zimbabwe and Zambia, and copper in Zaire reduce not only vegetation but also create unaesthetic sites (United Nations Industrial Yearbook, 1991). They also help the growth of dust storms in these African countries. Mining and quarrying are also known to destroy the natural habitats of wildlife.

There is another angle from which to view industrialization. Growth in industrialization has afforded humans with more biodegradable though complex and exotic products to consume. Solid wastes like garbage, trash and plastic result from such consumption. Other non-plastic industrial products are ashes, beer cans, sewage, sludge, building rubble and auto hulks. Rose (1972: 279) calls them 'third pollution' because they are solid wastes normally regarded as nuisance or worse. They are resources that degrade the environment because they are out of place. De Long *et al.* (1995: 79) argue that industrial treatment deposits which use air-borne pesticides degrade the environment by stunting the growth of vegetation, especially in areas close to the treatment plant. Chemistry has made man to live better by providing man with chemical plastic and aluminium products. But these degrade by existing long after their users are gone. The rural African communities frequently use bush burning, both for cultivation and hunting. The smoke haze from such fires screens off the sun at the same time that it emits million tons of CO_2 and particulate into the air. Furthermore, bush burning degrades the environment because it destroys millions of hectares of farmland and opens it to wind and flood erosion. This is true of the dry West African economies of Chad, Mali and Niger. In these economies, farmers have to cultivate a lot of land to produce more food for the increasing population. It is also known that bush burning destroys biodiversity. It has to be pointed out that this chapter does not argue that industrial activities per se contribute to environmental degradation. What it seeks to convey is that unplanned, unco-ordinated and uncontrolled industrial activities degrade the environment. Even though the benefits of industrial activities include rising employment, economic growth and improvement in the balance of payments position, one must remember that badly conceived industrial activities degrade the environment as well as reduce the quality of life. It, therefore, seems reasonable to argue that there is need for cost-benefit analysis of any industry so as to help the decision-makers in establishing the industry. As argued in Chapter 1 by Darkoh and Rwomire, the 'polluter pays' principle should be enforced in African countries.

The third contributor to environmental degradation is rising income or affluence. Affluence per se is not bad. What is bad is its improper use. Generally rising income positively correlates with such activities as increases in energy consumption, purchases of vehicles and pursuit of recreational activities. But these, when wrongly used, degrade the environment. Yoskihara and Tanaka (1995: 771) argue that diesel engine exhaust degrades the environment with its discharge of water-soluble methylamine into the air. African urban centres are blighted by clutters of junky vehicles. Furthermore, rising income enables people in many of the African countries to acquire and use the troublesome non-biodegradable plastic products rather than biodegradable leaves and papers in packaging. Even though affluence as shown above contributes to environmental degradation, one wonders if it is ethically and morally right to reduce affluence.

The author sees it both unreasonable and foolhardy to do so because we cannot have material things that money can buy without having affluence. Therefore, our goal should be to minimise the distortions on our environment caused by unwise use of affluence.

It is true that serious environmental degradation from population, industry and income is not yet common in the African countries. Though this is true, African countries should not relax in their efforts to minimise the current growth of environmental degradation. This is because African countries, like many other countries, are likely to be adversely affected by rapid growth in population, industry and affluence in the not distant future. What the countries have now in the urban and industrial areas are isolated cases of environmental degradation, which can intensify in future.

Generally, large population and urbanisation go together. Where urbanisation occurs without adequate plans to increase facilities in use, the latter become scarce and expensive. This seems to be the reason why tap water is expensive in some of the urban centres in Africa. Urban hardship that results from expensive water supply is a form of environmental degradation currently taking place in many of the African cities. Dasgupta (1993: 85) puts it that high population with its inevitable urbanisation that puts pressure on resources is a sign of environmental degradation. An example of an African city, which is facing such problems, and whose planning needs to be considered seriously is Lagos in Nigeria. The United Nations (1995: 28) shows that even though it was not one of the 15 largest cities in the world in 1994, it is projected to become the third largest with a population of 24.4 million by the year 2015. It is most likely that the demand for resources by its inhabitants will outstrip its capacity to provide them. This will lead to environmental degradation. Obi, in Chapter 3 of this book, offers more discussion on environmental crisis in Nigeria. Industrial activities, both directly and indirectly, degrade the African environment. For example, coffee processing in Kenya discharges toxic wastes that are harmful to aquatic life (Aringo, 1982: 45); the same is true of sisal in Tanzania. Liniger (1995: 37) shows that apart from wastes discharged into African land, air and waters by industries, the water used for cooling engines in textiles, beer, paper, and tannery industries either sinks into the soil or finds its way into the rivers and lakes as contaminants. These discharges cause negative effects to both the microfauna and the chemical characteristics of the rivers and lakes. Liniger (1995: 23) argues that human interference with the River Zambezi by damming for irrigation alters water flows and therefore adversely affects the downstream and the river basin. The same can be said of Aswan Dam in Egypt, Akosombo Dam in Ghana and Kianji Dam in Nigeria, to mention but a few. Aina and Akingbade (1994: 43) observe that off and onshore petrol exploration in Nigeria has endangered soil, plant and marine life close to the oilfields. This is because the area is noted for occasional spills of oil. The process of pumping oil in Nigeria involves flaring of the accompanying gas around Warri and Sapele on the coast. Such flaring of gas by the oil wells discharges sulphur oxide, nitrogen oxide and carbon dioxide into the atmosphere. When these mix with the moisture-laden wind, there occurs acid rain that affects plant life. Furthermore, smoke and fume from diesel engines reduce driving visibility on the roads. In brief there are traces of environmental degradation caused by growth in population, industry and affluence. Such degradation has economic and social repercussions on the nations of Africa.

Economic Effects of Environmental Degradation

There are three main sources of environmental degradation in Africa, namely, population explosion, unplanned industrialisation and licentious affluent consumption. Each of these causes generates its peculiar economic effects. Population explosion leads to increasing use of all resources. It erodes the standard of living of the economy and at the same time increases the level of a country's unemployment and housing needs. In Africa today, urban growth is generally associated with environmental decay. Population increase, by contributing to pollution, increases user costs or taxes imposed so as to collect the litter generated. Such costs are commonly known in economics as negative externalities or spillover costs.

Private and corporate enterprises engage in industrial activities in Africa. Though increasing industrial production of food and fibre boosts employment and income, it spurns pessimism because it discharges wastes into the ecosystem and reduces aquatic plant and animal life. Beer and textile industries close to rivers and lakes discharge soap and dyestuff, which destroy both plant and marine life. Furthermore, industrial discharges are a cause of reduced agricultural development, tourism and air purity. There have been declines in residential, commercial, and industrial properties occasioned by environmental degradation. Particulates destroy crops and plants. It is true that industry boosts agriculture by providing machinery and other farm inputs. Nevertheless, it discharges smoke and carbon which are a health hazard. Industrial exploitation deteriorates mineral, soil, animal and plant life. There are irreversible damages to farm lands by irrigation for extensive farming. These come in form of submergence of land, relocation of displaced population, increased evaporation rate of open water, reduced water supply downstream and increased accumulation of sediments at the bed of the river. Irrigation occasionally leads to waterlogging and inundation of agricultural and forest lands when there are excessive rains. Herbicides and pesticides, which are occasionally used by farmers, are toxic to man, birds and fish.

Corporate enterprises have argued that industrialisation boosts income and transfer of technology into Africa from other continents. However, these enterprises equally have claimed innocence of all wrongdoing and no responsibility to anyone except their owners. Anyone who questions these tenets is perceived as an enemy of the free market, capitalism and private property itself. By their pollution and therefore degrading activities, they have imposed financial costs on the public.

African economies are mostly skewed in income and wealth distribution. Democracy does not really exist in many of the countries. A small percentage of the population of each country is made up of plutocrats. With their wealth, they purchase luxury and polluting goods like tractors, lorries, aeroplanes and ships, which discharge fumes into the air, land and water. Though these goods pollute the environment to the detriment of the population, these plutocrats, for fear of the adverse use of their wealth against the rest of society, are not called upon to clean the environment that they have degraded. Income and affluence are therefore used to inflict economic damages on society. In brief, these are some of the economic effects of environmental degradations in African countries.

Social Effects of Environmental Degradation

Environmental degradation has negative social consequences. Some Nigerian politicians, for example, are aware enough to deplore such consequences for bidding for re-election. Even though environmental degradation has not reached catastrophic levels in many African countries, it is important that measures are introduced to minimise its adverse effects before the situation gets worse. There are pessimists who hold the view that environmental degradation cannot be avoided, especially in the urban centres and that if we want income to grow, the 'town must stink'. The rational objective to be pursued by planners in African countries should be to achieve growth without stink. Macrae (1972: 262) argues that high population is associated with urbanisation, dilapidated tenement and high rate of crimes. Generally, criminals live in a culture of grievance with the society and police. They are mostly youths with poor or no parental guidance. Crime is now a major problem in African cities such as Nairobi, Lagos and Johannesburg.

Industries contribute to environmental pollution and decay. Both private and corporate enterprises produce pollutants that devalue life and property. The socio-cultural impacts of environmental degradation from industries are evident in adequacy of health facilities and protection from hazards to industrial workers. Poor industrial practices lead to jammed highways, overcrowded recreational areas, reduced aquatic, vegetable, and animal foods for consumption, all of which militate against social and community life. This then calls for adherence to environmental guidelines by industrialists. Guidelines to evaluate the environmental impacts of industrial projects should be applied in all cases. Based on such guidelines, it will be possible to approve industries that refrain from discharging toxic and corrosive chemicals, harmful particulate, excessive and intensive heat, noise and noxious odour. There is no doubt that dumped contaminants that contaminate ground water may cause serious ailment and death to man. Industries which pollute land, air and water affect communities that use such resources.

Possible Measures to Reduce Environmental Degradation

African planners should not wait for the problems of their respective countries' environmental degradation to become acute before they begin to seek for remedial measures. They should heed Miller (1972: 261) who warns that unless we mend our ecological behaviour, we will be poisoned by the food we eat, choked by the air we breathe, and surrounded by rivers and oceans, which have become sewers. African countries must learn from the experiences of the already developed countries the need to avoid and/or reduce environmental degradation even if there are only sporadic cases on the continent now. The goal of every nation should be not to allow wastes of all kinds to accumulate before they become too big to eliminate. The economic plight of African nations typified by low per capita income is such that dismal environmental conditions should not be allowed to emerge at all. This chapter has already outlined evidence or indicators of environmental degradation in Africa. Implicit in the indicators presented are possible measures of environmental degradation control. We have already indicated that high population growth and urbanisation accelerate environmental degradation. The recommendation in this chapter is for pursuit of quasi population growth control. We must reduce population growth not because population growth is the villain that causes the trouble but because such a reduction releases resources, which can be used to improve the quality of life. The expected solution from this perspective is to hold a country's population growth rate down so as to reduce the pressure on the environment. This is necessary to make the use of other control mechanisms effective. In addition, hyper-urbanisation should be avoided. It is suggested that measures to reduce overconcentration of population be pursued on all fronts. Encouraging rural development can do this.

Furthermore, since man is a polluter, it is suggested that proper economic and legal measures in form of taxes on waste disposal be incorporated into the policies of African countries. Industry boosts employment in the African nations and contributes significantly to the growth of the GNP. Nevertheless, this chapter has shown that industry, by its relentless pollution, degrades the African environment. It is suggested that proposed industries be screened and selected on the basis of economic and environmental relevance. African governments should establish ambient standards for industries. Industries, through tax rebates, must be encouraged to incorporate industrial pollution technologies in their production processes. It is suggested that monetary evaluation of the contributions of an industry to environmental degradation be done on a cost-benefit basis. Benefits must be discounted at an officially accepted rate while all spin-off costs be recognised and collected. The best point to minimise industrial pollution is where the marginal cost of pollution equals the marginal benefit. At this point, environmental degradation is minimised. Governments are custodians of the society's welfare and should therefore play an active role in making the environment attractive. The government of Niger in West Africa has shown how it carries out its custodianship by its collaboration with industry, USAID, and German Technical Assistance to remove and destroy thousands litres of old pesticides in the country (UNEP, 1993: 72). Governments should require that vehicles be equipped with pollution control mechanisms. Industries must be made to direct wastes to positive uses such as irrigation, landfill, energy and manure production and recycling. Both for control and decentralisation mechanisms, industries must be encouraged to locate in rural areas. All these measures are likely to fight pollution and the consequent environmental degradation.

Affluence contributes to environmental degradation. This author does not argue for reduction of affluence but rather for reduction of the adverse effects of affluence which degrade the environment. Ad valorem tax is a good measure to fight adverse use of affluence.

Conclusion

First, at present the population growth rate of most African nations is above 2.5 per cent per year, and it has been growing at that rate for the last few years. Most demographers believe that in the absence of a major disaster or drastic changes in socio-economic conditions, this rate may rise during the twenty-first century. Since high population is associated with increasing pollution and, therefore, urban degradation, a slower rate of population growth would be desirable to abate environmental degradation.

Second, the geographical concentration of population is a greater threat to the environment than the expected population growth. Rural development is key to reducing population concentration in urban areas.

Third, pollution and environmental degradation are linked to high crime rates, transportation problems and social ills. Africa should seek ways and means to effectively tackle these problems as a way to reduce environmental degradation.

Fourth, project planning and implementation in Africa have to take the environment very seriously. Such must be cognisant of the costs and benefits of any actions in the economies. This requires the evaluation of the positive and negative effects of any programme or project before it is implemented. Definitely, proper planning and implementation will help to minimise environmental problems in the economies.

Finally, the government is the custodian of societal values and good. It must, therefore, use both economic incentives and regulatory measures to achieve a high level of environmental quality. While the economic incentives take the form of environment-degradation-abatement technologies, the regulatory measures take the form of disposal taxes whereby the polluter bears the cost of cleaning their pollution. It is hoped that the above suggested measures would help to minimise environmental degradation in African countries.

References

Aina, A. and Akingbade, T. 1994. 'The environmental price of oil: Petroleum resource development and the Nigerian environment', *African Technology Forum,* 1, 17-23.

Aringo, P. O. 1982. 'Issues and Trends in Industry and Environment in Kenya', *Industry and Environment,* 3, 45-48.

Dasgupta, P. 1993. ' Poverty, resources, and fertility: The household as a reproductive partnership', *In Environmental Economics and Natural Resource Management in Developing Countries* Mohan Munasinghe (Ed), Washington DC: World Bank Committee of International Development Institutions on Environment.

De Long, F.M.W., Van der Voet, E. and Canters, K.J. 1995. *Ecotoxico Environment* SAF, 30, 1, 15-25.

Goldman, M. I. 1970. *Controlling Pollution: The Economics of Cleaner America,* Englewood Cliff, New Jersey: Prentice Hall.

Heller, W. W. 1975. 'Should government share its tax intake?' *In Current Economic Problems: A Book of Readings,* Brandis Royall and Steven Cox (Eds), Homewood, Illinois: Richard D. Irwin Inc.

IBRD, 1994. *World Tables.* Baltimore, Maryland: The Johns Hopkins Press.

Liaskos, J. 1994. 'Rubber tyre recycling', *Industrial Environment,* 17, 37-50, July-September.

Liniger, H. 1995. *Endangered Water: A Global Overview of Degradation Conflicts and Strategies for Improvement,* Berne, Switzerland: Group for Development and Environment.

Markandya, A. and Munoz, C. 1993. *Environmental Economics and Natural Resource Management: The African Development Experience,* Mohan Munasinghe (Ed), Washington DC: World Bank Committee on International Development Institutions on the Environment.

Miller, H.P. 1972. 'Population, pollution, and affluence', *Current Economic Problems : A Book of Readings,* Brandis Royall and Steven R. Cox (Eds) Homewood, Illinois: Richard D. Irwin Inc.

Ngugi G. and Mbaria, M. 1995. 'Sustainability in agricultural development : evaluation of environmental impacts of large scale coffee enterprise in Kiambaa division of central Kenya', *International Journal of Environmental Health Resources,* 5, 1, 20-31.

Omoregie, E. and Okpanachi, M. A. 1992. 'Growth of *tilapia zilli* exposed to sublethal concentration of crude extracts of Azadirachta Indica', *ACTAHYDROBIOL,* 34, 3, 26-40.

Onianwa, P.C. 1995. 'Petroleum hydrocarbon pollution in urban top soil in Ibadan City Nigeria', *Environment International,* 21, 3, 31-37.

Rose, Sanford 1972. 'Population and affluence', *Current Economic Problems: A Book of Readings,* Brandis Royall and Steven R. Cox (Eds), Homewood, Illinois: Richard D. Irwin Inc.

Sal, V. S., Mishra, M. P. and Mishra G. P. 1995. 'Diseases affecting rural population along distant land at Kaymore Cement Works in India', *Environmental Ecology,* Vol. 13, No. 2.

Singh, J., Agrawal, M., and Narayan, D. 1995. ' Relative Susceptibility of Two Species ofTropical Fruit Trees to Thermal power Plant Emission', *Tropical Ecology.* 35, 1, 15–23.

United Nations, 1990. *Demographic Yearbook,* New York.

United Nations, 1991. *Industrial Yearbook,* New York.

United Nations Environmental Protection, 1993. 'Government and partnership solves African pesticide problem, *Industry and Environment,* January-June.

United Nations, 1995. *Concise Report on the World Population Situation in 1995,* New York: United Nations ST/ESA/SER.A/153.

Van Nostrand, J. 1989. *Scientific Encyclopedia,* Nostrand, New York.

Yoskihara, T. and Tanaka, T. 1995. 'Reduction of Oxides of Nitrogen in diesel exhaust with additional Methylamine', *Transport Japan Social Mechanical Engineering Service Bulletin,* 61, 582.

Chapter 3

Political and Social Dimensions of Environmental Crisis in Nigeria

C. Obi

Introduction

In the past decade, growing concerns have been expressed about Nigeria's threatened environment (NEST, 1991; Olaore, 1986). At the heart of this environmental crisis are issues of distributive inequity which denies the majority access to a fair share of resources and the authoritarian mould of state power and power relations in society. Other issues include the persistence of environmentally non-sustainable forms of production and consumption. Although there have been efforts by the government to erect institutional structures to contain the crisis, these have not succeeded in the face of complex political and social problems.

The environmental crisis in which Nigeria is immersed is quite deep. Its varied manifestations include the population explosion, rapid urbanisation which outstrips the capacity of planning and management institutions to cope, rural decay, rising crime rates, environmental stress and conflict. Others include improper waste management, industrial pollution, inefficient land use and the economic crisis.

It is apposite to point out that the foregoing is the outcome of historical factors, the nature of the state and society in Nigeria and the scant attention paid to environmental matters in development planning and social policy. This underscores the importance of the political and social dimensions of the problem, and informs strategies for its resolution.

This chapter is organised into five parts. The introduction identifies the critical issues in the environmental crisis confronting Nigeria. It is followed by a brief background of the institutional framework for environmental management in Nigeria. The third part deals with the major conceptual issues in the treatment of the political and social dimensions of the environmental crises. It dovetails into the core of the chapter - the political and social dimensions of environmental crisis in Nigeria. The fifth and conclusive part sums up the discourse and offers some suggestions towards resolving the crisis.

An Overview of the Institutional Framework for Environmental Management in Nigeria

Since the colonial times, and even after independence, it is possible to discern a low level of institutionalization of public management of environment in Nigeria' (Aina, and Salau, 1991: 120; Sada and Odemerho, 1988). During the colonial period, the environment was perceived in purely sanitary or health related terms. Unfortunately, this narrow perspective has persisted, in spite of the recognition that the environment is a much larger issue linked to sustainable development. It was only as recent as 1988, that the first coordinated effort was made at the national level to regulate and manage the environment in Nigeria (Adewale, 1991: 642). Setting up the Federal Environmental Protection Agency (FEPA) and the Natural Resources Conservation Council (NRCC) facilitated this. Also, the government came up with a National Policy on the environment, 'recognizing the link between developmental process, environmental factors as well as its ultimate goal for the achievement of sustainable development in the country' (Aina and Salau, 1991: 126).

Before the coming of FEPA, the tasks of monitoring and managing the environment were dispersed between several ministries (Health, Works and Housing, Youth, Sports and Social Development, Science and Technology) and the three tiers of government at the Federal State and local levels (Obi 1994: 68). Apart from the issues of dispersal, there is the problem of confusing, overlapping and conflicting jurisdictions which has further weakened incapacities of these institutions.

This could be traced to the little attention given to environmental considerations in development planning (Adeniyi, 1986). Although the third development plan (1975-1980) noted the link between environment and development, it continued the over-emphasis on economic growth and gross under-funding of the environmental sector (Aina and Salau, 1991: 64).

From a constitutional perspective, Akande, in her examination of the 1989 Nigerian constitution (now suspended) notes that 'it does not deal specifically with the environment, although it has sections that are relevant to the environment' (Akande, 1994: 38). Apart from creating a major lacuna in the area of establishing a coherent policy framework for environmental management, this omission poses a problem for the enforcement of environmental laws (Adewale, 1991: 650). Thus, an institutional and policy centralisation that negates the spirit of federalism has emerged. As Akande again points out, 'it is quite ironic, that the level of government which is closest to the people, that is then local government, has the least constitutional responsibility to the management of the environment' (Akande, 1994: 38). If the 1989 constitution failed to address the specifics of environmental management, then its suspension by military fiat has complicated matters. Alongside existing structures have emerged military-led Special Task Forces to enforce a 'sanitary' perspective of environmental management. Usually the victims of such sanitation drives are the poor and vulnerable social groups.

From the foregoing, it is clear that the crisis is partly the outcome of low level of public participation in policy making as it affects the environment. Furthermore, the retention of powers and resources at the federal centre divert the energies or competing jurisdictions and levels of government towards the struggle for resources and space. This further weakens their capacities, and they end up reproducing costly top-heavy structures that neither address nor integrate adequately the needs of the majority into the processes of environmental management.

Some Conceptual Issues

The major conceptual issues in the environmental crises in Nigeria revolve around a predominantly managerial approach to the environment, the absence of social equity in environmental management and the character of state and society in Nigeria.

A managerial approach towards the environment traps itself within narrow physical terms. Such an approach to environmental management as 'a specialized, technical and highly professional endeavour' routinely excludes certain social categories from its preview (Aina, 1992). More importantly, its orientation favours the use of expensive, capital-intensive, import-dependent strategies of environmental management which are difficult to cope with (given the scarcity of resources), and tend to favour the rich and powerful against the poor (Obi, 1994: 70). Thus, environmental planners are faced with the problems of high costs of importing and maintaining equipment, and the inappropriateness of imported ideas in tackling local problems.

Closely related to the problem of the managerial approach, is that of the absence of social equity in environmental management and by extension, the enforcement process. This comes out clearly in the enforcement of environmental laws that punish the poor more, while the large and powerful corporate polluters go virtually unpunished. The argument in some quarters is that these large industries bring in foreign capital and contribute to the development of the economy and should be given more latitude to comply. Also, these companies are smart enough to exploit loopholes in the law and employ the services of experienced corporate lawyers. This way they escape sanctions, while the urban poor and rural inhabitants bear both the full brunt of the law, and the effects of environmental pollution. In the oil industry, the fine for gas-flaring is so low that the companies find it more economical to flare the gas and pay the fine rather than re-inject it into the ground, while the villages around the flare-sites continue to suffer the atmospheric pollution. Yet, it is necessary to point out that both issues raised are linked to the nature of state and society in Nigeria.

The Nigerian State is a product of British colonialism. The process of its creation involved the forcible bringing together of diverse people with equally diverse culture, ethnicity, production systems and political structures. The impact of colonialism was threefold. First, its administrative policies created divisions within the diverse people it ruled. Second, it subordinated the Nigerian economy to the demands of the world market for cheap primary products. Third, the elite it created lacked a strong economic base, which was to raise the stakes in political contests in the nationalist and post-colonial periods. At independence, the nascent political elite thus faced the problem of a weak economic base and an externally oriented economy.

The attempt at nation-building was beset by two problems: the winner-take-all attitude of the political elite which deployed regionalism, religion and ethnicity in the quest to capture and retain power, leading to political instability, and the domination of political life by the military.

Looking at the current status of state and society in Nigeria, the role of petroleum is significant. Since 1970, petroleum has become the mainstay of the economy and state power. It accounts for over 95 per cent of foreign exchange earnings and 80 per cent of all government revenues. Unfortunately, the sector has not been able to forge strong linkages with other sectors of the economy (Obi and Soremekun, 1995: 1-31; Khan, 1994: 148). The control of oil is therefore of strategic importance to the state, and its distribution, to the stability of the society. Unfortunately, the wealth from oil has not laid a firm economic base, nor has it promoted stability. Rather it has fed into corruption, undermining the ethics of efficiency and productivity (Khan, 1994: 10).

From the discussion so far, it is not difficult to grasp the link between authoritarianism and the oil-renter economy. Those in power do not want to leave because of the enormous resources at their disposal, while those outside seek entry by all means. In this zero-sum context, democracy and social cohesion are severely undermined.

Related to the environment, the state operates from an economic perspective that dwells on the need to optimally exploit nature for profit. Environmental degradation is therefore not perceived as a problem except where it blocks the realisation of profit. This approach has led to environmental stress, degradation, and conflict between those who plunder the environment, and those whose lives depend on its sustainability.

Thus, in a context of economic crisis, when the oil revenues are on the decline, political and social tensions remain high as a result of struggles to control power, or to ensure a change in the distribution of oil revenues. Hardly any attention is paid to the environment or sustainable development.

From the foregoing, only a holistic treatment of the political and social dimensions can provide a proper understanding of the environmental crises. For the most part, the power relations support the optimal exploitation of the environment for the benefit of the power elite and foreign economic interests while, the people are left out, or struggle over the crumbs. In the face of shrinking resources and growing demands, the growing gap between the rich and the poor, and the stretching of institutional capacities beyond breaking point, the crisis is bound to deepen.

The Political and Social Dimensions of Environmental Crisis in Nigeria

The power relations between social groups is strategic to the understanding of the Nigerian crisis. Yet one must not ignore the political framework. Theoretically, Nigeria is a federation made up of three tiers of government: federal, state and local, with distinct and equally distributed powers and responsibilities. In practice, however, power has been concentrated at the federal centre. Two factors are responsible for this: the imposition by the military of their command structure on federal governance, and the centralisation of the collection and distribution of all oil revenues by the federal government, thereby rendering the other tiers of government (which generate very little revenue on their own) dependent on the federal government. Some of the problems arising from this centralised fiscal federalism is duplication of institutions across the tiers of government to capture resources, and the lack of resources at the local tier to enable it effectively manage the environment.

Another problem is the conflict, confusion and institutional overlaps that exist between the various tiers of government. As noted elsewhere, such conflicting relations have made the enforcement of environmental regulations difficult (Obi, 1994: 72). An example of such was the disagreement between FEFA and the Lagos State government over whose duty it was to enforce appropriate waste disposal methods (*ibid*). Akande (1994) also draws attention to the lack of uniformity in the setting up and defining of duties of state environmental enforcement agencies as another instance of the confusion that existed in the implementation of environmental policies.

The relevance of the preceding observations is to give the example of Lagos with a population of over six million inhabitants, that, appropriate waste disposal methods have remained a problem. Thus the city continues to suffer from the following environmental problems: pollution, careless dumping of untreated sewage and industrial effluents in drainage, open spaces and water bodies, water and energy shortages, unemployment and the growth of slums. These do not only lead to environmental degradation and pose health risks, they provide a breeding ground for social stress, crime and violence.

A critical look at the structure of power relations in Nigeria shows a wide gulf between the ruling elite, made up of the military top brass, highly placed bureaucrats in the public and private sectors, prominent traditional rulers, and the rest of the populace (Obi, 1994: 73). Due to the strong link between political and economic power, this elite controls the bulk of resources and ultimately decides 'who gets what'. Being so critical to access and distribution, the elite are constantly immersed in struggles over 'who gets what' and how much. Thus, it is highly fractionated and unstable, living off the spoils of the oil economy. It also routinely excludes the rest of society from a fair share of resources. Furthermore, the power elite manipulates institutions and defies laws that do not favour its interests.

Within the context described above, several problems arise. Political instability feeds into regime instability and a rapid turnover of governments, administrators, and technocrats. This often leads to discontinuities in policy, confusion and institutional crises. The capacity for effective environmental management is therefore undermined, while enforcement becomes problematic. Worse still, some 'enforcers' become the violators of environmental laws. Examples include building on land reserved for other purposes (play grounds, parks, drainage systems and canals) without paying attention to environmental considerations, and the inability to prosecute industrial polluters. This has not only led to environmental degradation, but it has also promoted public cynicism and conflict.

A critical dimension of environmental crisis in Nigeria relates to the strategic oil industry. Apart from the immense contribution of oil to the Nigerian economy, the major issue in this context is the inability of the Nigerian State to effectively regulate the oil industry and enforce environmental regulations, therein. The limited institutional capacities of the Petroleum Inspectorate Commission, the body charged with regulating the oil industry, has been discussed elsewhere (Obi, 1994: 75). It is important to note that the inspectorate's limited capacity is partly linked to wrangling with some officials of the Ministry of Petroleum Resources and the shortage of skilled manpower. The outcome of this is the virtual non-regulation of the industry outside the sense of responsibility and conscience of the operating companies. Oil pollution has therefore become an explosive issue leading to violent conflict between the people of the oil-producing areas and the oil companies and government. A lot of people have lost their lives in the conflict, while the environment has been severely polluted (Obi, 1995).

The consequence of the state-environment nexus is that development and environment have been immersed in crisis. Political and social instability has further diverted resources towards the maintenance of state security, at the expense of development, the protection of the environment and social rights.

The economic crisis in which Nigeria has been immersed since the 1980's has had an adverse impact on the Nigerian environment and its management. In that decade, Nigeria slipped from being a medium-income country to a low-income one in the World Bank's ratings, in the face of a drop in global oil prices. According to Khan (1994), export earnings from oil fell from $5.2 billion in 1980 to $9.9 billion in 1983, and went down further to $5.7 billion in 1986, leading to a balance of payments crisis in the import-dependent economy. The severe shortfall in revenue also led to a domestic economic crisis and the adoption of a programme of structural adjustment with its harsh social consequences. Since 1986, the crisis has persisted with dire consequences.

The environment has been subjected to more pressures to provide more money for economic recovery. At the same time, the poor have been forced to put more pressure on the environment for their basic needs including food, shelter, water and energy. Thus have the state and the people come to pursue conflicting interest with respect to the environment, with adverse consequences for social harmony.

Nowhere is the crisis more evident than in Nigeria's cities and the oil producing areas of the Niger delta. The state and local governments have remained in the shadow of federal centralization and have been unable to stem the rate of urban and rural decay. Economic crisis has increased the influx from the villages to the cities, putting more strain on existing inadequate urban facilities, worsening living conditions and social stress. In its bid to stem the crisis, urban authorities tend to rely on compulsive methods. Mass arrests of street traders and the demolition of illegal settlements are some of short-run measures that have been adopted. Yet these have largely remained ineffective as the problems reappear just a few weeks after highly publicized campaigns. Worse still, the environment is exposed to more degradation from the major polluters.

Perhaps more dangerous, politically, is the environment-related conflict in the oil producing areas. Since the oil is located largely in areas inhabited by ethnic minorities, oil pollution compensation and equitable access and distribution have become explosive issues (Soremekun and Obi, 1993). Since the mid-1980s, there have been violent protests against the activities of oil multinationals operating in the Niger delta, and the federal government. Recently, these struggles were merged with those for democracy and self-determination of minorities in Nigeria. Within the context of existing ethnic and class divisions in Nigeria, and the determination of the power elite and the military to control oil and secure its continued supply, the politics of oil is bound to turn into a national crisis.

Yet, while a lot of energy and lives are being wasted on oil politics, the ecology of the Niger delta bears the brunt of ecological degradation. Apart from toxic wastes introduced into the environment by refineries, the operators of the industry pollute and degrade the environment. This strikes at the means of existence and livelihood of millions of people whose water, land and air are daily polluted and degraded. It is this situation that feeds social and political conflict and instability (see Chapter 19).

Environmental crises in Nigeria have thus been exacerbated by political and social factors, which are themselves linked to authoritarian structures of power and social inequalities. It therefore stands to reason that the resolution of the crises should take place in a context of political and social change.

Conclusion

It is clear from the discourse that political and social factors play a critical role in Nigeria's environmental crisis. The challenge therefore, is to devise a framework for crisis-resolution and transcendence. The first step revolves around the democratisation of the Nigerian state and breaking the hold of the power elite over it and would provide a basis for reforming Nigerian federalism on a more equitable basis that gives more autonomy and resources to tiers of government closest to the people. It would also open up political space for popular empowerment, debates, and participation in decision-making in matters that affect the lives and livelihoods of people. In this way, stakeholders can input substantially into the policy and planning process.

The issue of accountability is quite important. Unless accountability is institutionalised, the connection between power, office and corruption cannot be broken. It would also enhance the exposure of corporate actors who take advantage of corrupt and weak institutions to exploit the economy and pollute the environment. And would provide less incentives for the manipulation of institutions to selfish ends.

Another important point is environmental awareness. This has to be integrated into development planning as a step towards sustainable development and environmental protection. The popularisation of environmental education would also promote environment-friendly practices amongst the populace and feed into the policy-making process.

The limitations of the existing legal framework can be transcended in several ways. The framework should be transformed emphasising social justice and equity, both in making laws and enforcing them. Lawmakers should be sensitised to the importance of the environment and its protection as a means of getting constitutions to respond to the challenges of efficient environment management.

Social values that promote extreme wealth and poverty breed a culture of aggressive acquisitiveness that has been responsible for most of the contradictions besetting African societies today. It is thus imperative to move away from such disabling values towards those that uphold the welfare of the people, fairness, and the rights of vulnerable groups and minorities. These would form the fundamental step towards an alternative model of development centred on popular participation and the survival of present and future generations.

References

Adeniyi, E. O. 1986. 'Environmental management and development in Nigeria', In Adeniyi, E.O. and Bello-Imam, I.B. (Eds), *Developmental and the Environment: proceedings of a National Conference*. Ibadan: NISER.

Adewale, O. 1990. *Sabotage in the Petroleum Industry: Some Socio-Legal Perspectives*, Lagos: NIALS.

Adewale, O. 1991. 'The right of the individual to environmental protection: a case study of Nigeria', *Revista Guiridica Dell Ambiente*, 4, 26-45.

Agobola, T. 1988. 'A review of environmental components in Nigeria's National Development Plans', In Sada, P.O. and Odemerho, F. O. (Eds), *Environmental Issues and Management in Nigeria's Development*. Evans, Ibadan.

Aina, A.T. 1990. 'The politics of sustainable Third World urban development', In Cadman, D. and Payne, G. (Eds), *The Living City: Towards a Sustainable Future*, London and New York, Routledge.

Aina, A.T. 1990a. *Health, Habitat and Underdevelopment in Nigeria.* London: IIED.

Aina, A.T. and Salau, A.T. 1991(Eds). *The Challenges of Sustainable Development in Nigeria.* Ibadan, NEST.

Aina, E. 1992. *Transition to Sustainable Development through Environmental Protection,* Annual FEPA Press Briefing, Abuja.

Aina, A.T. 1992. 'Environmental management and social equity', paper presented at the International Conference on Sustainable Environmental and Resource Management Futures for sub-Saharan Africa, organised by the United Nations University, Tokyo, held in Accra, Ghana.

Akande, J.O. 1994. 'Allocation of environmental responsibilities under the Nigeria Constitution: Institutional challenges and conflicts', In Ajomo, A. and Adewale, O. (Eds), *Environmental Law and sustainable development in Nigeria,* Lagos: NIALS and the British Council.

Khan, A.S. 1994. *Nigeria: The Political Economy of Oil,* Oxford: Oxford University Press, Oxford.

Nigeria Environmental Study/Action Team (NEST). 1991. *Nigeria's Threatened Environment: A national profile,* Ibadan, NEST.

Obi, C.L. 1992. 'State and Oil in Nigeria', Unpublished Ph.D. Dissertation, University of Lagos, Nigeria.

Obi, C. I. 1994. 'Political and social consideration in the enforcement of environmental laws', In Ajomo, A.M. and Adewale, O. (Eds), *Environmental Laws and Sustainable Development in Nigeria.* Lagos: NIALS and the British Council.

Obi, C. I. 1995. 'Oil minority rights versus the Nigerian State : Environmental conflict, its implications and transcendence', Paper presented at CODERSA's 8th General Assembly and Conference, Dakar, Senegal.

Obi, C. I. and Soremekun, K. 1995. 'Oil and the Nigerian state: An overview', in Soremekun, K. (Ed), *Perspectives in the Nigerian Oil Industry,* Lagos, Amkra Books.

Olaore, G. O. 1986. 'The cultural basis of environmental crisis in Nigeria',In Sada, P. O. and Odemerho, F.O. (Eds), *Environmental Issues and Management in Nigeria's Development,* Ibadan: Evans.

Soremekun, K. and Obi, C. 1993. 'Oil and the national question', in Onah, F. (Ed.), *The National Question and Economic Development in Nigeria,* Ibadan: Nigerian Economics Society.

Chapter 4

Environmental Security in the Drylands of Africa: The Role of Science and Technology

M. B. K. Darkoh

Introduction

Everywhere in the African drylands, there is clear evidence that the environment is not secure. The environment is failing to offer opportunities for sustainable livelihoods to rural and urban populations dependent upon it for their survival. For example, long spells of drought, desertification and floods are only a few examples of the serious environmental problems that make it impossible for rural people to co-exist in harmony with nature and/or, at times, with each other. The famines and food shortages that have severely hit people in the 1970s and 1980s resulting in mass starvation, death, disease and the exodus of millions of environmental refugees moving to urban areas or to more humid lands in desperate search for survival, are the clearest manifestations of the unfolding environmental insecurity in the drylands of Africa. This chapter reviews the extent and significance of environmental insecurity in the drylands of Africa and the role science and technology can play in any effort to find solutions.

Environmental security is viewed from two related perspectives: first, in terms of those environmental conditions that enable individuals and groups to satisfy their basic needs and those that offer opportunities for sustainable livelihoods, and second, as a pre-condition for peace and conflict resolution. In societies where the reproduction of resource endowments (farms, herds, trees etc.)

depends, to a large extent, on nature's bounty, environmental and human security are intertwined and cannot be treated in isolation from each other (Mohamed Salih, 1992). This line of thought has been forthcoming in recent literature (Frankie and Chaisen, 1980; Redclift, 1984; Spooner and Mann, 1982; Hjort, 1982, 1989 and Mohamed Salih, 1992). The linkage between people and environmental security has also reached a new height in the social science agenda, particularly when an association between political conflicts and environmental stress is evident in many parts of African drylands (Mohamed Salih, *ibid.*). In essence, this means that environmental security implies the elimination of the factors which may threaten people's survival such as soil erosion, depletion of natural resources and their accompanying food shortages. In other words, to enjoy environmental security is to satisfy the basic needs of human existence and overcome those environmental conditions that threaten the safety, health and well-being of human populations.

The second perspective which sees security as a pre-condition for peace assumes that peace cannot be attained if the physical, social and political environments are in a situation of imbalance (Buzan, 1983; Mohamed Salih *ibid.*). Thus, the depletion of natural resources or their appropriation by a few wealthy and powerful individuals or groups, jeopardises peace and undermines security. The security of one group of actors in their respective relationship to natural resources or the harsh conditions of the drylands will automatically affect the security of other groups of actors in the same environment.

Considering the above salient features of what is 'environmental security', it seems that the whole logic of the concept is skewed when applied to the drylands (Mohamed Salih, *ibid.*). It appears as if we mean insecurity when we refer to environmental security in the drylands of Africa.

Major Environmental Security Issues

Climate

Dryland environments in Africa are undergoing rapid environmental change. Major environmental security problems stem from the effects of drought, desiccation, climatic variability, climate change, population and human activities. Drought denotes a regional deficiency in soil moisture, which may be caused by a combination of lower than normal precipitation and higher than average evapotranspiration. Desiccation is a process of aridification resulting from a dry period lasting maybe decades. Climatic variability denotes shorter-term climatic variability including interannual variations, and climate change refers to secular trends or shifts in climate resulting from natural processes or human activity.

Every year, drought occurs in some part or other of the arid and semi-arid lands (ASAL) of Africa. In fact, major droughts regularly affect large portions of these drylands. Examples include the most severe drought in recorded history in 1968-73 and other droughts in 1982-85 and 1990-92. In the Sahel, during the three decades from 1961 to 1991, annual rainfall had been between 20 and 40 per cent less than it was during the three previous decades (Kelly and Hulme, 1993). Over the last quarter of a century, the Sahel has undergone severe desiccation and increasing deterioration of the soil quality and vegetative cover. Some studies have pointed out that, within contiguous Africa, there has been a net shift of land area towards aridity, especially toward hyper-aridity and a consequent net loss of semi-arid and dry sub-humid areas (Hulme and Kelly, 1993; Kelly and Hulme, 1993; Hulme *et al.*, 1992).

Drought, desiccation, climatic variability and other dryland climate characteristics greatly influence human use of drylands. These factors influence vegetation productivity, carrying capacity of the land, susceptibility of the land to erosion, surface water availability and aquifer recharge (Williams and Balling, 1994). As such they impose considerable constraints on the human use of drylands. They also tend to complicate the management of drylands and impose limitations on food security and the quality of life of people. For example, drought, when it occurs, creates a shortage of available moisture for plant growth and limits total rangeland productivity. Since the use of supplementary feeds is limited, lack of nutritious forage results in high rates of livestock mortality. As a result of livestock loss and the total impoverishment of nomadic families, dependence on famine relief food by a population which has traditionally been self-reliant has increased (Oba, 1992). Recovery of livestock herds during post-drought periods is low. Following the severe weakening of the traditional security system of reciprocity and opportunistic lifestyles, a majority of former pastoral families have become economically marginalised and live outside the pastoral system. According to UNSO (1992), the effects of desiccation on pastoral communities have been even more serious than droughts. In the Sahel, many peasant and pastoral communities have 'simply ceased to exist after the desiccation of the last 20 years'.

Water

Closely related to climate is the issue of water. All life forms on earth depend upon water and water issues pose many challenges to environmental security in Africa. These challenges relate to water availability, shortages and quality. Because of aridity, water is a scarce resource in arid and semi-arid lands and in many areas, it is not possible to obtain a ready source of sufficiently pure drinking water. Rapid population growth and urbanisation pose threats to water utilisation and depletion. Surface water that is used for food production and other domestic purposes is often the medium of pollution and transmission of diseases. Ground water is not easily accessible and technologies for underground water detection, acquisition and monitoring are often lacking or underdeveloped and generally have to be imported.

Because of its scarcity, water may be a source of strategic local and regional conflicts in the drylands of Africa. In Southern Africa, for example, the members of the Southern African Development Community (SADC) have responded to the issue of limited water resources in the region with a protocol on shared river basins like the Zambezi. However, this creates conflict in the region. As will be discussed elsewhere in this chapter, areas such as river basins and wetlands which are more naturally endowed with water have become focuses of land use conflicts in African drylands and sometimes even of international conflicts.

Population Growth

A matter closely related to the problem of environmental security in Africa is the scarcity of natural resources vis-à-vis human carrying capacity. For instance, Africa's population growth rate of 2.8 per cent p.a. is putting pressure on an already scarce supply of water and other natural resources. As *Our Common Future* (WCED, 1987: 1) reminds us: 'The issue is not just numbers but how those numbers relate to available resources'. The ever-growing demand for natural resources for human use or application against the background of an ever-growing population, not only intensifies competition over these resources, but has a number of other important consequences. First, there are problems created by the rapid depletion of non-renewable resources and associated with extracting more diffuse and inaccessible deposits. Second, the exploitation of renewable resources exceeds their natural ability to renew themselves. This threatens their availability. Third, the possibilities of expansion into remaining natural areas and resources are limited and could lead to a growing competition for natural resources.

Human Activities

Agriculture and pastoralism are the predominant human activities in the drylands of Africa. They have both had direct negative impacts on the biophysical environment. Soil erosion, soil and water pollution, water supply and species extinction have all become important environmental issues. While impacts were initially small scale and local, they increased in extent and severity as agricultural systems became more industrialised and less in tune with natural ecosystems, processes and cycles.

Modern agriculture of the type common in high income nations is characterised by high levels of investments in technology, machinery, chemicals and genetics, a very extensive scale of operation, and constant change (Aplin *et al.*, 1995). Indeed, change is so much a part of modern farming that it is difficult to describe a typical farming enterprise. Animal and human labour have been replaced by machines and fossil fuels; natural cycles of soil fertility have been replaced by the use of chemical fertilisers, monocultures have been replaced by complex polycultures; and weed and pest control are dominated by commercial chemicals. The geography of modern agriculture reflects climate and soil quality modified by human ingenuity and technology. Among the most intense application of modern science and technology are biotechnology, genetic engineering and information technology (Mannion, 1995). These innovations have been developed as scientific and technological response to management problems.

In Africa, on the other hand, attempts at agricultural transformation have concentrated mainly on the export sector, on which most African countries depend primarily for employment, income, foreign exchange, and government revenue, while the traditional sector has been largely neglected.

The technological content of African agriculture is among the lowest in the world, which means that agricultural production is largely dependent on the exploitation of the natural resource base without much replacement.

Land Degradation

Unprecedented population increase and the problem of feeding people with the use of technologies in farming that have shown little change and agricultural policies that are encouraging farmers to grow higher priced non-food cash crops, have led to excessive pressures in arable lands, rangelands, forests and other landed resources, leading to accelerated land degradation.

Land degradation is not just soil erosion; the term includes any change in the condition of the land which reduces its productive potential. In the drylands of Africa, the primary agents of land degradation are overcultivation, overgrazing, deforestation and poor irrigation practices (UNEP, 1992). Of the four main threats, three (deforestation, overcultivation and overgrazing) pose the principal threat to the livelihood of the peoples of the drylands.

In the woodland savannas and forests of Africa, rapid population growth has shortened the fallow periods, and the high demand for agricultural land has caused an expansion of cultivation into water catchment areas with serious ecological consequences as the soils of these lands degrade rapidly and become depleted of major plant nutrients. In a natural forest ecosystem, the nutrients released by plant litter are rapidly used by living plant tissues and become locked up in the biomass for most of the forest's life. When a forest canopy is cleared and burnt, the nutrients are released, but the soil fertility is short-lived. Loss of vegetation cover therefore accelerates soil loss. Furthermore, the removal of forest cover from a water catchment increases flood risks.

The present trend of over-exploitation and mismanagement of forest resources can be reversed if the remaining forestlands are protected and the forest cover replaced where possible. Although afforestation programmes have been attempted in several countries, the introduction of exotic species has not restored the ecological stability that existed when there was natural forest cover. Deforestation in Africa is proceeding 30 times faster than reforestation and in the early 1980s, an estimated 3.7 million hectares was being lost every year through farming activities and cutting down of trees to meet energy and other needs including lumbering by an increasing population. Partly as a result of tree cropping, wildlife resources are being depleted. Fuelwood crisis is manifesting itself in severe fuelwood deficits in the savanna regions of West, Central and East Africa and in the arid areas of the Sahel, South Eastern and South Western Africa.

Most of the fragile arid and semi-arid lands do not have the capability to support large-scale arable farming for non-food cash crops. Their real potential lies in livestock production and wildlife management. However, with rapid population growth and use of increasing areas of land for cash cropping rather than subsistence agriculture, rangelands in eastern and southern Africa are fast disappearing.

Many large-scale mechanised schemes, irrigation projects, resettlement schemes and big dam projects have taken away lands used by pastoralists during droughts to alleviate pressure on the fragile drylands environment. The food security policies of African nations which see the key to agricultural development as the expansion of large-scale projects rather than in improvement of the production conditions for nomadic animal husbandry and smallholder farmers are contributing to the insecurity of herders and small farming communities (Darkoh, 1994). The rights or needs of the indigenous populations and of small-scale producers and mobile livestock keepers in African drylands and river basins are often not given sufficient attention in development planning, a fact which often aggravates conflicts.

Because of their role as key production areas in drylands, African river basins and wetlands have become seedbeds of land use conflicts. As elsewhere, conflicts arise between the various production sectors or stakeholders, of which the major ones are agriculture, livestock, wildlife and urban development. Other conflicts are caused by the polarised population interests (e.g. pastoralists versus peasants) over river basin and wetland resources. In broader terms, the land use conflicts reflect the lack of land reforms and co-ordinated national land use policies for drylands. Each sector views the key production areas as optimal for their particular activity. What ensues is intense competition for the same land. Without a mechanism to prioritise competing demands, the resultant land use is not necessarily the most appropriate in ecological terms. For example, in many countries, large chunks of riverine forests and other wetlands used normally by herders in the drylands as fall-back areas for dry-season grazing have been designated as National Parks or Forest Reserves, which has marginalised the weakest sector, namely, pastoralism.

As a result of this endemic marginalisation, the pastoralists have had little alternative but to utilise more intensively and at times over-exploit the range and water resources available to them. The decrease in available pasturage, with simultaneous increase in livestock, results in overgrazing. This in turn causes the destruction of vegetation, soil erosion, desertification and so on, and thus the destabilisation of the ecological balance in the river valleys and their adjoining upland areas.

Commercialisation of the pastoral economy and land resources and sedentarisation policies of African countries have also increased pressure on natural resources in many areas. The main causes are the reduced distances which livestock now roam and the heavy concentration of men and animals around new high capacity watering places. Also, sedentarisation or villagisation without improved lifestyles and management practices have done little to prevent land degradation in the region.

Africa has 57 out of the 200 major river basins in the world. Twelve of these river basins are shared by four or more countries. African river basins have an international character and as such many countries cannot embark on large-scale river basin projects without affecting the interest of other countries. The construction of dams for irrigation and hydropower upstream usually impedes similar developments downstream. Some cases have shown that such projects may precipitate interstate conflict as has happened between the Niger and Nigeria over the two dams on the Lamido and river Maggiya, and between Nigeria and Cameroon over the Ladoo Dam on the Benue river. Because of their international character and the polarised national interests over river basin resources, river basins are good examples of areas in African drylands where national political and economic interests can transcend ecological considerations to trigger major social and political disasters. Witness, for example, the conflict between Senegal and Mauritania in the 1980s over the utilisation of resources in the Senegal basin, and the potential conflict between Botswana and Namibia over the Okavango Swamps.

Indeed, natural resources and environmental issues underlie most forms of conflicts in the drylands of Africa. Military conflicts between states and between groups within a state frequently occur over access to and control of natural resources. Military conflicts do also frequently occur within and between states when environmental degradation and natural resource scarcities undermine social, political and economic viability of societies and lead to political instability, a breakdown of law and order and increased flows of refugees. The increased flows of refugees across international borders have further aggravated the political and environmental security situation in several African countries (see Chapter 7 for further information on refugees, security and environmental impact.)

Improper irrigation practices are also turning large areas of precious irrigated farmlands into wet salty wasteland, causing declining yields and often loss of productivity. The problem is usually inadequate drainage. Where excessive volumes of water are used, groundwater levels rise, creating the twin problems of waterlogging and salinisation. Salinisation and alkalinisation are serious problems in some irrigated areas in North Africa, the Sahel, Eastern and Southern Africa.

All causes of land degradation discussed so far lead to soil erosion of one form or other and desertification, two other major environmental problems in the drylands. The rates of soil erosion in the drylands of Africa are amongst the highest in the world. The continent's precious topsoils are being lost at incredible rates. Data provided by GLASOD (UNEP, 1992) shows that some 332.3 million hectares or 27.6 per cent of the susceptible drylands in Africa, are already affected by soil degradation at different degrees. FAO studies indicate that either erosion or salinity affects more than 35 per cent of land north of the Equator. Zimbabwe alone could be losing, on average, 1.6 million tonnes of nitrogen and 240,000 tonnes of phosphorus each year through erosion. FAO's findings show that sediment in many African rivers is increasing roughly 1.5 times as fast as the population in the catchment areas. Thus, sediment levels are increasing at approximately 5 per cent in Nigeria, Tanzania and Zimbabwe, 4.3 per cent in the Sudan and 3.2 per cent in Tunisia. (FAO, 1990: 10).

Desertification, defined simply as dryland degradation (see Chapter 10), is affecting millions of hectares of arable land and grazing land as pressure from humans and animals increases, thereby removing the vegetative cover through farming, overgrazing and bush fires as the net shift of land area towards aridity in the ASAL environments continues to gain momentum. The exposure of soil to wind erosion leads to an extension of desert-like areas outside the climatic desert.

Urbanisation and Environmental Security

The growth of human settlements in Africa today is marked by an unprecedented urban explosion which is reversing the rural trends which had characterised the continent until recently and is compounding the problems of land degradation and environmental security. There are certain common characteristics in the general geography of human settlements and pattern of urbanisation. At the macro-level, there tends to be an imbalance in the relationship between people, resources, environment and development which has led to a geography of disparities, including sharp disparities between the urban and the rural areas and disparities between the towns and the villages which are reflected in the sharp drop in environmental quality and services (Saha, 1992). Over-concentration of wealth in the big cities is contributing to the abandonment of the rural areas and the small towns. Attempts to redress the situation are beset with difficulties both in view of the low level of investment that has been available to inject into the rural areas to raise them from their existing level of deprivation or resource degradation, and due to such structural issues as institutional capacity and absence of meaningful intersectoral linkages. This has led to depletion of human resources and inability to maintain infrastructure and service provision.

In most African human settlements, large masses of people live in environments that are neither life-sustaining nor conducive to positive development trends. Areas are poorly serviced and the inhabitants are susceptible to both man-made and natural hazards. These environmental hazards can be identified at different spatial scales of the settlement hierarchy.

At the very local level, the neighbourhood or village, millions of Africans are exposed to pathogens, with serious impacts on human health; e.g. pathogens from human excreta in water supplies, and smoke fumes from stoves which cause respiratory problems. There is also the problem of unsafe housing sites subject to flooding and exposed to other risks of natural disasters. These problems are compounded at the next level, the small or medium town, where there are additional risks from toxic chemicals used in the workplace without adequate safeguards; household, and on occasion, industrial, solid wastes which are dumped around houses, attracting disease-causing agents like rats and flies, that feed on the uncontrolled garbage dumps; and dirty water pools which occur on account of inadequate sewage disposal facilities and drainage.

At the third level, the city or conurbation, problems centre on inadequate and overcrowded housing; unplanned peri-urban development and proliferation of squatters and shanties; high levels of air pollution caused by poorly maintained vehicles on congested streets and high levels of lead additive in the petrol used; high levels of water pollution on account of inadequate and strained systems, absence of drains and sewage treatment plants, inappropriate dumping practices and lack of enforcement, and inadequate environmental control on industry (HABITAT, 1990; Saha, 1992).

In the context of the potential impacts of industrialisation, it is worth noting that the contribution of cities in the developing African countries to global environmental problems remains relatively small - it is mainly the cities of the Developed World that are emitting greenhouse gases. Thus, the main threat from industry located in African cities is to the local population. Even so, African cities are threatened by global warming which may lead to more severe storms as well as rising sea levels.

Finally, at the sub-national level, problems arise from the interaction between cities and their rural hinterlands. Urban-based decisions have a powerful influence on rural resource exploitation and rural areas suffer from the impact of city-based activities or city-generated wastes. Among the most common examples are the degradation of coastal and estuarine fisheries as a result of water pollution from city-based enterprises; and air pollution from city-based industries, damaging vegetation through acid rain and disrupting natural systems. Other resource degradation would occur from over-intensified demand, say, for fuelwood; and the dumping of solid wastes from city enterprises; from poorly prepared and maintained landfill sites, which often contaminate water used by rural households because of the high incidence of run-off and seepage during periods of heavy rain.

Some Way Forward: Science and Technology

Most of the rural and urban environmental problems outlined above are linked to the failure or inability of African countries to provide appropriate scientific and technical knowledge, effective regulations and planning frameworks to address the problems of the poor citizens and protect the rural and urban environments from contamination or degradation. Rapid population increase, urban explosion, recurrent droughts, increasing scarcity of land and increasing pressure of exploitation of the limited resources of soil, water and vegetation are some of the major factors behind the repeated famines and food shortages occurring in the continent. The cures for the problems of land degradation, low biological productivity and food security in the drylands of Africa can generally be formulated as planned and improved land use on a sound ecological basis (Rapp and Hellden, 1979). This is true both for small-scale traditional farming or herding as well as for large-scale mechanised projects.

Understanding and Knowledge

To be successful, planning for better land use in African drylands must be based on a sound knowledge of both biophysical and social sciences; neither the biophysical nor the social sciences alone will suffice, but nor will a reliance on technology while neglecting basic science. We desperately need greater depth and breadth in our collective knowledge of dryland environments. We need, above all, to understand the processes operating within the dryland ecosystems and the complex web of interconnections between these component parts. Much of the background data for planning a better land use can be provided by natural and social scientists.

Combating land degradation requires scientific knowledge of available natural resources, their conditions and ecologically viable management methods. This, in turn, requires inventories of natural as well as human resources, appropriate counter measures (cures) and environmental monitoring (repeated, comparative surveys or registrations). The repeated famines and food shortages show that traditional knowledge of land use in drylands, though highly useful (Darkoh, 1996b), is not sufficient to meet the challenge and changing situations of pressure from growing populations and annually fluctuating rainfall. Research is needed on improved land use and alternative land management in drylands.

We need to gain so much knowledge so rapidly that we cannot afford to work in competition and replicate the work of others unnecessarily, though that is not to say that a high degree of specialisation is the answer. African developing countries are marginal producers of new scientific knowledge and technology. Their share of the world research and development is less than 0.5 per cent and the research work being done does not produce much scientific and technological innovation (ECA, 1995). Indeed, the research and development effort in Africa has made little contribution to technological and economic development during the last 30 years and it is not expected to make much contribution either in the years ahead (*ibid.*; Jugessur and Hammel, 1995). Under these circumstances, what is needed is international co-operation in research and the easy and unselfish exchange of ideas, data banks and other information. This has become increasingly easy with the advent of regional and international computer-based networks.

Technology

There is no universally accepted definition of technology or technology transfer. Technology can be defined simply as practical knowledge applied to satisfy human needs and technological transfer can be defined as the process of assimilation and exploitation of technological resources (ECA, 1995). Technology has a crucial role to play in any attempt at finding solutions, but we should note that technology by itself will not fix anything. Technology is a social construct responding to social, cultural, political and economic demands and priorities. These factors determine not only whether technology is used positively or negatively, but which forms of technology are developed in the first place. Technology, then, must be seen in a social, economic and political context.

There are at least five separate but closely related ways in which changes in technology can assist in avoiding environmental crises in drylands (Aplin *et al.*, 1995): by reducing our dependence on diminishing resources; by helping develop more effective ways of managing resources such as soil and water; by avoiding or reducing pollution of the environment; by giving us better data on our activities, the environment, and the interactions between the two; and by providing more powerful means of analysing that data. In the long run, the last two of these may turn out to be the most important, as one very real barrier to managing environmental issues in Africa has been inadequate data and the inability to analyse it, build models and develop management plans.

Many types of instrumentation will help to increase the amount of information available. Satellites provide the capacity to record enormous amounts of information for every point on the Earth's surface. Information can also be updated at very frequent intervals, providing a unique opportunity to analyse processes and to heed early warning signals of impending problems. However, we have to be extremely careful in interpreting the data. Pretty maps on a screen are one thing; making practical use of them is another thing altogether. We still need to supplement information gathered from satellite data with information collected from the ground or ocean surface. Environmental monitoring should utilise information from three levels: ground, air and space. Ground checks of sample areas should be combined with analysis of air photographs and satellite images of Landsat or other available types.

Geographic Information Systems (GIS) is a young science which is related to remote sensing, surveying, geodesy and photogrammetry, all of which deal with spatial data handling. GIS is basically a tool for urban and regional research, policy analysis, policy simulation and planning. In addition to GIS, which holds a variety of applications for planning arid land development, two other Spatial Information Systems need to be mentioned: Land Use Information System (LIS) and Computer-aided Design (CAD). LIS is primarily a tool for the legal, administrative and economic management of land resources. It consists of two main components: a database containing spatially referenced, land-related data and a suite of procedures and techniques for systematic collection, updating and querying of data. The differences between GIS and LIS from the hardware and software perspective are the fact that LIS are used primarily for storage and retrieval of spatial data while GIS are used essentially for more complex spatial analysis (Scholten and Vlugt, 1990). CAD refers to graphical systems which support the work of architectural or industrial designers. It provides facilities to perform calculations and to produce technical drawings and three-dimensional displays of designs.

In order to utilise and share the existing knowledge on drylands using some of the powerful and sophisticated computer-based tools such as those described above, we recommend that the high-income countries (the USA, the countries of the European Union, Australia and so on) should support research and development of methods in environmental monitoring and control, particularly, of desertification in the drylands of Africa. They should identify and work closely with local African research institutes, universities and other research organisations.

Transfer of Information and Technology

Sound technological transfer and acquisition must play a bigger role in African development than they have up to now and must receive greater attention from policy makers than hitherto. In particular, human, institutional and legal capacity in technology transfer must be enhanced.

Chapter 34 of *Agenda 21* deals at length with the international transfer of basic data, scientific knowledge and environmentally sound technology. It states that the basic goal is to place all nations in a position where they can make informed choices as to which technology is most appropriate in their particular circumstances. This will then strengthen each nation's own technological capabilities, especially in the case of low-income nations. Both information and technology are vital if global environmental catastrophe is to be avoided. As Aplin *et al.,* (1995) have noted, both information and technology are vital if global environmental catastrophe is to be avoided. However, information and technology transfer from high-income to low-income nations raises questions on what terms such transfers will occur and the kind of information and technology that should be transferred.

Transfers of technology and information frequently involve questions of ownership, patent rights and commercial advantage. Much of the technology needed by developing countries or low-income nations to ameliorate or overcome serious environmental problems is too expensive for them to buy at market rates. Some is certainly in the public domain and therefore freely or relatively cheaply available, but what of emerging 'state-of-the-art' technology being developed commercially. Their owners have definite rights to profit from its sale, so ways must be found to subsidise costs to poorer nations. There is an obvious role for aid in this area, as both parties can benefit; the low-income nations obtain subsidised access to desired technology, and the commercial developers are appropriately rewarded.

Equally important is the question of the appropriateness of technology transferred to low-income countries. The high-income world is not the citadel of all technological wisdom, nor are excellent technological solutions to environmental problems in the high-income nations necessarily appropriate for similar problems in poorer countries. The belief that they are, could only be maintained if one believed that all other circumstances were also the same, including levels of education, technological expertise, ability to pay, social and cultural traditions, and so on. But the context clearly differs. High-income nations and their transnational corporations may well try to impose their solutions because they see profit in doing so, but such attempts must be resisted and more appropriate local solutions sought wherever possible.

Agenda 21 stresses the need for co-operative research and the sharing of information on an international scale. Again, national and commercial interests can be at odds with the need to assist poorer nations. But it is also essential that as much relevant information as possible, is available so that each nation can make its own informed decisions on environmental issues.

The Challenges

There are many challenges facing African countries in acquiring and gaining access to appropriate and cost-effective information systems and technologies in support of the proper management of physical environmental resources. Some of the specific areas posing the greatest challenge for scientific and technical co-operation in arid lands development in the developing countries of Africa are:

- climate change and global warming in arid and semi-arid lands;
- synoptic meteorological observations of dustorms and other conditions of high dustiness;
- changes occuring in chemical cycling during a disturbance of arid ecosystems;
- physical and social indicators of desertification and land degradation;
- drylands management systems, avoiding the risk of desertification;
- land use conflicts in river basins and wetlands;
- land use changes, sand dune fixation and sediment control;
- conservation of biodiversity;
- coastal zones and wetlands management systems for enhanced long-term aggregate economic productivity;
- revegetation of eroded and saline soils in rain-fed and irrigated areas;
- dynamics of the contamination of soils and underground

aquifers by agro-chemicals;
- waste reduction and pollution abatement;
- the use of solid and liquid urban wastes in soil improvement in peri-urban agriculture;
- integrated water management, focusing on water treatment and recycling, water conservation, particularly in agriculture and industry, water pollution risk assessment and prevention, improved water supply systems enhancement, and water harvesting technologies;
- ground water monitoring;
- management systems of mountain environments;
- improved food systems in marginal lands and fragile environments;
- biotechnology and genetic engineering for agriculture, medicine and pollution control;
- irrigation and water resource potential;
- conservation and technologies targeted at reduction of post-harvest losses or quality deterioration;
- technologies for reducing post harvest damages and spoilage;
- improved rangeland management for multipurpose livestock and wildlife production;
- endangered species and habitat monitoring;
- rangeland rehabilitation; and
- animal health systems and resistance to infectious diseases; and
- animal feeding strategies based on the use of food crops products and by-products, specialised forages or pastures; and aquaculture.

The European Union has recently identified areas of scientific and technological co-operation in developing countries, which include several of the areas outlined above (*Official Journal of European Communities*, No. C117/29, 1997).

Conclusion

Finding solutions to environmental problems in the drylands of Africa involves more than simply gathering facts but understanding scientific issues of particular problems and finding types of actions to be taken. This calls for research and technology development and transfer. Co-operation with developed countries and international organisations would appear to us to be a foremost priority if success is to be achieved. Also, solutions of our environmental problems have much to do with our systems of values and issues of social justice (Botkin and Keller, 1995). To solve our environmental problems, we must understand what are our values and which potential solutions are socially just. Then, we can apply scientific knowledge about specific problems and environmentally sound technology to find acceptable solutions.

Acknowledgement

This chapter was first presented as a paper at the International Workshop on Science and Technology for Drylands Security, Albuquerque, New Mexico, USA, July 8-11, 1997. I am indebted to the sponsors, and especially, to W. Franklin G. Cardy of UNEP and Sheila Nelson of Sandia National Laboratories, Albuquerque for their helpful comments on the original draft. This article has also appeared in the *Desertification Control Bulletin*, 34, 1999, pp. 6-14 and it has been modified and reproduced here with the permission of UNEP.

References

Aplin G., P. Mitchell, H. Clough, A. Pitman and D. Rich, 1995. *Global Environmental Crises: An Australian Perspective*, Oxford University Press, Melbourne and New York.

Botkin, D.B. and E.A. Keller, 1995. *Environmental Science*, John Wiley and Sons, New York.

Buzan, B., 1983. *People, States and Fear: The National Security Problem of International Relations*, University of North Carolina Press, Chapel Hill.

Darkoh M.B.K. 1994. 'The deterioration of the environment in Africa's drylands and river basins', *Desertification Control Bulletin*, 24, 35-41.

Darkoh, M.B.K. 1996a. 'Desertification: its human costs', *Forum for Applied Research and Public Policy*, 11(3), 12-17.

Darkoh, M.B.K. 1996b. 'Towards an adaptive and community-based approach to the management of natural resources in the drylands of Sub-Saharan Africa', in A. Hjort af Ornäs (Ed), *Approaching Nature from Local Communities, Security Perceived and Achieved*, EPOS, Research Programme on Environmental Policy and Society, Institute of Tema Research, Linkoping University, Sweden.

Darkoh, M.B.K. 1998, 'The nature, causes and consequences of desertification in the drylands of Africa', *Land Degradation and Development*, 9, 1-20.

ECA, 1995. *Technology Transfer in Africa*, Economic Commission for Africa, NRD/STS/PUB/2/95, Addis Ababa.

FAO, 1990. *The Conservation and Rehabilitation of African Lands*, FAO, Rome.

Frankie R.W. and B. Chaisen, 1980. *Seeds of Famine: Ecological Destruction and the Development Dilemma in West Africa Sahel*, Allanheld Osmun, Monclair, New Jersey.

Hjort af Ornäs, A. 1982. 'A critique of ecological models for pastoral land use', *Ethnos*, 3, 4.

Hjort af Ornäs, A. 1989. 'Environmental security of dryland herders in eastern Africa', in A. Hjort af Ornäs and M.A. Mohamed Salih (Eds), *Ecology and Politics: Environmental Stress and Security in Africa*, Scandinavian Institute of African Studies, Uppsala.

HABITAT, 1990. *People, Settlements, Environments and Development*, United Nations Centre for Human Settlements, Nairobi.

Hulme, M. and M. Kelly, (1993) 'Exploring the links between desertification and climate Change', *Environment*, 35, 6, 1-45.

Hulme, M., R. Marsch, and P. D. Jones, 1992 'Global changes in humidity index between 1931-60 and 1961-90', *Climate Research*, 2, 1-22.

Janzen, J. 1991. 'Mobile livestock keeping - a survival strategy for the countries of the Sahel, the case of Somalia', *Applied Geography and Development*, 38, 53-65.

Jugessur, S. and J.L. Hammel, 1995. *The Science and Technology Protocol of the African Economic Community*, Economic Commission for Africa, ECA/NRD/STS2 (A), Addis Ababa.

Kelly, M. and M. Hulme, 1993. 'Desertification and climate change', *Tiempo*, 8, 1-7.

Mannion, A.M. 1995. *Agriculture and Environmental Change*, John Wiley and Sons, Chichester.

Mohamed Salih, M. A .1992. 'Environmental and social insecurity in the drylands of the Sudan', in A. Hjort af Ornäs (Ed), *Security in African Drylands*, Scandinavian Institute of African Studies, Uppsala.

Oba, G. 1992. 'Sedentarisation of pastoralists: right or wrong?', *Resources*, 31, 1, 15-23.

Rapp, A. and U. Hellden, 1979. *Research on Environmental Monitoring Methods for Land Use Planning in African Drylands*, Department of Physical Geography, University of Lund, Sweden.

Redclift, M. 1984. *Development and the Environmental Crisis: Red or Green Alternative*. Methuen, London and New York.

Ringrose, S., C. Vanderpost, R. M. Kwerepe, and M.I. Mulalu, 1997. Comparison of range degradation in Botswana using satellite imagery, Unpublished, ODA, commissioned report, University of Botswana, Gaborone.

Saha, V.L. 1992. *Towards Sustainable Development of Human Settlements in Africa*, Consultancy prepared for United Nations Economic Commission for Africa, Background Paper, Addis Ababa.

Scholten, H. and M. van der Vlugt, 1990. 'A review of geographic information systems applications in Europe', in L. Worrall (Ed) *Geographic Information Systems, Developments and Applications*, Belhaven Press, New York.

Spooner P. and H.S Mann, 1982. *Desertification and Development in Dry Ecology in Social Perspective*, Academic Press, New York.

UNEP, 1992. *Status of Desertification and Implementation of the United Nations Plan of Action to Combat Desertification*, United Nations Environment Programme, Nairobi.

UNSO, 1992. *Assessment of Desertification and Drought in the Sudano-Sahelian Region*, United Nations Sudano-Sahelian Office, New York.

Williams A.J. and R.C. Balling Jr., 1994. *Interactions of Desertification and Climate*, WMO and UNEP, Nairobi and Geneva.

World Commission on Environment and Development, 1987. *Our Common Future*, Oxford University Press, New York.

Chapter 5

Environmental Security in South Africa: A Prognosis for Regional Security

J. K. Van Wyk

Introduction

African states are institutionally weak and vulnerable to internal and specifically non-military pressures such as the environment. States in Southern Africa are no exception in this regard. Environmental degradation in Southern Africa has been taking place for decades, if not centuries. As elsewhere on the continent, environmental degradation and the competition for natural resources are prompting massive migrations of people and putting new pressures on state boundaries. These forces interlock with political and economic centrifugal forces on these states. The threats to state institutions are specific and measurable. These are, amongst others, water shortages, deforestation and the relentless advance of the Kalahari desert, soil erosion, the scarcity of agricultural soil, and disease (*Africa Confidential*, 6 January 1995: 1-4).

Environmental resource scarcities contribute to social impacts such as population movements, economic decline, and the weakening of states. These impacts contribute significantly to violent conflicts in many developing societies. The rate and extent of such conflicts increase as natural resource scarcities worsen, and these are more likely to occur in areas where communal access to natural resources is more critical to survival (Gleick, 1993: 83). Scarcity sharply increases the demands on key institutions within a state, while it concurrently reduces their capacities to meet those demands. The result is often large population flows and the inability to comply with international agreements on collective security and global environmental protection when a country is under such stress. It is suggested that countries facing chronic internal conflict because of environmental stress will either fragment or become more authoritarian. Recent literature on national security emphasises the impact of environmental degradation and resource scarcity on the viability and stability of democracy in countries that are in transition from authoritarian rule.

The end of the Cold War has contributed significantly to the imperative to redefine approaches to national, regional and international security to take cognisance of environmental issues. This trend relates directly to the relationship between environmental issues and traditional security concerns, such as their potential to add to stresses that make societies ungovernable. In addition, environmental issues contribute to 'traditional' sources of conflict[1] and instability by their increasing importance in factors underlying border conflicts, populations pressures, transboundary migration, immigration, access to strategic natural resources, inter-communal conflict and cultural and ethnic clashes (Rodal, 1994: 1). This is illustrated in the studies by, amongst others, Durham (1979) on the 1969 'Soccer War' between El Salvador and Honduras; Homer-Dixon (1991, 1994) on environmental conflicts in the Jordan and Nile river basins, Bangladesh, Nicaragua, the Philippines and China, and Gleick (1993) on water-related conflicts in the Middle East.

Baldwin (1995: 118) suggests three emerging themes for post-Cold War security studies. These are a decline in the importance of military power, a re-examination of the links between international relations and national security and, thirdly, a broader view of national security which makes provision for the inclusion of non-military threats. Against this background, this chapter seeks to locate environmental issues in South Africa's post-apartheid, long-term security policy concerns. This chapter is organised as follows. First, a brief overview of the state of the debate to include the environment in security policy is presented. The second part of the paper presents a survey of environmental threats to security in South Africa and how policymakers respond to these issues. Lastly, the prospects for regional security are addressed.

The Environment and Security

The end of the Cold War era called for a dramatic review of the concept of security. This led to a plethora of alternative concepts of security. A critical overview of this debate is beyond the scope of this chapter.[2] Traditionally, however, security was defined in terms of 'coping with potential threats to and the establishment of conditions conducive to the promotion of core values cherished by individuals or communities' (Stern, 1995: 213-214). These 'core values' relate to survival, autonomy, prosperity and health. Yet, for the purpose of this contribution, the definition of, amongst others, Ullman (1983:133) is particularly helpful as it takes cognisance of the social aspect of security:

> A threat to national security is an action or a sequence of events that (1) threatens drastically and over a relatively brief period of time to degrade the quality of life of the inhabitants of a state, or (2) threatens significantly to narrow the range of options available to a state or to private, non-governmental entities (persons, groups, corporations) within a state.

Ullman (1983: 134) continues and specifically identifies 'a drastic deterioration of environmental quality caused by sources from either within or outside a territorial state' as falling within the scope of the above-mentioned definition. Also writing in 1983, Buzan also suggested an inclusion of environmental aspects in the reconceptualisation of national security. He writes (1983: 82) that threats to national security might appear in ecological forms 'in the sense that environmental events, like military and economic ones, can damage the physical base of the state, perhaps to a sufficient extent to threaten its idea and institution.' Therefore, environmental problems could be seen as security threats. Not only do they threaten the physical foundation of the state and its population, but also the territorial, natural and man-made wealth contained within its area (Buzan, 1983: 62).

The environment is understood as including *all* natural resources. Tennberg (1995: 144-245) divides natural resources into living (such as agricultural crops, wood and fish) and non-living resources (such as land, water, minerals and fossil fuels). Furthermore, the environment consists of different kinds of resources which can be divided into two subclasses. These are:

- Non-extractive resources (such as land, soil, water and

atmosphere) and
- Extractive resources (renewable such as certain minerals and non-renewable resources such as a mineral such as gold).

The geographic sources from which natural resources are obtained can be divided into two categories. These are:

- National territories (such as territorial areas, seas and possessions) and
- Extra-national territories (such as oceans beyond national jurisdiction and the seabed. It is also possible for a natural resource to extend into two or more national territories and extra-national territories.)

On the question of what constitutes an environmental issue, the definition of Levy (1995:38) is particularly helpful:

> An issue is an environmental issue if it involves physical or biological systems to a significant extent.

The question now arises: When does the environment become a security issue? According to Levy (1995: 43) there must be some 'demonstrable connection to some vital national interest.' To elucidate his point, he mentions the cases of ozone depletion and migration. For ozone depletion to be an environmental threat it needs to impact on public health; and for migration to be a threat there needs to be an impact on regional political stability and environmental quality. Hence, an environmental issue becomes a security issue if:

- Change in the environment is extensive, rapid and sustained over a period of time.
- There is a high degree of compatibility between attainment of ecological and other important issues.
- The real and monetary costs of attaining an acceptable trade off are increasing.
- This change in the environment interacts negatively with other structural or political weaknesses within a state that leads to social turbulence (Maddock, 1995: 25).

The conflict/insecurity-environment nexus in developing

countries has been studied by Homer-Dixon (1991a, 1991b, 1994). He addresses large-scale human-induced environmental issues with long-term irreversible consequences (like climatic change, deforestation and the degradation of agricultural soil). According to him, the total effect of human activity on the environment in a region is the function of the product of the total population in that region and the physical activity per capita (Homer-Dixon, 1991: 85). A second variable is the vulnerability of the ecosystem in this specific region to these physical activities. Activity per capita derives from a combination of the availability of physical resources (which includes non-renewable resources) and societal factors (such as policies, institutions, relations and beliefs).

He also addresses the connection between the environment and social effects (Homer-Dixon, 1991: 90-98) showing that environmental degradation leads to production, economic decline, population displacement and the disruption of institutions within a state. A number of intervening factors occur in the way environmental issues affect human relations. Among these are patterns of land and wealth distribution, family and communication structures, economic and legal incentives, patterns of trade, the strength of institutions within a state and beliefs in government, industry and the cost of environmental degradation. These factors influence the manner in which environmental effects of human activity will be transformed into social effects and conflicts.

Another issue addressed by Homer-Dixon (1991a and 1991b) relates to the types of conflict and eventually insecurity that are most likely to result from environmental degradation. These are simple scarcity conflicts, group identity conflicts and relative deprivation conflicts. Simple scarcity conflicts are expected when actors are in a zero sum or negative sum situation which could arise from resource scarcity. Three types of resources – river water, fish and agriculturally productive land – are likely to result in simple scarcity conflicts.

Group identity conflicts arise from, amongst others, large-scale movements of populations caused by environmental change. Such conflict is expected to occur under circumstances of deprivation and stress. The objective of the conflict is the protection and re-enforcement of group identity (Homer-Dixon, 1991a and 1991b). Relative deprivation conflicts can be expected when widening gaps develop between groups' perception on what they actually should achieve and what they actually experience. As in the case with group identity conflicts, the scope of the conflict can be either domestic or international.

Environmental scarcity, that is the combination of resource base, population development and consumption patterns, has cumulative social impacts. Among these are population movement, economic decline and the weakening of state institutions. Environmental resource scarcity can act as a long-term stressor, but environmental scarcity per se could be an important cause of change in politics and economics.

Traditional security concerns in Southern Africa ignored the impact of environmental resource scarcity (Van Aardt, 1993). Although environmentally linked threats are not likely to dominate the security agenda in South Africa and the region (Nathan, 1994) – already environmentally degraded to an extent – Southern African security policy-makers cannot afford not to take notice of this new emerging security issue. The next section addresses the question of environmental issues that pose a threat to security in South and Southern Africa.

Southern African Environmental Issues and Security

Environmentally induced scarcity in South Africa is a combination of three factors. These are:

- Resource scarcity is no longer an issue of a single natural resource, but of multiple resource scarcity,
- These shortages may grow rapidly to crisis proportions (Tennberg, 1995: 246) and
- The commitment to specific resource consumption patterns make it difficult to address the crisis (Homer-Dixon, 1991a: 30-33).

It is clearly not possible in a chapter with limited scope to address all the environmental issues and their 'demonstrable connection with some vital national interest' (Levy, 1995: 43), so a judicious selection is required. Among the environmental issues that pose a threat to security are ozone depletion, global climate change, water availability and erosion (Table 5.1). A number of social problems are related to these environmental issues. They are population growth and migration. The question of boundaries also has an impact on the environmental and social problems associated with it. This chapter will only limit itself to the following interrelated issues within the South and Southern African context:

- Global climatic change,
- Water,
- Soil erosion, and
- Boundaries.

Table 5.1: South African Environment Security Risks

Environmental Threats	Impact
Ozone depletion	Public health
Climatic change • Air pollution	Economic disruption
Population growth	Increased stress on natural resources
Water • Availability • Quality • Distribution • Also marine life	Scarcity conflicts
Erosion • Afforestation • Deforestation • Desertification	Scarcity conflicts
Migration • Environmental refugees	Regional political stability
Boundaries	Regional political stability

Source: Levy (1995: 48).

Climatic Change

The South African economy is not only extremely energy-intensive, but also 'carbon'-intensive. This is evident in the following statistics:

- It produces 4.8 per cent of the world's coal.

- Up to 90 per cent of all South African electricity is generated by coal fired power stations.
- South Africa emits 350 million tonnes of carbon dioxide and
- In sum, it contributes 1.2 per cent to the global change phenomenon (Rowlands, 1996: 165).

South Africa does not escape the impact of global climatic change. In their assessment on the impact of global climatic change on South Africa, Preston-Whyte and Holden (1994: 175) apply atmospheric scales that range from the meso and macroscale through to the global scale. On the mesoscale, their analysis focuses on the impact climatic change has on hailstorms. Although the occurrence of hail is often limited to a specific area its impact on agricultural production is in most cases severe.

On the macroscale, droughts and air pollution impacts on the environment. Preston-Whyte and Holden (1994: 188) describe droughts as 'acts of neglect'. The absence of predictability of the climatic impact contributes to the fact that political decision-makers respond with great difficulty to early drought warning. These occurrences contribute to global scale impacts. Globally, the effect is droughts and below average rainfall. This impacts severely on agricultural and food production (Preston-Whyte and Holden, 1994: 181) and increases the potential for conflict in Southern African societies.

Water as a Security Issue

The complexity of the environment-security debate is well illustrated in the issue of water. Water is likely to be a source of strategic rivalry depending on the degree of scarcity, the extent to which water supply is shared by more than one state or region, the relative power relationship of the sharing states, and the accessibility to alternative fresh water resources (Gleick, 1993: 84-88). This natural resource has often been described as a *casus belli* – particularly in the Middle East (Beschorner, 1992 and Gleick, 1993). Water is one of the key elements in the path to national prosperity as it relates to irrigation and food production, a form of cheap energy, and one of the important structural controls in flood and drought control schemes (Hassan, 1991: 37). Furthermore, as water flows from one area to another, one country or region's access can be affected by another's actions.

Water issues pose three challenges to security policy-makers in

South and Southern Africa. These relate to water shortage and its availability and quality. Rapid population growth and urbanisation also pose threats. Threats to water quality are uncontrolled urbanisation, urbanisation and rapid population growth (Coetzee, 1994: 142). Southern Africa is vulnerable to endemic droughts as a result of the El Nino/Southern Oscillation phenomenon that causes dry conditions in the region (Preston-Whyte and Holden, 1994: 176).

The influence of the past policy of apartheid is also present in the access to and distribution of natural resources.[3] This is particularly evident in the case of water supply in areas of South Africa: individual white suburban residents have daily access to 200 to 350 litres of water. In black townships more often up to 40 000 share a single tap. The Eastern Cape rural areas are the most hard hit with a daily use limited to 9 litres per day. In these areas, there exists a relationship between water, poverty and wealth. Water that is important for food production is often the carrier of disease and the transporter of pollution (Coetzee, 1994: 144-145).

To give an indication of the importance of water and the problems related to it, Falkenmark's (1991) study is particularly helpful. To quantify the present water demand and relate it to the pressure placed on it by, amongst others, increasing population, Falkenmark rated water scarcity on a scale of 1– 5, using the number of persons per million cubic metres of water per day. According to this model (Table 5.2), a country can survive with 1 million cubic metres per 2000 people. Falkenmark refers to this as the 'water barrier'. However, in terms of this scale, projections for South and Southern Africa are not too optimistic. By the year 2000, South Africa, Zimbabwe and Malawi will experience scarcity to absolute scarcity. The projection for 2025 is far bleaker: absolute water scarcity will be experienced by South Africa, Zimbabwe, Malawi and Lesotho. Mozambique will experience water stress (Coetzee, 1994: 144).

Table 5.2: Levels of Water Scarcity from Increasing Pressures on Availability of Water Resources

Persons per million cubic metre	Level of scarcity	Problems
< 100	1	Water quality
100-600	2	
600-1000	3	Water stress
1000-2000	4	Absolute water scarcity
>2000	5	

Source: Falkenmark (1991: 81-94).

Recent forecasts made by the United Nations Intergovernmental Panel on Climatic Change predicted that the overall trend for the region will continue to become drier and hotter. This will undoubtedly result in problems related to agricultural food production. It will also hinder the already limited industrial and urban development in countries such as Zimbabwe, Namibia, Botswana and South Africa (*The Times*, 18 October 1995).

The politics of water in South and Southern Africa is clearly illustrated in the following. South Africa is dependent on Lesotho for most of its water supply. In the past, South Africa negotiated in vain with Lesotho to divert water to what is now Gauteng, South Africa's industrial complex. South Africa's support in 1986 to a successful military coup against the Lesotho government turned out to be a decisive moment in these negotiations. It justified its involvement in this coup, as Lesotho was known to give sanctuary to members of the ANC (at that stage still involved in the armed struggle). Although this was a motivation, it seemed of more importance as the South African government and the new government of Lesotho reached within months of the coup an agreement to construct the Lesotho Highlands Water Project (LHWP) to meet South Africa's water needs. South Africa's desire for water seems likely to have been an ulterior motive behind this country's support for the coup.[4]

To meet the increasing demand for water in the Gauteng area and South Africa's northern provinces, a series of dams is under construction as part of the Lesotho Highlands Water Project. The demands for water in these areas alone are expected to increase from 3.5 per cent to 4 per cent per annum. The first supply of water from the Katse Dam was delivered in 1999 (SAIRR, 1994: 53, 67). However, the Lesotho Highlands Water Project remains controversial. One issue involves a legal battle between the Maseru government and the Lesotho registered Swissborough Diamond Mines company on the site of the Katse Dam. Since 1992 the company has been involved in a series of court cases centred on disputed mining lease areas expropriated in favour of the water project. The company is also suing the South African government (Guy, 1995:1).

Another issue surrounding the LHWP is the fact that the South African government feels that the 1986 Water Treaty between the two countries is outdated because it was engineered during the apartheid era and it was signed at a time when South Africa had limited access to foreign loans (*Cape Times*, 22 July 1996).

For Southern African water is undoubtedly *the* limited natural resource. However, the issues of South Africa's water needs remain unresolved. Regionally, members of the Southern African Development Community (SADC) responded to the issue of limited fresh water resources with a protocol on shared water course systems in the region that makes provision for equitable utilisation of water in shared river basins. However, this creates conflict in the region. SADC's project on the Zambezi aimed at meeting the water needs of Zimbabwe, Botswana and South Africa is heading for a serious confrontation with local inhabitants (Bwalya, 1995: 17).

Boundaries and the Environment

Boundaries are responses to shifting power and security relations (Griggs, 1995: 89). South Africa is no exception in this regard. The recent political changes in South and Southern Africa had a profound impact on the boundaries within South and Southern Africa. Not only were state boundaries changed with the return of Walvis Bay to Namibia in February 1994, but also the boundaries of regions, the delimitation of nine new provinces (with only the boundaries of the Free State remaining the same), the incorporation of ten former homelands and the reorganisation of local authorities. Furthermore, South Africa became a member of SADC in August 1994. This incorporation into SADC opened its doors to illegal immigrants and environmental refugees. The question now arises:

how much concern for the environment was in all these changes? The issue remains unresolved, but still a matter of urgency due to the question of migration in Southern Africa.

Soil Erosion

Apartheid policies concentrated millions of black South Africans in some of the country's least productive and ecologically sensitive areas, where population densities were worsened by high birth rates. Some rural areas have a population density of 82 persons per square kilometre. Poverty and a lack of alternatives caused wide areas to be completely stripped of trees for fuel wood and this resulted in topsoil erosion.

One third of South Africa's population rely on wood for fuel. People who rely on wood as their source of energy burn an average of 3 kg of wood per day. This amounts to 12 tonnes of wood per annum and contributes significantly to the rate of deforestation in Southern Africa (SAIRR, 1994/95: 44). Bad agricultural and town-planning practices contributes to the problem of soil erosion in South Africa. The country loses approximately 3000 tonnes of topsoil annually through drainage into rivers and oceans. Soil was lost at 30 times the rate at which it was formed and costs about R900m a year to replace (SAIRR 1994/95: 50). The effects of soil erosion are a lower per capita availability of productive agricultural land, which results in food insecurity. The most significant social effects are migration, poverty and natural resource scarcity related unrest (Homer-Dixon, 1994: 32).

Environmental degradation in South Africa has crossed the threshold of irreversibility. Even if the government, through ambitious reconstruction and development programmes, removes all the original political, economic and cultural causes of the degradation, it will continue to burden the society. Environmental resource scarcities cause conflict. This conflict, once it occurs, tends to be persistent, diffuse and beyond national boundaries. Countries experiencing chronic internal conflict as a result of environmental degradation will either fragment or become more authoritarian (Homer-Dixon, 1991a, 1991b, 1994). Fragmented countries in Southern Africa – like Angola and Mozambique – are already the source of large out-migrations, mostly to South Africa.

Discussion

Over the past five years the political landscape in Southern Africa has undergone dramatic changes. The major historical conflicts have been solved or are in the process of being solved. Most of these conflicts, in Angola, Namibia, South Africa and Mozambique related directly to the Cold War. However, the region remains insecure. Although events at Cuito Cuanavale in Angola signalled the end of superpower rivalry in Southern Africa, conflicts in the region are far from over. Ching'ambo (1992: 36) mentions some of the sources of regional insecurity that need to be addressed. These are:

- Ethnic conflict,
- Border disputes, or territorial claims, and
- Conflicts relating to inequitable access to and distribution of natural resources between or within states.

The region is characterised by abject poverty and underdevelopment. The situation poses a direct challenge to the traditional role of the military in restoring and maintaining long-term security. Under these circumstances in Southern Africa, the broadening of the concept of security to include political, economic, social military as well as environmental dimensions seems more than imperative. Human security is after all more important than conventional state security (Buzan, 1991: 26-28). Against this background, the next section will attempt to address the prospects for environmental security in post-apartheid Southern Africa.

Prognosis for Regional Security

Traditionally, security concerns in Southern Africa related to Cold War concerns - more often with devastating effect on the environment. Southern African states and their relations with one another offer a focal point for an examination of an environmentally linked redefinition of security as the subcontinent is currently in search of a post-Cold War, post-apartheid security agenda. This section addresses the question of what considerations, if any have been given to the environmental dimension of regional security.

There has been a traditional reluctance to include the environment on the security agenda in Southern Africa. As Vale

(1994: xvii) states, the environment has been relegated to the realm of 'low politics'. It has also been considered to exclude key questions relating to the interests of states or groupings within these states. Yet in spite of the preoccupation with 'high politics', a number of significant decisions were taken by the leadership in Southern Africa during the 1950s and 1960s which included such 'low political' considerations. These decisions related to the hydroelectric schemes at Kariba, Cahora Bassa and Calueque.

Southern Africa's security policy undoubtedly should include the reduction and, if possible, the elimination of environmental insecurities. The latter is caused by the presence of threats and vulnerabilities (Buzan, 1983: 73). A threat is present when developments are possible that would be harmful, should they occur at a future date (Ullman, 1983: 133). The presence of vulnerability is evident when a society lacks the means to limit the impacts of threatening actions, developments or events (Soroos, 1994: 321). Therefore, the most successful strategy for achieving and maintaining security must be orientated toward the reduction of threats or toward the reduction of vulnerabilities to threatening actions, developments or events (Buzan, 1992: 5). However, it is feasible to reduce both threats and vulnerabilities in order to achieve a higher level of security although limited natural resources often prescribe environmental security policy choices.

Strategies to achieve environmental security can also be undertaken unilaterally or collectively. Whereas the unilateral approach involves an individual state applying the means at its disposal to reduce threats, the collective approach involves co-operation among a number of states to address the issue of mutual or shared insecurity. To this end Soroos's (1994) model is particularly helpful. This model suggests the application of combinations of these two-paired approaches - reducing threats and vulnerabilities, and the unilateral and collective approaches to achieve security (Table 5. 3).

Table 5.3: Policy Options for Enhancing Environmental Security in Southern Africa

	Unilateral	Collective
Reduce threat	I Self-prevention	II Collective prevention
Reduce vulnerability	III Self-defence	VI Collective defence

Source: Soroos (1994: 322).

The first strategy involves *self-preservation.* An individual state applies the means at its disposal to reduce threats unilaterally.

The second strategy is *collective protection/mutual restraint, which* involves a group of states agreeing to moderate behaviours that contribute to one another's insecurities. This is considered to be by far the most preferred strategy.

Self-defence, or strategy III, involves actors' reliance on their own devices to reduce the harmful impact of potentially damaging actions, developments and/or events.

The fourth strategy, *collective defence,* involves a co-operative effort to reduce shared vulnerabilities.

Environmental security in Southern Africa can be pursued by applying anyone, or a combination of the types of strategies outlined above. This section explores the applicability of each of these strategies for coping with environmental insecurities in Southern Africa.

Environmental self-preservation States are more willing to act on their own to reduce environmental threats than has been the case with military threats. States in Southern Africa have in the past acted unilaterally to reduce environmental threats. Examples include legislation and strict border control. The primary motivation is to reduce the extent to which a state is a source of its own environmental insecurities, notwithstanding the fact that neighbouring states might benefit from this action.

Environmental collective prevention In this strategy, in order to succeed, it is imperative that all countries primarily responsible for an environmental threat participate fully. Although several international agreements (such as the one between South Africa and Mozambique on the Cahora Bassa and between South Africa and Lesotho on the Lesotho Highlands Water Project) have been signed, only one (the SADC Treaty) addresses the issue of environmental security. Negotiating environmental treaties is a cumbersome process and it may take years before a treaty is ratified.

Environmental self-defence States among themselves may decide to do what they can on their own to reduce their environmental vulnerabilities. This approach, however, has a number of disadvantages: comprehensive defence against a number of environmental threats may not be possible or feasible. Also, unexpected natural events ('acts of God') can expose inadequacies of such an approach. One manifestation of this approach – although limited – is evident in the activities of SADC.

Environmental collective defence Only a limited number of collective strategies have been undertaken to reduce environmental vulnerabilities. The reason for this is that this strategy has a number of inherent limitations. States perceive themselves to be less susceptible to certain environmental insecurities and may therefore be reluctant to invest in risk-sharing arrangements (Soroos, 1994: 324-325). However, Westing (1989: 11) suggests a regional approach to environmental security. It can involve the prevention of transboundary pollution, sustainable use of shared resources and the establishment of equitable means of allocating shared natural resources. The emphasis on regional security is preferable for a number of reasons. These are: regional efforts are more likely to succeed than global ones, because problems are more clearly definable, the need for joint action is more readily apparent and the potential partners are more easily dealt with.

On environmental collective defence as a policy option, one of the earliest indications of such a redefinition of 'national' security in Southern Africa to include the environment was at the Windhoek Summit of SADC in August 1992. Here SADC members adopted the Treaty of the Southern African Development Community, which provided for the negotiation of protocols of co-operation in, amongst others, food security, natural resources and the environment. These developments were followed by a SADC theme document tabled at the consultative conference of January 1993 which suggested a number of ways to promote co-operation relating to security matters. However, no specific reference was made to environmental issues (Davis, 1994: 7-10).

Regional co-operation in Southern Africa is restricted by institutional weakness. This is further complicated by several multilateral organisations that aspire to serve as the institutional foundation for a post-apartheid security agenda. Organisations in the region can be divided into those centred around and dominated by South Africa (such as the Southern African Customs Union and the Common Monetary Area) and those formed as a response to the South African policy of apartheid and destabilisation. These are SADCC (SADC's predecessor) and the eastern and southern African Preferential Trade Area (PTA) (Davis, 1994: 2). The focus of these organisations is mainly on economic co-operation and integration.

It has been suggested that the adoption of the Conference on Security and Co-operation in Europe (CSCE) model for Southern Africa will resolve the question with how best to address non-military security issues (Breytenbach, 1994; Cilliers, 1994). Negotiations for the CSCE were initiated in 1972 and the body was established in 1975. At the 1990 Paris Summit the leaders of 34 eastern and western European governments signed the most comprehensive disarmament treaty ever. The question remains: how appropriate is the CSCE model for the post-apartheid Southern African security landscape? The CSCE has been applied to three fields, or 'baskets'. These are security, economy and the environment and human rights. These issues remain particularly relevant to the region. However, 'security' in Southern Africa is still circumscribed by military overspending, armed exiled groups and arm smuggling (Breytenbach, 1994: 58).

It is expected that the role of the military in Southern Africa will decline as is already evident in regionwide reductions in defence budgets. Against this background, as Breytenbach (1994: 53) states, it is fair to assume that Southern African states will change their definition of 'national' security to include areas of a non-military nature allowing for non-state interests and actors. Except for the situation in Angola, military threats have disappeared from the subcontinent. Non-military sources and types of conflict will continue to threaten the security in the region. These sources and types of conflict relate to problems of a transnational and regional nature. In this regard, Breytenbach (1994: 55) refers to a natural resource that has the potential to cause insecurity: water. A second potential source of non-military insecurity relates to resource-based conflict arising from droughts, failed agricultural crops, poverty and environmental refugees as a result of these.

Concluding Remarks

This chapter has dealt with the impact of environmental resource scarcities on security within the region. It discussed the redefinition of security to include an environmental dimension as well as a selection of environmental issues in the region - of which water is considered to be the most crucial issue. The chapter concludes with a prognosis for regional security by presenting a number of policy options of which environmental collective defence seems to be the solution.

Security in a society is security in the widest possible sense. In a developing region such as Southern Africa, survival is acutely dependent on the environment. As Buzan (1983 and 1992) indicates, a national security issue must be considered at all three levels of analysis; i.e. individual, national and international. The concept of national security refers to the phenomenon on the national level, but the connections between the individual and international levels are too numerous to deny or ignore.

The challenge to policy-makers in South and Southern Africa will continue to be to balance conflicting interests for the need for the creation of employment opportunities and industrialisation with the need to conserve natural resources. Environmental induced threats are not likely to dominate the security agenda in the short or medium term. However, the social problems caused by scarce resources will remain within the region unless co-operation is strengthened.

Acknowledgements

The author writes in her personal capacity and wishes to express her gratitude to two anonymous reviewers for their valuable comments on a previous draft.

Notes

1. What in the past has been regarded as 'traditional' sources of conflict is not so far removed from so-called 'new' sources of conflict. See, in this regard, the work of Westing (1986: 204-209) who lists a selection of major twentieth century wars and conflicts and their natural resource aspects.

2. For a discussion on the debate on the environment as a security issue, see, amongst others, McNamara (1986), Ullman (1983), Buzan (1983; 1991), Tuchman-Mathews (1989; 1994), Hassan (1991), Rodal (1994), Romm (1993), Levy (1995), Maddock (1995), Tennberg (1995) and Stern (1995). For a discussion on the debate *against* linking the environment and security, see Deudney (1990).

3. For a discussion on the environmental impact of apartheid, see, amongst others, Timberlake (1988), Durning (1990), Ramphele (1991), Cock and Koch (1992) and Chenje and Johnson (1994).

4. See *Africa Report* (March-April 1986, January-February 1987 and May-June 1988).

References

'A new Lesotho?' *Africa Report*, January-February, 1987.

Baldwin, D.A. 1995. 'Security studies and the end of the Cold War. A review article', *World Politics*, 48, 1, 118.

Beschorner, N. 1992. 'Water and instability in the Middle East', *Adelphi Paper*,273, IISS, London.

Breytenbach, W.J. 1994. 'Types of conflicts in Southern Africa and mechanisms for their resolution', in M. Venter (Ed), *Prospects for progress, Critical Choices for Southern Africa*, Maskew Miller Longman, Cape Town.

Buzan, B. 1983. *People, States and Fear: The National Security Problem of International Relations*, University of North Carolina Press, Chapel Hill.

Buzan, B. 1991. *People, States and Fear: An Agenda for International Security in the Post-Cold War Era*, Lynne Rienner, Boulder.

Bwalya, J. 1995. 'Zambia River project heading for choppy waters', *The Argus* (Cape Town), 15 September and *Cape Times* (Cape Town), 22 July 1996.

Chenje, M. and Johnson, P. (Eds) 1994. *State of the Environment in Southern Africa*, SARDC, IUCN and SADC Harare.

Ching'ambo, L.J. 1992. 'Towards a defence alliance in Southern Africa?' *Southern Africa Political and Economic Monthly*, 5, 8, 36-38.

Cilliers, J.K. 1994. 'National and regional stability: expectations versus reality', in M. Venter (Ed), *Prospects for progress. Critical choices for Southern Africa*, Maskew Miller Longman, Cape Town.

Coetzee, H. 1994. 'Southern African water issues', in M. Venter, (Ed), *Prospects for progress. Critical choices for Southern Africa*, Maskew Miller Longman, Cape Town.

Cock, J. and Koch E. (Eds), 1992. *Going Green, People, Politics and The Environment in South Africa*, Oxford University Press, Cape Town.

Davis, R. 1994. 'Creating an appropriate institutional framework', in M. Venter (Ed), *Prospects for Progress. Critical Choices for Southern Africa*, Maskew Miller Longman, Cape Town.

Deudney, D. 1990. 'The case against linking environmental degradation and national security', *Millennium*, 19, 3, 474-475.

Durham, W. 1979. *Scarcity and Survival in Central America: The Ecological Origins of the Soccer War*, Stanford University Press, Stanford.

Durning, A.B. 1990. 'Apartheid's environmental toll',*Worldwatch Paper* 95, Worldwide Institute, Washington.

Falkenmark, M. 1991. 'Rapid population growth and water scarcity: the predicament of tomorrow's Africa', in K. Davis and M.S. Bernstam (Eds), *Resources, Environment and Population: Present Knowledge and Future Options*, Oxford University Press, New York, pp 81-94.

Gleick, P.H. 1993. 'Water and conflict. Fresh water resources and international security', *International Security*, 18, 1, 83-88.

Griggs, R.A. 1995. 'The boundaries of a new South Africa', *Boundary and Security Bulletin*, 2, 4, 85-89.

Gurr, T.R. 1985. 'On the political consequences of scarcity and economic decline', *International Studies Quarterly*, 29, 45-53.

Guy, D. 1995. 'Katse Dam court battle looms', *The Star and SA International*, 1, November, 1.

Hassan, S. 1991. 'Environmental issues and security in South Asia', *Adelphi Papers* 262, IISS, London.

Homer-Dixon, T. 1991a. 'On the threshold. Environmental change as acute causes of conflict', *International Security*, 16, 2, 36-41.

Homer-Dixon, T. 1991b. 'Environmental change and economic decline in developing countries', *International Studies Notes*, 16, 1, 90-111.

Homer-Dixon, T. 1994. 'Environmental scarcities and violent conflict. Evidence from cases', *International Security*, 19, 1, 39-40.

'Lesotho water project gets under way', *Africa Report*, May-June, 1988.

Levy, M.A. 1995. 'Is the environment a national security issue?' *International Security*, 20, 40-48.

Maddock, R.T. 1995. 'Environmental security in East-Asia', *Contemporary Southeast Asia*, 17, 1, 25-30.

McNamara, R. 1968. *The Essence of Security*, Hodder and Stoughton, New York.

Preston-Whyte, R. and Holden, M. 1994. 'Climate and the regional economy', in M. Venter (Ed), *Prospects for progress. Critical Choices for Southern Africa*, Maskew Miller Longman, Cape Town, pp 175-188.

'Pretoria has its way in Lesotho', *Africa Report*, March-April 1986.

Ramphele, M. (Ed), 1991. *Restoring the Land. Environment and Change in Post-apartheid South Africa*, Panos, London.

Rodal, B. 1994. 'The environment and changing concepts of security'. *Commentary,* 47, (A Canadian Security Intelligence Service Publication).

Romm, J.J. 1993. *Defining National Security. The Non-military Aspects,* Council on Foreign Relations Press, New York.

Rowlands, I.H. 1996. 'South Africa and global climate change', *Journal of Modern African Studies,* 34, 1, 163-178.

Soroos, M.S. 1994. 'Global change, environmental security and the prisoner's dilemma', *Journal of Peace Research,* 31, 3, 321-325.

'South Africa: Nuclear reactions', *Africa Confidential, 1994,* 35, 5, 1–17.

South African Institute of Race Relations (SAIRR). 1994. *Race Relations Survey 1994/95,* SAIRR, Johannesburg.

Stern, E.K. 1995. 'Bringing the environment in: The case for comprehensive security', *Cooperation and Conflict,* 30, 3, 213-216.

Suhrke, A. 1994. 'Environmental degradation and population flows', *Journal of International Affairs,* 2, 47–55.

Tennberg, M. 1995. 'Risky business: defining the concept of environmental security', *Cooperation and Conflict,* 30, 3, 244-245.

The Times (London), 18 October 1995.

Timberlake, L. 1988. *Africa in Crisis. The Causes, the Cures of Environmental Bankruptcy,* Earthscan, London.

Tuchman-Mathews, J.C. 1989. 'Redefining security', *Foreign Affairs,* 68, 2, 12-22.

Tuchman-Mathews, J.C. 1994. 'The environment and international security', in Klare, M. and Thomas D. (Eds), *World security: Challenges for a New Century,* St Martin's Press, New York.

Ullman, R. 1983. 'Redefining security', *International Security,* 8, 1, 133-138. 'Unmaking the state', (6 January 1995). *Africa Confidential,* 36, 1, 1-26.

Vale, P. 1994. 'Fashioning choice in Southern Africa', in M. Venter (Ed), *Prospects for Progress. Critical choices for Southern Africa,* Maskew Miller Longman, Cape Town.

Van Aardt, M. 1993. 'In search of a more adequate concept of security for Southern Africa', *South African Journal of International Affairs,* 1, 1, 216-225.

Westing, A.H. 1986. 'Wars and skirmishes involving natural resources: a selection from the twentieth century', In A.H. Westing (Ed), *Global Resources and International Conflict. Environmental Factors in Strategic Policy and Action,* Oxford University Press, Oxford.

Westing, A.H. 1989. 'Environmental approaches to regional security', in A.H. Westing (Ed), *Comprehensive Security for the Baltic: An Environmental Approach,* Sage, London.

Chapter 6

Population-Environment Interface in Botswana: Issues, Challenges and Prospects

G. Letamo and O. Totolo

Introduction

There is no simple correlation between population and environment as population, environment and development interact in many complex ways. Human populations require resources for their survival and the environment is the primary source of these resources. Development provides the means by which these resources are converted and delivered for human use. Development encompasses use and conservation of the environment in which human population resides. It considers issues of economic growth, distribution of wealth, poverty alleviation and human welfare. By environment we refer to an ecosystem which both supplies resources and serves as a repository for wastes (UN, 1984). Environment is often used to describe the planet's life-support systems for human and other forms of life. Resources refer to anything obtained from the living and non-living environment to meet human needs and wants. They are not limited to material resources drawn directly from nature but also include capital goods and human resources.

Resources become useful only with human effort and technological ingenuity. For instance, petroleum was an unknown fluid until humankind learned how to find it, extract it and refine it into gasoline, heating oil and other products. Resources are normally divided into renewable and non-renewable resources. Non-renewable resources are those that exist in fixed amounts in the earth's crust and thus can be used up e.g. coal and oil. Also, resources can be considered non-renewable if the rate of exploitation is slower than the rate of use.

Over millions of years, geological processes can renew them. Renewable resources, on the other hand, are those that can be replenished relatively quickly by natural or artificial processes if managed wisely, e.g. forests. It should also be noted that there is need to distinguish between non-critical zone and critical zone renewable resources. Non-critical zone resources such as solar energy do not have a user cost since their renewability is infinite whereas the critical zone resources such as wood have a user cost since they can be depleted if their rate of use is greater than their rate of replenishment. It should be noted, therefore, that just because a resource is renewable does not mean that it can never be exhausted.

Although human population can lead to environmental degradation, it is not the only cause of environmental change. Environments are constantly changing but human impact may accelerate the process of change. It is against this background that we need to understand the complex interaction between population, environment and development. Sustainable development cannot be attained unless the current generation uses the environment on a sustainable basis.

In developing countries, pressures created by rapid population growth, uneven distribution, and inadequate socio-economic development are contributing to widespread environmental degradation. In the past, low densities and low consumption of early societies had little environmental impacts. The populations were adapting to their environments rather than altering them. However, since 1960, there have been dramatic increases in population, technology, consumption and human-induced environmental changes. The world's population has doubled over the past four decades to reach 5 billion (1989). According to UN (1991) projections, it will rise to 8 billion by 2025, levelling off at 10 billion in the second half of the next century. This projected exponential growth in population has important implications for resource use and conservation.

Most of the increase will take place in developing countries, several of which are already hard-pressed to feed their existing populations. The rapid growth has also far-reaching implications for human life, economy and environment, particularly in Africa where populations are growing fastest. Africa's 500 million people in 1987 had become 850 million in 2000, a 70 per cent increase in only 13 years. The problems induced by this rapid population growth are readily visible and include soil erosion, deforestation, desertification and massive influx of the rural poor to urban slums and shantytowns. There is therefore need to understand the complex relationships between population and environment if both the expected short-term and long-term implications of these trends are to be properly addressed. The role of population in environmental degradation and how the environment influences population dynamics need to be spelt out.

Botswana has experienced rapid economic growth and this has led to enormous pressure on the natural resources. An annual population growth of 3.5 per cent, means that Botswana's fragile semi-arid environment is placed under more stress from rangeland degradation, woodland depletion, veld product depletion and reduction of wildlife resources. As Botswana's economy improves, more non-biodegradable waste is generated which makes it difficult to improve the quality of the environment. Drilling of more boreholes for water in the rural areas has also led to rangeland degradation. The Government of Botswana has recognised the major environmental problems in the country and has introduced policies and strategies to address them. Examples include the National Conservation Strategy, the Accelerated Development Programme, the Financial Assistance Policy, the Water Resources Policy and the Industrial/Urban Pollution Policy. If well implemented, these policies will go a long way towards solving environmental problems.

This chapter examines interrelationships between population and environment in Botswana. The chapter basically reviews the impact of population growth on the environment and the prospects for mitigating these impacts through government policies and programmes. Population-environment relationships are also examined from the perspective of a theoretical framework, the findings of which reveal that the I = PAT model is relevant to Botswana. Botswana's population continues to increase and this has placed more demand on the use of natural resources.

Theories of Population Growth and Resource Interactions

Determination of Population-Resource Relationship

There are a number of measures used to investigate how population influences resource availability and vice versa. The means of identifying population-resource relationships can be properly dealt with under a well-defined conceptual framework. It is therefore necessary to develop a framework that can be used to comprehend the complex interaction of population and environment.

Theoretical Framework of Population-Environment Nexus

Figure 6.1 illustrates the theoretical framework, which helps us to identify the relationships between population dynamics and the environment. This model clearly indicates that human institutions modify the impacts of population variables upon the environment. The framework shows that population and environment are not directly linked. Figure 6.1 shows the relationship between population variables, development and environment. Development is seen as the intermediary between population growth and environmental change. This approach was considered the most appropriate at more local levels, from household to nation than at global levels. This is mainly because of intervening factors of globalisation such as world trade, international technological transfers, and other global issues that transcend local or national borders (Clarke, 1995: 24).

Generally, the population-environment nexus has generated heated debates. In some circles, it is argued that population growth leads to environmental change whereas in other circles population growth is seen as an advantage that generates wealth and therefore needs to be tapped. The first school of thought is often referred to as 'Alarmists' or 'Pessimists' school of thought whereas the latter is referred to as 'Non-alarmists' or 'Optimists' school.

The 'Alarmists' or 'Pessimists' School of Thought

Malthusian Theory

Malthus argued that population increases geometrically whilst food output at best increases arithmetically. He argued that, in order to maintain a balance between population and food supply, population growth must inevitably be checked by what he termed 'positive' and 'preventive' or negative measures. The preventive or negative checks of limiting population growth include celibacy or postponement of marriage, which would lower the reproductive rate. This is what Malthus calls a moral constraint. Positive checks are ways in which population size can be reduced by such events as war, famine and disease. Malthus has argued that rising income per head would lead to early marriage and hence an increase in population. As this population increases geometrically, it would finally outrun the food supply, leading to falling standards of living and an increased death rate. Thus, Malthus believes that population increase cannot go unchecked, particularly in traditional societies. The theory simply states that there is a direct link between population growth and environmental degradation, that is, more people will consume more resources and produce more waste (Lowe and Bowlby, 1992). Malthus assumes that there is a finite resource base and a fixed carrying capacity, which cannot be exceeded without disastrous effects.

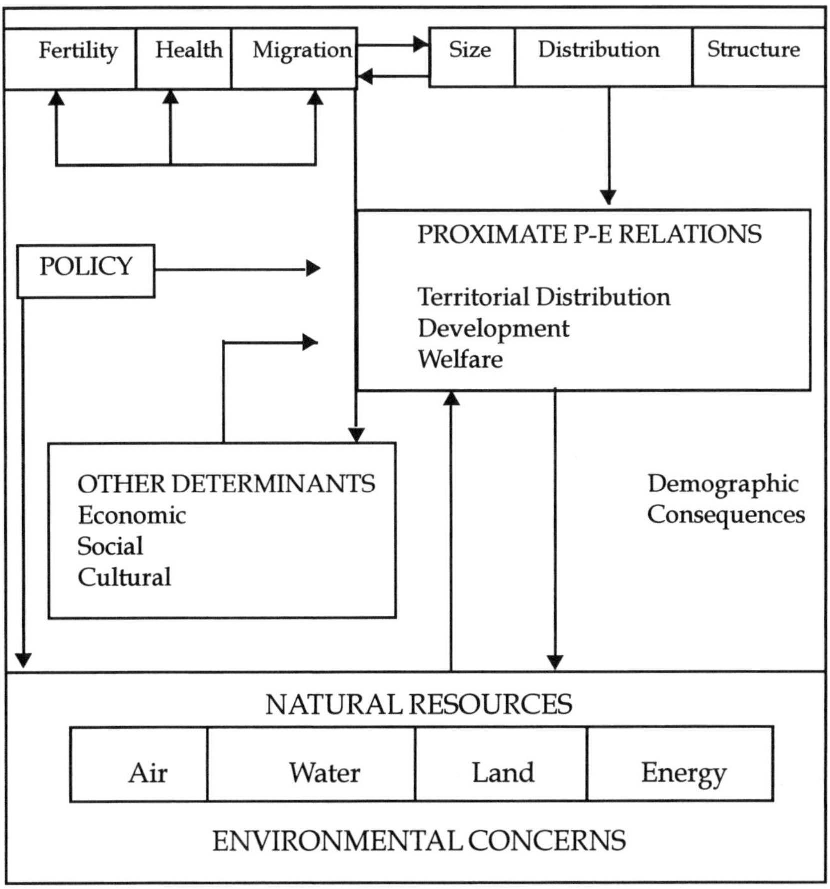

Source: Clarke, 1995.

Figure. 6.1: Model Illustrating Relations between Population Dynamics and Environment

Changing technology, which increases food production and allows human beings to live at greater densities under fixed landmass, violates the assumption of this theory. For instance, it has been noted that as population pressure increases, people adopt certain measures such as high-yielding crops, the use of fertilisers and irrigation in order to increase their yields. To a certain degree, however, Malthus ideas are applicable to some pre-industrial societies.

The Optimum Population Theory

The concept of optimum population is important in that planning for human beings given limited resources, requires us to obtain maximum satisfaction with minimum effort (the concept of rational economic man). In any planning exercise, population size is usually related to the available resources in order to attain optimal resource use. Both the conceptualisation and measurement of the optimum population is difficult. Despite the fact that this concept has been criticised for a number of reasons, it survives even today simply because it represents a fundamental problem that must be confronted in one way or another. Writers on optimum population theory argue that there is an optimum population, which maximises output per head at a certain level in economic production. Before or after this point exist underpopulation and overpopulation, respectively. The theory assumes a fixed ceiling of population growth vis-à-vis resources. That is, while population increases, other factors of production remain constant. For instance, the theory assumes that there is no technical change, available land is constant and the distribution of income is also constant.

How is overpopulation defined? Here are some examples of the definitions of overpopulation by Ehrlich *et al.* (1977). A population that is too large and too poor to be adequately supplied by resources has exceeded the optimum. Similarly a population so large that it can be supplied with physical necessities only by the rapid consumption of the non-renewable resources or activities that irreversibly degrade the environment has also exceeded the optimum because it is reducing the earth's carrying capacity for future generations. If an increase in the population decreases the well being of a substantial number of people in terms of necessities while increasing that of others in terms of luxuries, the population has exceeded an optimum.

The concept of population density is usually used to measure the relationship between population and resources, particularly land. In its simple form, it is the number of persons per square kilometre, mile, hectare, etc. A high population density is sometimes taken as a proxy for overpopulation whereas a low population is taken to reflect underpopulation. Population density is, however, full of weaknesses. For instance, shortages of many resources other than physical space may make it difficult to supply even a sparse population in a given area with the necessities of existence (Ehrlich *et al.* 1977). This is because over large areas, people may lack sufficient dependable water, fertile soil etcetera. How then do we define optimum population? Optimum population is defined as that level of population beyond which the average product in an economy falls. Thus optimum population is the maximum population that can be supported with the existing resources (Thirwal, 1983). Ehrlich *et al.* (1977) and Thirwal (1983) argue that overpopulation is a condition whereby population exceeds the optimum for a country with given resources and technology. For every land area to which a given system of land usage is applied, there is a population limit which cannot be exceeded without setting in motion the process of land degradation (Allan, 1972). This is where the concept of carrying capacity comes in. The carrying capacity refers to the maximum number of people a particular habitat can sustain indefinitely without deterioration.

The Ehrlich and Holdren Model

The model takes the form: **I = PAT**
Where

I = Impact on environment
P = Population (size, distribution and rate of growth)
A = Affluence (per capita consumption, determined by income and lifestyle)
T = Technology (level of impact of consumption)

This model states that population is not the only cause of environmental degradation. The consumption of materials and energy per person are also important factors. So is the type of technology that people use to make consumption possible as well as the economic, social and political forces that influence decision-making (Ehrlich *et al.*, 1977: 719). All these factors are intricately linked by an array of cause-and-effect connections. Population, consumption per person and the careless use of technology have been increasing rapidly at the same time and the ability of individuals and governments to adopt the behaviour to ever changing conditions and problems have not kept pace. To properly address environmental problems, it becomes imperative to investigate all factors that interact to cause these problems and the nature of their interactions. Hence, the need to develop models that seek to investigate the nature of the factors that cause environmental degradation. The current model shows that in order to reduce environmental degradation, there are various ways that can be used. One way is to reduce population growth. Another is to limit consumption. The other option is to reduce the impact of technology on the environment. Usually in developing countries like Botswana, the environment is damaged by rapid population growth with high levels of consumption and destructive technologies having a lesser impact.

Affluence is measured by consumption per person (GNP). Gross National Product (GNP) is a measure of economic growth. Unlike the gross domestic product (GDP), which is the total market value of all goods and services produced within a country despite where their owners live, GNP (in current dollars) is the market value of all goods and services produced by the residents of a country within a year despite their place of production. Foreigners may own some of the goods and services and this leads to a flow of income out of the country whereas some of the goods and services owned by the residents may be located abroad leading to a corresponding flow of income from abroad into a country. The net difference between these flows is called net property income from abroad. If this item is added to GDP, we get the GNP figure. Thus, GNP = GDP + net property income from abroad.

If we adjust for GNP inflation (any increase in the average price level of final goods and services), we get real GNP. Dividing the real GNP by the total population gives us real per capita GNP. If population expands faster than economic growth, then real per capita GNP falls. GNP indicators are poor measures of human welfare because:

- They tell us nothing about how the resources are distributed among the people in a country.
- They don't include the depletion and degradation of natural resources on which all economies ultimately depend e.g.

exhausting mineral resources, eroding soils and pollution.

- They hide or underestimate some of the positive effects on society. For example, GNP indicators exclude labour we put in volunteer work, the food we grow for ourselves, the cooking and cleaning, etc.

The model shows that developing countries, with their large populations but limited economic advancement, can have a major impact on the environment, hence prospects for sustainable development. The equation also makes clear that developed nations also generate environmental problems through their high consumption of goods and services per person. The I = PAT equation was initially formulated by Ehrlich and Holden in 1971 to emphasise the fact that it is necessary to deal simultaneously with overpopulation, excessive affluence and faulty technology (Ehrlich and Holdren, 1972, cited in Martine, 1995). The model argues that the negative impacts of resource use on the environment increase as a result of:

- an increasing human population, which requires more resource use;
- an increasing standard of living which leads to more resource consumption per person; and
- technology can have both a negative and a mitigating effect, depending upon its environmental friendliness (VanderPost, 1995).

The model has been criticised for a number of reasons: 1) the statistical association between the variables is assumed as causation; 2) the deeds of the poor and rich are lumped together; 3) it assumes population as independent of affluence and technology and; 4) it overlooks the fact that all the elements in the equation are influenced by the social, economic, institutional and political factors (Martine, 1995).

The 'Non-Alarmists' or 'Optimists' School of Thought

Ester Boserup's Theory

Boserup states that in pre-industrial societies, population increase stimulates agricultural technology so that food can be increased to match the increasing populations. She studied different land-use systems classified according to their intensity of production and found a close connection between agricultural technology used and the type of land-use system. For instance, she found that axe and digging stick were associated with forest fallow, the hoe with bush fallow, and the plough with short fallow. From these findings, she concluded that:

- any increase in intensity of production by the adoption of new technology was unlikely unless population increased, and
- if population increased beyond a certain point and no extra land was available, the length of fallow was reduced in order to maintain the same level of consumption per capita and this leads to a decline in soil fertility and per capita output.

Boserup acknowledges the importance of the population factor in development. Her belief is that 'an increasing population size may make life easier because there will be more people to share the burden of collective investments but may also make life more difficult because the ratio of natural resources to population decreases (Boserup, 1981: 5). The reduced ratio of resources to population motivates a society to import or develop new technologies to reduce or eliminate these disadvantages. She argues that population pressure is necessary for the development of new agricultural methods and land use. She argues that population increase actually stimulates innovation and technological development. Hence, the carrying capacity of the area is expanded to accommodate the additional population.

Boserup, like Malthus, does not specifically address population-environment relations but rather the narrow topic of land use and population. In this chapter, implications on general linkages between population and environment are inferred from her work as her viewpoint represents issues relevant the theme under discussion. The primary focus on land precludes the role of other natural resources in economic development (Arntzen, 1989). Another limitation of Boserup's theory is that it neglects the comparative advantages of land for specific forms of land use (for instance, semi-arid environments are more suitable for extensive livestock production) (Arntzen, 1989). Other relevant aspects such as culture and economic aspects, although not necessarily left out, are marginalised. The final criticism is that although Boserup's theory is based on a large number of detailed empirical studies, other studies do not confirm her theory (e.g. Brookfield, 1975 and Blaikie and Brookfield, 1987) and the data used is rather old (Arntzen, 1989).

Julian Simon's Theory

One proponent of the optimist theory is Julian Simon (1981). In *The Ultimate Resource*, he argues that population growth leads to increasing living standards and economic development through an increase in the invention of means to cope with the population increase because there are more potential inventors and because the pressures of growth create the impetus for invention. He argues that 'more people need not cause scarcities or environmental decay in the long run . . . population growth has positive economic effects in the long run, though there are costs in the short run' (Simon, 1981: 9). He believes that 'the increased demand for copper leads to increased pressure for new discoveries. Second, and perhaps even more important, a larger population implies more people to think and imagine, be ingenious, and finally make these discoveries' (Simon, 1981: 216).

Davis' Theory of Demographic Change and Response

Kingley Davis, a demographer, argues that when mortality declines and children survive through adulthood putting pressure on the resources, people reorganise their lives so as to relieve that pressure, that is, people respond to the demographic change that is occurring. Davis argues that people respond to personal goals rather than national goals. He further argues that the response that individuals make to the population pressure created by additional members joining their ranks is determined by the means available to them (Weeks, 1981). The first non-demographic response is said to be an attempt to increase the resources by working harder (longer hours, a second job etc). If insufficient resources are accumulated by the first response, then emigration of some family members, especially of unmarried sons or daughters takes place. The response of the second generation is to take advantage of social or economic opportunities by avoiding the large families that caused problems for their parents (Weeks, 1981). Davis concludes that the most powerful motive for family limitation is not so much fear of poverty or avoidance of pain as suggested by Malthus but it is the prospect of rising prosperity.

In short, Davis' central thesis is that land use responses to population growth are essentially 'multiphasic'. He attempted to explain how rural households in Japan and Western Europe responded to threats to their living standards in the nineteenth and twentieth centuries. He postulated that they would change their demographic behaviour by postponing marriage, reducing fertility or outmigration. Davis thus saw the responses as 'multiphasic', that is, several responses could occur simultaneously. But he did not consider the possibility of other non-demographic responses.

Population Growth and Environment in Botswana

The Demographic Profile

Census taking has been conducted in Botswana since 1904. However, the censuses carried out between 1904 and 1956 were irregular and inaccurate in that the colonial administration was not so much interested in the total population but the number of people available for colonial needs such as taxation and military recruitment. The headmen of settlements, who were more likely to have not been able to count properly, did this type of counting. It is therefore not surprising that the figures were rather grossly under-estimated. The first census based on house-to-house canvassing was the 1964 census which was a modified *de jure* counting based on the village of allegiance of the respondents (Central Statistics Office, 1987). The subsequent censuses of 1971, 1981 and 1991 were based on the *de facto* method of enumeration.

Population Size and Growth

Based on the last three censuses, it is evident that the population of Botswana has been growing rapidly over the past few decades. The *de facto* population was counted to be 596,944 in 1971, 941,027 in 1981 and 1,326,796 in 1991. These figures imply an annual growth rate of 4.7 per cent between 1971 and 1981 and 3.5 per cent between 1981 and 1991 (Central Statistics Office, 1984 and 1993). Assuming an annual intercensal growth rate of 3.5 per cent, the population will double in about 20 years' time. The population growth rates were higher in urban areas compared to rural areas. Such rapid population growth has serious ramifications as will be shown later in the chapter.

Fertility

Although fertility in Botswana is high, it has been declining since the late 1970s. The overall fertility rate declined from 6.5 children per woman in 1971 to about 5 children per woman in 1991. The fertility decline is explained by increasing female labour force participation rates, increased female school enrolment rates, increasing modern contraceptive use, declining infant and child mortality and declining marriage rates.

Mortality

Infant mortality rate declined from 97.1 deaths per 1,000 births in 1971 to 48.0 deaths per 1,000 births in 1991. Life expectancy at birth increased from 55.5 years in 1971 to 65.3 years in 1991 (Central Statistics Office, 1995).

Population Structure

Given the relatively high level of fertility and low level of mortality, the population of Botswana is young and has a disproportionately large number of females. It is worth noting that the proportion of the population under age 15 went down between 1971 and 1991. The percentage of population of those under 15 years declined from 47.5 per cent in 1971 to 43.2 per cent in 1991. The sex ratio also improved during the corresponding period from 84 males per 100 females in 1971 to 92 males per 100 females in 1991 (Central Statistics Office, 1994). The improved sex ratios are a result of reduced male labour migration to South Africa. The predominance of the female population could be explained by the low sex ratios at birth and as a result of high mortality among males as well as male emigration to South Africa.

Population Distribution

The percentage of the population residing in urban areas increased from 9 per cent in 1971 to 45.7 per cent in 1991. The increase in the urban population may be explained by the high rate of rural to urban migration and the reclassification of some major villages to urban status. The population density has also increased from 1.0 person per square kilometre in 1971 to 2.3 persons per square kilometre in 1991.

Population density varies by district, with the urban districts having the largest densities (Table 6.1). Table 6.2 shows some examples of population growth, distribution and resource access in both urban and rural areas.

Table 6.1: Botswana's Population Density by District, 1971-1991

District	Area (km²)	Density (persons/km²) 1971	1981	1991
Gaborone	169	104.8	353.0	789.8
Franscistown	79	235.6	393.2	825.9
Lobatse	42	284.2	453.2	620.3
Selibe-Phikwe	50	98.8	589.4	795.4
Orapa	17	71.1	307.6	519.2
Jwaneng	100	–	55.7	111.9
Sowa Town	159	–	–	14.0
*Ngwaketse	26,467	2.6	3.9	4.9
*Barolong	2,003	5.5	7.7	9.2
South East	1,780	11.3	17.2	24.5
Kweneng	35,890	1.8	3.3	4.7
Kgatleng	7,960	3.9	5.6	7.3
C. Serowe	31,319	–	2.6	4.1
C. Mahalapye	16,463	–	5.0	5.8
C. Bobonong	14,204	–	3.3	3.8
C. Boteti	33,716	–	0.7	1.1
C. Tutume	46,018	–	1.6	2.2
North East	5,120	5.0	7.2	8.5
Ngamiland	109,130	0.4	0.6	0.9
Chobe	20,800	0.2	0.4	0.7
Ghanzi	11,500	1.0	1.7	2.1
Kgalagadi	105,200	0.13	0.2	0.3
National	**582,000**	**1.0**	**1.6**	**2.3**

Sources: Population and Housing Census (1971, 1981, 1991), Central Statistics Office (1994), Rakgoasi and Gaisie (1999).
* Ngwaketse and Barolong form the Southern District.

Table 6.2: Population Growth, Distribution and Resource Access: Some Examples

SETTLEMENT	POP 1981 Absolute Nos	POP 1981 in %	POP 1981 Absolute Nos	POP 1981 in %
Urban	159 614	17.0	606 239	45.7
Rural:				
over 5000 inhabitants	167 577	17.8	12 482	0.9
1000-4999 inhabitants	148 607	15.8	225 660	17.0
500-999	103 959	11.0	116 437	8.8
100-499	239 500	25.5	234 252	17.7
less than 100 inhabitants	121 770	12.9	131 726	9.9

Source: Population Census (1991).
Central Statistics Office, Gaborone Botswana.

History of Environmental Issues

Worldwide, population increase is blamed for many environmental problems and Botswana is no exception (Arntzen, 1995). Population increase exerts pressure on the environment. According to Cooke (1991), Botswana's population was 120,000 in 1902. By the year 2000 it is estimated that it will have reached about 2,000,000, a sixteen-fold increase in just 100 years. This has inevitably involved a massive increase in human use of environmental resources. Livestock raising has been the dominant activity and as people have increased in number, so in parallel have livestock numbers. These animals make increasing demands on available grazing, and cattle posts have spread into areas previously used only on a temporary basis after good rains have made water available in a normally waterless country. Modern technology in the form of deep borehole drilling has now, however, made permanent occupation of marginal grazing land possible. Towns and villages demand even more water. More people need more fuel for cooking and warming in winter, land for ploughing, and slowly woodlands become depleted. More land is ploughed, and the soil, which is overused and overexposed, becomes progressively degraded. The flexibility of the past, when communities moved easily, leaving old settlements and establishing new ones elsewhere, and when livestock could be easily and rapidly shifted around to match grazing and water availability, is becoming more and more difficult. Some urban areas such as Gaborone have become large, rapidly expanding agglomerations. They demand ever increasing supplies of water, food, power and land and create large amounts of non-degradable wastes which have to be disposed off if they are not to pollute, clutter and disfigure the places in which people have to live and work.

All these changes create unrelenting pressure on a finite and fragile environment. Unfortunately, traditional patterns of behaviour have not changed fast enough in time and space with the rapid growth of population and its increasing demands for natural resources. Most Batswana still cultivate the land and raise their livestock in much the same way as their forefathers did. Too many head of cattle exhaust the grazing resource, displace wildlife and interfere with their migration routes. People cut down woodland but do not replant it. Modern rubbish and waste, mostly non-degradable and unlike that generated by more traditional modes of life, accumulates, and methods of disposal have not changed to cope with it. As a result, the inhabited areas have become dirty and badly polluted, and an ugly, unhealthy wasteland results. This whole spectrum of maladjustment and disturbance of the human-environment balance rises in intensity as population increase continues to accelerate. Thus, the rapid population growth will cause substantial extra resource demand in the near future (Molutsi, 1992; Arntzen, 1995).

Table 6.3: Population Concentration in Southeastern Botswana (South-East, Kgatleng and Kweneng)

	1981	1991
Population in settlements of over 5000 inhabitants	118 137	260 356
Total Population	251 894	405 259

Source: Population Census (1991), Central Statistics Office, Gaborone, Botswana.

In Botswana, uneven population distribution is a result of environmental conditions. Two thirds of Botswana is semi-arid and covered by the Kgalagadi sands and therefore most settlements and arable agricultural activities are concentrated in the eastern part of the country. These spatial disparities in population causes substantial pressure on local natural resources around an increasing number of large settlements and at the same time underutilisation of resources elsewhere. Resource pressure is particularly evident in south-eastern Botswana (Table 6.3).

The inhabitants of small districts, in particular, such as those of the South East district, face rangeland constraints and therefore keep fewer livestock (Table 6.4).

Table 6.4: Population Density and Resource Access / Use

Region	Population Density Persons/km²	Per Capita Cropland (ha)	Per Capita Cattle	Per Capita Livestock Units
Small district: South East	69.1*	0.31	0.57	0.54
Medium sized district: Kgatleng	7.5	0.37	1.97	1.60
Large district: Central District	4.0	0.24	2.21	1.75
Botswana	2.2	0.25	2.11	1.66

*excludes Gaborone.
Sources: Population Census (1991), Central Statistics Office, Gaborone, Botswana.

The Impact of Population Growth on the Environment

Environmental Degradation

Scholars disagree about the environmental impacts of population. Regarding human adaptability, some are pessimistic (e.g. Malthus, Ehrlich), others are optimistic (Boserup, Simon) and yet say that the situation needs to be assessed on a case by case basis (revisionists). In Botswana, there have been many debates on land degradation, especially desertification and rangeland degradation. Arntzen (1989) and Ringrose (1987) believe that most rangelands in the country are seriously degraded. Some aver that soil erosion in arable areas is also considerable in extent (Arntzen and Veenendaal, 1986), while others believe that there is no land degradation. The major cause of land degradation is generally considered to be overstocking and the subsequent overgrazing of land (Williamson, 1987). It has also been alleged that Botswana's development has led to widespread environmental degradation (Arntzen, 1989).

There have been quite a number of studies on environmental degradation in Botswana, for example, Ringrose (1987), Arntzen (1989) and Kwerepe (1997). Examples of bush encroachment, decrease in grass density as well as in the number of grass species, and increase of patches of bare soil have been documented by, among others, Van Rensburg (1971), Field (1978), Cooke (1981) and Van Vegten (1981). Regarding soil erosion, Van der Poel (1980) has recorded high net soil losses from a plot of grazing land near Mahalapye. Ringrose (1987), by employing Landsat imagery and bare soil as an indicator of the extent of land degradation, has shown that land degradation is widespread throughout the country, especially Western Botswana and in the Okavango area. However, several authors have strongly challenged these findings and instead argued that many of the 'truths' about land degradation in Botswana are exaggerated and based on insufficient data (Abel and Stocking, 1987; Abel and Blaikie, 1989; Biot 1990; White, 1992; Perkins and Thomas, 1993). It is suggested that rates of soil loss from a small test plot are much higher than from a large catchment area where the soil is usually not lost but simply redistributed (Abel and Stocking, 1987; Biot 1990). Biot (1990) modelled soil loss in a heavily used and, by Botswana standards, steeply sloping area and estimated that the ability of the land to sustain present cattle densities can be maintained for another 400-500 years (Abel and Blaikie, 1990). Biot (1990) also argues that bare soil is a common feature during periods of drought and around areas of high population density and does not, therefore, necessarily indicate land degradation. Bush encroachment does not have to be an indicator of overgrazing nor does it necessarily have to be interpreted as being environmentally detrimental (Perkins and Thomas, 1993; Dahlberg, 1994, 1996). Queiroz (1993) has argued that range degradation is not a myth but a reality in Botswana. Queiroz (1993) dispels allegations by Biot (1992) and White (1992) that there is no strong evidence of range degradation in Botswana.

Ringrose (1987), using satellite imagery, showed the North-East (NE) District to be under serious environmental degradation. The

NE District plan summarises the situation as follows: 'The single greatest development problem in the NE District is shortage of grazing and arable land, which is compounded by high population densities, especially in the northern part of the district. The land shortage has resulted in the overcrowding and deterioration of land quality' (Botswana, 1986: 10).

Despite the arguments by Perkins (1991) and White (1992), amongst others, that Botswana's arid and semi-arid ecosystems are extremely resilient, and that there is thus no desertification or acute dryland degradation in Botswana, the view expressed by Darkoh (1999) that 'the combined effects of increasing pressure from conflicting land uses (including intensive grazing, wood harvesting, the gathering of veld products and crop production), and recurrent disturbances such as drought, have led to dryland degradation in the rangelands throughout Botswana', is more plausible. Botswana's environment is undergoing some significant changes and therefore it is important that the changes are acknowledged by research scientists, policy makers, the general public and the government in order to come up with appropriate policies and programmes to combat environmental degradation.

According to Darkoh (1999), environmental degradation is widespread in Botswana. However, whether the degradation in Botswana is temporary (slight) or severe (that is, permanent or irreversible) is a matter for further research. The mid-Boteti study by Sefe *et al.* (1996) has demonstrated that once an ecosystem is degraded, it is very difficult to recover. Therefore, precautionary measures have to be taken against environmental degradation which includes desertification.

Population and Resources

Woodland and Forest Resources

In Botswana, people lived in harmony with trees but this delicate balance is changing as a result of population growth. Batswana have deep respect for trees and have detailed knowledge of their names and uses. The deep respect is expressed in mythology, traditional religions and ceremonies.

The felling of trees around settlements is very common (Kgathi, 1982). The distance to collect firewood has increased with time. This is a result of overharvesting of woodland resources around settlements. Fuelwood is used for heating homes, lighting and cooking. Wood is popular for construction of houses, kraals and for the carving of sculptures for selling, mainly to the tourists. In Botswana, *colophospermum mopane* (Mopane) provides poles for construction of houses and homesteads. Mopane poles are highly resistant to faunal activity like termites.

In order to uplift their living standards, some Batswana, especially in Ngamiland and Chobe districts, have resorted to basket making. They produce various baskets from forest resources and these baskets include cereal containers, fishing baskets and tunnel nets. Most are made from young leaves of fan palms.

In Setswana culture, most traditional ceremonies like Bogwera and Bojale are carried out in secluded areas in forests and woodlands and therefore it could be argued that conservation of trees is fundamental to the continuation of the cultural inheritance like initiation ceremonies.

Population increase could also result in deforestation, which will inevitably result in the reduction of biodiversity, which entails ecosystem diversity, genetic diversity and species diversity. The response to deforestation has always been to start single species plantations and reforestation programmes. While it might be acknowledged that such a measure might arrest the problem of soil erosion and may be the alternative to harvest firewood, the natural diversity of wildlife will never be replaced by such measures. As more trees are removed from the watershed, accelerated soil erosion will dominate the landscape, leading to siltation of dams downstream. Siltation of dams reduces their water storing capacity. In Botswana, there are a few good sites for damming and therefore the country cannot afford to mismanage its water catchment areas since this will lead to dams like Shashe, Letsibogo, Nywane and Gaborone experiencing a short lifespan.

According to Arntzen (1995), a population increase of 3.3 per cent in Botswana implies an annual population increment of around 50, 000. Using average per capita resource figures, these are the likely environmental implications (Arntzen, *ibid.*):

- An annual increase in wood consumption of 50 000 tonnes of biomass;
- An annual increase in cultivated land of approximately 10-25 000 ha;

- An annual increase in water consumption of 4.6^{3};
- An annual increase of 85 000 livestock units requiring at least 1 million ha (assuming an average carrying capacity of 12 ha/lsu); and
- The expansion of the human influence sphere reducing the natural wildlife habitat.

Energy Resources

One of the reasons for the complex relationship between population growth and resources is that resources acquire economic significance due to technological advancement. Increased technology has led to great expansion of non-renewable and renewable resources, which have been brought into use over time. This is exemplified by change from the use of fuelwood to wind and water power, coal, hydro-electric power, oil, solar power and nuclear power (Clarke, 1995). According to Arntzen (1995), a comparison of energy figures for the period 1981 and 1991, show that wood is still the most important urban and rural energy source for cooking. However, its share has decreased from 85.8 per cent to 64.5 per cent of the households in 1991 (see Table 6.3). A significant group has switched from wood to gas. For lighting purposes, paraffin is most important in the rural areas and the low-income areas of urban areas (Table 6.5). A growing minority has electricity (10.1 per cent in 1991). The use of candles and wood is declining. Even though the proportion of households who used fuelwood declined from 86 per cent in 1981 to 64 per cent in 1991, mainly because of switching to petroleum-based fuels, there has been an increase in the absolute number of households using fuelwood during this period due to the rapid increase in human population, estimated at 3.5 per cent per year (Table 6.6) (Government of Botswana, 1993, cited in Kgathi 1997).

Table 6.5: Energy Sources for Cooking

Energy Source	1981 Total	1991 Total	1981 Urban	1991 Urban	1981 Rural	1991 Rural
Wood	85.8	64.5	48.8	40.6	95.4	90.6
Electricity	1.8	2.7	7.7	4.8	0.2	0.2
Gas	5.4	21.7	18.9	2.0	2.5	2.0
Paraffin	6.4	10.8	23.3	18.2	2.0	2.5
Coal	0.3	0.1	1.1	0.1	0.1	0.1
Other	0.2	0.3	0.1	0.4	0.2	0.2

Source: Population Census (1991), Central Statistics Office, Gaborone, Botswana.

Table 6.6: Energy Sources for Lighting

Energy Source	1981 Total	1991 Total	1981 Urban	1991 Urban	1981 Rural	1991 Rural
Candle	24.5	11.8	1.4	14.1	30.5	9.3
Electricity	5.4	10.2	21.7	17.5	1.2	2.0
Gas	0.6	0.8	1.4	1.2	0.4	0.4
Paraffin	53.8	64.6	34.5	65.7	58.8	63.2
Wood	14.8	11.4	40.6	0.8	8.1	23.2
Other	0.8	1.2	0.3	0.5	0.9	1.9

Source: Population Census (1991), Central Statistics Office, Gaborone, Botswana.

According to Mogwe (1992), with the high rate of unemployment, low salaries, little electrification, and limited access to participation in development, the poor in the urban areas have little choice other than to use fuelwood unsustainably. Because of their marginalised position in development, their access to adequate living conditions and a healthy environment is greatly hindered.

Building Materials

The use of thatch is rapidly declining and corrugated iron is common (41.7 per cent in 1981 and 49.6 per cent in 1991). Similarly, there has been a marked increase in brick walls (49.3 per cent in 1993) (Table 6.7).

Table 6.7: Building Material for Houses (1981-1991)

A: Walls

Materials	1981 Average	1991 Average	1981 Urban	1991 Urban
Bricks	29.3	49.3	74.3	17.8
Metal	1.5	1.4	1.9	1.4
Mud/Poles/ Reed	68.4	45.8	23.6	79.8
Others	0.9	3.5	0.2	1.1

B: Roofing

Materials	1981 Average	1991 Average	1981 Urban	1991 Urban
Thatch	64.7	41.2	13.2	77.9
Iron/Zinc/Tin	34.4	49.6	86.4	21.5
Others	0.9	9.2	0.4	1.0

Source: Population Census (1991), Central Statistics Office, Gaborone, Botswana.

Water

Table 6.8 shows that around 70 per cent of the households have access to safe, piped water, in most cases, groundwater. In urban areas, the entire population has access to piped water. In smaller settlements and at the lands or cattleposts a variety of locally available water sources, some of poor quality, are used (boreholes, pans and hand-dug wells).

Table 6.8: Water Sources by Household (1981-1991)

Water Source	1981 Total	1991 Total	1981 Urban	1991 Urban	1981 Rural	1991 Rural
Pipedwater	56.1	77.0	97.2	100.0	45.5	51.6
Borehole/well	21.2	11.7	0.8	0	26.6	24.6
River Flow	5.0	1.5	0.5	0	6.1	3.1
Dry river	8.5	2.3	1.0	0	10.5	4.8
Dam /lake	5.5	2.8	0.1	0	6.9	5.9
Other	3.8	4.6	0.7	0	4.5	9.7

Source: Population Census (1991), Central Statistics Office, Gaborone, Botswana.

Agricultural Land and Livestock

It should be noted that the agricultural felling of trees rather than fuelwood collection is the cause of deforestation (Kgathi, 1997). Therefore, where arable agriculture is the dominant economic activity, one would expect to observe deforestation. Fuelwood is collected mainly from dead trees. From Table 6.9, it could be observed that there has been a decline in agricultural involvement. Agricultural activity in Botswana is mostly pronounced in the rural areas. The decline in agricultural participation is mostly pronounced in the arable sector. A substantial decline in arable agriculture would probably reduce deforestation as more and more people engage in non-agricultural activity. Government is blamed for encouraging deforestation through its policy of requiring people who have been allocated agricultural fields to clear their fields within five years from the date of allocation to avoid forfeiture (Sekhwela, 1997).

Table 6.9: Popular Involvement in Agriculture (as % of Households / Holdings)

Families/ Holdings with:	Households 1971	Households 1991	Agricultural Holdings 1981	Agricultural Holdings 1986	Agricultural Holdings 1990
Cattle	56.5	51.3	68.6	63.7	62.5
Goats	68.4	60.3	56.7	69.4	76.6
Cultivated arable land	74.9	31.8	84.1	74.5	77.3

Source: Population Census (1991), Central Statistics Office, Gaborone, Botswana.

According to Abucar and Molutsi (1992), the degradation of land is a human-made situation. If we are only concerned with the physical aspects of the land without understanding the socio-cultural fact behind man's use of the environment, we will not be able to resolve environmental problems in Botswana.

Poverty and the Environment

The Role of Poverty

In the developing countries, a substantial amount of environmental degradation is a result of poor people searching for the basic essentials of life: food, water, fuel and fodder. The majority of these poor masses press upon the marginal lands, destroying the very resources upon which they depend and contributing to the processes of soil erosion, desertification, deforestation and water shortage. Poverty is an important factor in outmigration, especially in Sub-Saharan Africa (Clarke, 1995). 'Although the poor will continue to be mostly rural, the numbers living in cities are rising fast, ensuring that P-E problems in cities will also increase' (Clarke, 1995: 23). Botswana is no exception to the conditions described above.

It should be noted that the rate of urbanisation in Botswana is currently considered the highest in the world. Between 1981 and 1991, the country's urban population grew at about 14 per cent (Chanda, 1996). According to Molebatsi (1996), the high urban growth rates are mainly due to rural-urban migration. It is observed that the driving force behind rural-urban migration is the non-productivity of the rural sector and consequently, the migrants include people who want to escape rural poverty. The movements amount to the poor shifting from the rural areas to the urban areas. This inevitably exerts more pressure on the environment in terms of more energy requirements, clearing land for more houses, increased water needs and consequently the problems of waste management. High levels of urban poverty affect the physical environment not only within the urban area, but the hinterlands as well. In Botswana, and probably in other countries, the most outstanding urban environment, related problems are poor housing structures, inadequate infrastructure provision, overcrowding and the location of low-income neighbourhoods on marginal environments.

Poor Housing Structures

The problem of housing is mostly evident in urban areas where some people construct houses in ungazetted areas, leading to what is normally referred to as squatter settlements. Houses found in these areas are far below the minimum required safety standards and building materials, which include mud, poles, grass cement, concrete bricks and corrugated iron sheets, are organised in a haphazard fashion. Molebatsi (1996) observes that quite often included in the list of building materials is an assortment of materials such as plastic bags, sacks, old tyres and cardboard boxes salvaged from dumping sites.

Overcrowding

Overcrowding in urban areas is a prevalent problem in Botswana. This problem is mostly common in low-income residential areas where plots are overcrowded with a number of housing structures. This has resulted in a marked population increase as recorded in the 1991 population census. According to Rantau (1992), upgraded areas like Naledi in Gaborone, Botshabelo in Selebi-Phikwe and Peleng in Lobatse are known to be overcrowded as indicated by the 1991 population census which recorded the population of Naledi as 20652, Botshabelo 13851 and 9037 for Peleng.

According to Molebatsi (1997), overcrowding in these areas could also be deduced from the number of building structures in a plot. The Development Control Code of 1978 provides for one building structure per plot, but the majority of plots in these areas have more than one structure. Rasesigo (1989) came to the conclusion that in most cases the second or third structure on a plot is used for renting purposes. The high population densities and structures per plot have significant bearing on infrastructural services in the low-income areas.

Pollution and Waste Management

As already highlighted, Botswana is one of the fastest developing economies of the world. This development has hitherto brought some environmental problems, which are relatively new to the country, and therefore authorities have not been thoroughly prepared to deal with such problems. Before independence in 1966, most of the waste generated by the residents was mostly biodegradable. But with rapid industrialisation came new materials like plastic bags, tin cans and bottles, which have been thrown indiscriminately into the environment. Authorities suddenly were overwhelmed with environmental problems that they were not familiar with. This meant both the rural and urban areas were becoming dirty.

As the population increases in both urban and rural areas, more waste is generated and therefore the need arises to manage waste. The Government of Botswana, with the help of the German Technical Co-operation (GTZ), has produced a waste management act. The Department of Sanitation and Waste Management was established in 1999, with a mandate to advise the Government on waste management issues in the whole country. In most small villages in Botswana, there is no waste collection and people have been encouraged to dig rubbish pits in order to dispose off waste they generate. In bigger villages like Kanye, Molepolole, Maun and Serowe, rubbish bins are provided in most places within the village but the collection is highly irregular, which results in rubbish being scattered over the place by animals. This poses a serious health hazard. Waste disposal methods in Botswana are rudimentary and have resulted in ungazetted dumps throughout the country. According to Segosebe and VanderPost (1991), the methods of waste disposal in Gaborone and other towns in Botswana are crude and this refers to land tipping and burning. Some of the dumps on which the tipping takes place were poorly constructed. Segosebe and VanderPost (1991) also reported the casual conversion of sandpits into dumping sites. In most low-income areas, some open space left for recreational purpose have been used as dumping sites.

Waste-related Issues

Essentially human groups consume the available resources and produce waste, which in an ideal situation would be recycled back to the resource base. Sustainable use of resources is quite simply the consumption at or below the level at which the resources regenerate.

If the population increases rapidly, then there is increasing competition for finite resources, a situation which leads inevitably to the poverty cycle, with accompanying miseries of famine, disease and resource destruction so apparent in many parts of Africa. Alternatively, there can be increased consumption of resources per capita, a characteristic of economic development. Although quality of life improves with development, the cost in resources is high, particularly in terms of dealing with waste.

In practice, Botswana is dealing with both problems simultaneously. The population is growing at 3.5 per cent per annum with 50 per cent of the population below the economic return. Improvements in health care and life expectancy have long made obsolete the need for large families and the notion prevalent in many developing countries that 'A Populous Nation is a Mighty Nation' has been shown to be untenable in the face of current ecological realities.

Sanitation

There has been a marked increase in pit latrines (an increase from 3.2 per cent in 1981 to 34.65 per cent in 1991). A significant minority use flush toilets. Despite this progress, an estimated 45 per cent of households still have no sanitary facilities (Table 6.10). Pit latrines and poorly constructed septic tanks increase the risk of nitrate pollution of groundwater.

The public latrine services are inadequate and the uninhabited areas of the city become latrine facilities, affecting the immediate environment and creating a health hazard. Batswana have been plagued by a 'throw away syndrome' and this has resulted in indiscriminate dumping of waste along the roads and streets and other ungazetted areas.

Table 6.10: Type of Household Sanitation, 1981-1991

Sanitation	1981 Total	1991 Total	1981 Urban	1991 Urban	1981 Rural	1991 Rural
Pit latrine	3.2	34.6	5.2	55.1	2.8	17.0
Flush toilet	8.6	12.8	33.5	23.7	2.2	2.6
Communal Facility	16.8	2.3	36.8	2.4	11.6	2.3
Pail/Bucket	0.3	0.1	0.9	0.0	0.2	–
Other	7.0	4.1	3.2	5.4	7.9	4.2
None	64.0	45.0	20.7	19.2	75.2	74.0

Source: Population Census (1991), Central Statistics Office, Gaborone, Botswana.

Air Pollution

As Botswana is slowly being industrialised, there are emerging problems with air pollution. The extent of air pollution in Botswana is still minimal. However, the mining town of Selebi-Phikwe in the north east of the country, is experiencing some air pollution problems (Asare, 1999; Asare and Darkoh, 2000). During the processing of copper, one of the waste products is sulphur dioxide which is released directly into the atmosphere. Once it is in the atmosphere, it can combine with water vapour leading to the formation of sulphuric acid. Sulphuric acid is a component of acid rain and therefore during the rainy season, this acid falls down on trees, water resources and buildings. The environmental impacts, which are discernible, are yellowing of tree leaves and corrosion of buildings. Although this problem is not severe, caution has to be taken to ensure that sulphur dioxide emission is minimised.

It was recently reported that some hospitals have been treating 'patients from Selibe-Phikwe complaining of symptoms attributed to sulphur dioxide pollution' (Anneleng, 1997:16; Asare 1999). This observation indicates that even though air pollution is not yet considered to be a threat, it is important that pollution abatement strategies be put in place in time to avoid disastrous effects.

Other Possible Impacts

As the population increases, more sanitary facilities will be needed. However, in most rural areas, pit latrines are mostly used and if not properly constructed, could lead to groundwater pollution. Batswana have developed a 'throw away' culture where tin cans, bottle sand plastic bags are just dumped into the environment. This practice makes the environment very untidy and unhealthy to human beings. More people exert more demand on already diminished natural resources and this is mostly apparent around big settlements.

The Linkage Between the Conceptual Framework and the Theories of Population and Empirical Evidence from Botswana

Having elucidated the conceptual framework and the population theories and relating them to the data collected for this study, the I=PAT model has been deemed most relevant. This is explained by the high population increase (3.5 per cent per annum) in the country which has led to an increased demand on natural resources. The improvements in the living conditions of Batswana has resulted in lower mortality rates, high internal migration rates and a fairly high but declining fertility, resulting in high population growth rate. The high population growth rate has in turn resulted in rapidly increasing population, which is faster than the rate at which resources are generated. More people have meant more consumption of resources that are already in short supply. Natural resources like fuelwood are being depleted at an alarming rate around settlements. Villagers now travel longer distances in search of veld products like thatching grass.

Botswana has one of the fastest growing economies in the world, which has resulted in high consumption patterns. Because of this affluence, the country is turning into a wasteful society with supermarkets and other retailers over-packaging bought items. The consequence of this is that plastic materials are discarded indiscriminately into the environment. This 'throw away syndrome' has led to plastics strewn all over the environment, posing a health hazard to all living organisms, including domestic animals and young children.

Technology has also influenced the status of Botswana's environment. Botswana is a dryland country with most of its water resources being underground. To have access to these water resources, borehole technology has been used to exploit them. As a result, livestock population tends to be concentrated around boreholes, leading to overgrazing of range resources. Cattle posts have spread into areas previously used only on temporary basis after good rains. Unlike in the past, livestock can no longer be moved easily to match grazing and water availability.

Although there are policies and programmes in place to address the existing and future environmental problems, they do not sufficiently deal with the problems. For instance, the Government of Botswana, upon allocating agricultural land for ploughing, recommends that the allocated field be cleared within five years from the date of allocation (Sekhwela, 1997). The practice encourages land clearing that may not necessarily be followed by ploughing. Thus, the policy encourages people to clear their agricultural land, which exposes the land to soil erosion. The next section focuses, in detail, on policy issues.

Policy and Programme Responses to Population-Environment Interaction in Botswana

Because problems of population growth and environmental decline require immediate attention, both remedial and preventive measures need to be instituted. Solutions can only be provided if the population-environment linkages are properly understood. In order to tackle the complex nexus of population-environment, we need a policy package covering several sectors and action areas (Hansen, 1994). Thus, a multi-pronged and a multi-sectoral approach to population-environment relationship is more appropriate than a single sector approach. Policy measures adopted by the government of Botswana are outlined below.

Population Policy Formulation

All governments should develop the capacity to integrate population factors into development planning. This process requires a comprehensive assessment of the trends and levels in population growth and distribution, both present and future (UN, 1991). Population growth affects the environment through overuse of agricultural lands, depletion of water supplies, decline in forest cover, spread of deserts, atmospheric pollution, etc. Thus, all population factors should be evaluated as part of national strategies with respect to their environmental linkages. Botswana has developed an antinatalist population policy, which has been approved by Parliament, and this policy takes cognisance of the fact that 'resources are under pressure because of increasing population and development needs. This has given rise to the concern about the capacity of the environment to sustain the needs of future generations' (Government of Botswana, 1997: 18).

The Information, Education and Communication (IEC) programme is important to foster awareness of the critical linkages between population and environment. All sectors of society should become thoroughly informed about the importance of population planning and environmental safeguards as key components of sustainable development. Educational programmes can be designed in such a way as to address issues of environmental concerns. Exposing students to environmental issues at an earlier age creates a greater chance of making people to be environmentally aware. Informal education can also be used to reach people who cannot be reached through formal education. The IEC of the national population policy of Botswana recognises that population growth negatively impacts upon the environment and therefore aims at sensitising the communities about environmental problems such as unsustainable methods of harvesting natural resources; creating awareness on the causes of land degradation and motivate communities to develop local strategies that mitigate against these; and to create a forum for the collection, processing, exchange and dissemination of environmental information (Government of Botswana, 1997).

Family planning should be readily accessible to all who need the services to allow both men and women to avoid unwanted pregnancies; reducing the incidence of abortions, and promoting birth spacing. This will indirectly lead to reduced population growth, which in turn will have less impact on environment. This is particularly useful where antinatalist population policies are not in existence.

Population Distribution, Migration and Urbanisation

Governments need to formulate policies that would result in balanced population distribution and regulated rural-urban migration. Population pressures around rapidly growing areas are visible. Policies to develop rural areas, which help to reduce urban growth by reducing rural push, can be implemented. The government should provide support to increase agricultural production and increase agricultural producer prices, which will reduce rural push. This approach encourages increased agricultural production in both the arable and pastoral sectors. Instead of discouraging rural-urban migration, the improved agricultural producer prices may ease it. Regulatory measures need to be instituted to guard against encroaching on fragile environment that may promote environmental degradation.

The Government of Botswana has noted that 'rural poverty, high birth rates, drought and environmental degradation have driven a significant proportion of the rural poor to urban centres in search of job opportunities and better life' (Government of Botswana, 1997: 16). The sparse population distribution and the attendant dispersal of settlements in the rural areas make it difficult and more expensive for government to provide essential services such as education and health services. The government has therefore deliberately embarked on policies aimed at stimulating development in the rural areas by expanding and improving physical and socio-economic infrastructure, creating alternative growth points to achieve a more even population distribution, and creating employment opportunities in the rural areas (Government of Botswana, 1997: 29)

Rural development is usually a top priority for most African leaders, especially when they come to power. Rural areas are settlements with the largest portion of the population and supply the urban areas with food products and manpower. Despite rapid urban growth, the Government of Botswana has emphasised rural development through the implementation of her successive National Development Plans. Rural development is implemented based on improved agricultural production, raising income levels of the rural communities and the provision of social services and facilities (Government of Botswana, 1992).

Women's Status

Fertility is generally high among women who are denied the opportunity to enjoy productive life. Lower fertility rates can be achieved by availing opportunities to women such as education, health and employment. In addition, emancipated and better educated women are more able to make vital contributions to resource management and conservation. In Botswana, it has been noted that there are various socio-cultural values and laws which act as barriers to women's full participation in the development process, particularly in decision-making (Government of Botswana, 1997: 20). Because of this, the government has made deliberate efforts to develop appropriate institutional mechanisms and programmes for effective integration of women in development policy. The following measures have been adopted to improve women's status and empower them: review and enact laws to ensure gender equality in access to and control of productive resources; develop training programmes to provide women with the necessary skills for empowerment; encourage women to participate in male dominated occupations; and establish programmes both in formal and informal sectors for women and to provide appropriate resources to enable them start up business enterprises.

Poverty Alleviation

Policies that can change the intergenerational flow of wealth from children to parents so that wealth instead flows from parents to children will lead to low fertility. This means providing the poor with better access to markets and social services (schools, extension services and health centres). The policy calls for the development of rural social and physical infrastructure and public infrastructure in peri-urban areas. Providing better income-earning opportunities increases the costs of raising children, which in turn results in smaller families. Smaller families mean that better health, nutritional and educational services can be provided.

The Government of Botswana has a policy aimed at improving the standard of living of Batswana, which dates back to the early 1970s. The Accelerated Rural Development Programme (ARDP) was devised as a strategy to improve social services and infrastructure and create jobs and income generating activities for the rural majority (Government of Botswana, 1973; 1977; Ditlhong, 1997). Most of the policies discussed below do have a bearing on poverty alleviation. These policies include Arable Lands Development Programme, whose main objective is to increase arable production, enhance rural development, improve the welfare, and create productive and remunerative employment in rural areas (Ditlhong, 1997). The other policy is the Financial Assistance Policy (FAP), whose main objective is to stimulate sustainable manufacturing growth and productive employment. The hope was that through FAP, employment creation would be achieved. Other policies include basic education provision, labour based works, and destitute policy. Under the destitute policy, the poorest people in Botswana are provided with basic necessities such as food shelter, health care, and education areas (*ibid.*).

Environmental Policy

Within the national population policy, the Botswana government acknowledges the intricate relationships between population, environment and development. Policy measures that attempt to prevent or mitigate the negative impacts of population growth on the environment are clearly articulated. The policy measures outlined are: 'create public awareness on the relationships between population and the environment; safeguarding the quality of, and preventing over-exploitation of renewable and non-renewable resources; intensify afforestation programmes; strengthen the enforcement of existing laws, and enacting laws to enhance the protection of the environment including environment impact assessment legislation; incorporate population and environment education in the programmes of educational and training institutions; strengthen the capacity for the co-ordination and management of environmental issues; and promote the production of reusable, recyclable and environmentally friendly packaging materials' (Government of Botswana, 1997: 33).

Depletion of Forest Resources Policy

The Government of Botswana has proposed to preserve woodland areas through policy measures such as promotion of fuelwood substitutes by use of subsidies, promotion of planting trees and woodland management, to develop an economically viable Forestry Industry, and proper management of the existing resources (Government of Botswana, 1990).

Energy Policy

In Botswana, dependence on fuel wood as a source of energy for cooking has led to the emergence of deforested zones. Because of the high population growth and a rapidly modernising economy, demand for energy is likely to increase. In recognition of this likely problem, the government plans to 'provide technical support for the development of community based woodlot schemes; increase accessibility and affordability of electricity; and develop appropriate and affordable energy saving technologies and promote their safe use' (Government of Botswana, 1997: 33). These measures are expected to reduce the impact of increasing population on the environment.

Water Resources Policy

Two-thirds of Botswana's population depend upon underground water. The demand for water resources will increase with increasing population size. As demands for water increase, competition among different water users will emerge, particularly among towns, mining, irrigated agriculture, rural and major villages. Increased water demand has therefore necessitated the initiation by the Government of a long-term Water Master Plan to review existing water legislation, assess water demands and availability of water resources to meet this demand up to the year 2020 (Government of Botswana, 1997: 19). The policy suggests improved planning and administrative measures in the interests of both protecting water resources against pollution and improving multipurpose use; the extension of water tariff system into the rural areas; and an incentive to encourage the collection of rain water (Government of Botswana, 1990). The government further suggests enhancing capacity for early warning systems for monitoring, assessing and forecasting climatic and environmental changes and enhancing water exploration testing techniques and strengthening water treatment and recycling procedures (Government of Botswana, 1997).

Industrial/Urban Pollution Policy

The existing measures that will continue into the future are contained in the Atmospheric Pollution (Prevention) Act, the Water Act, and the Public Health Act. Amendments will cover all aspects of licensing, registration, reclamation, waste disposal, industrial discharge, the import and export of chemicals, emission/waste standards, etc. Anti-pollution initiatives such as the establishment of waste processing factories and the introduction of incentives to encourage recycling are to be investigated (Government of Botswana, 1990).

Conservation of Wildlife, Heritage and Cultural Resources

The proposed policy includes gazetting of the designated Wildlife Management Areas, the upgrading of some game reserves to protect all forms of wildlife, expanding National Park boundaries on account of their biotic interest, the preparation and implementation of management plans; the provision of increased anti-poaching controls, and the execution of wildlife cropping projects mainly for the benefit of local communities (Government of Botswana, 1990).

Policy on Rangeland Pasture Degradation Issues

The issue of pasture degradation is said to be the hardest to solve because the suggested measures run counter to traditional customs. The suggested policy measures are: strengthening the Range Ecology Unit within the Ministry of Agriculture; extending the zoning and gazetting of land for use of livestock and wildlife; developing a comprehensive water supply policy for all forms of livestock and wildlife; improving the provision of information to livestock farmers concerning carrying capacity; and improving livestock marketing infrastructure to improve offtake (Government of Botswana, 1990).

Conclusion

In this chapter, the development of a clear conceptual framework and an understanding of the interactions between population factors and the environment are needed to best deal with environmental issues. There is clearly lack of data to argue that rapid population growth causes environmental degradation. However, there is evidence that the clearance of forests for agricultural household purposes, poor waste management practices, the increased demand for energy resources, the large numbers of livestock and poverty are some of the causes of environmental problems in Botswana. Poverty and large family sizes create demands that force a substantial number of people to exploit the environment on an unsustainable basis. Large families mean that more resources are needed to sustain such families. Because these people are poor, it implies that most children are unlikely to receive good education and health care services. Lack of these services leads to poor human resource development which in turn leads to poor economic performance. Poor economic performance results in poverty. Thus, the vicious poverty-population-environment circle that reinforces poverty, sustains high population growth and leads to increased pressures on fragile natural resource systems. Soil degradation, the depletion of forest cover and the lack of agricultural land ultimately force the rural poor to migrate to urban centres, creating spontaneous squatter settlements that are evident in most of the developing countries, including Botswana.

In order to lower the rate of population growth, policies geared toward: the reduction of fertility levels through good maternal and child care, and family planning services, formal education, particularlyof women, and employment opportunities for women outside thehome are required to promote environmental conservation. Creation of employment opportunities in the rural areas is likely to reducerural-urban exodus. The policies and programmes created by the Government of Botswana in tackling the impact of population on the environment are a testimony to the fact that the government is concerned about environmental degradation and is trying its best to resolve emerging conflicts. How successful these policies and programmes are remains to be seen and is beyond the scope of this chapter. All that can be concluded from the evidence presented in the chapter is that rapid population increase negatively affects the environment and proper policy and programme measures are required to address these negative impacts.

References

Abel, N.O.J. and P.M. Blaikie. 1990. *Land Degradation, Stocking Rates and Conservation Polices in the Communal Rangelands of Botswana and Zimbabwe*, Pastoral Development Network, Paper 29a., ODI, London.

Allan, W.A. 1972. 'How much land does a man require?' In M. Prothero (Ed) *People and Land in Africa South of the Sahara*, Oxford University Press, New York.

Anneleng, D. 1997. 'Norway may help Phikwe solve air pollution problem', in *The Botswana Gazette*, Wednesday 26, November, Gaborone.

Arntzen, J.W. and Veenendal, E.M. 1986. *A Profile of Environment and Development in Botswana*. IES, Free University: Amsterdam and NIR: University of Botswana.

Arntzen, J. 1989. *Environmental Pressure and Adaptation in Rural Botswana*, PhD Dissertation, Virje University, Amsterdam, Holland.

Asare, B. 1999. 'Socio-economic and Environmental Impacts of Mining in Botswana: The Case Study of the Selebi-Pikwe Copper-Nickel Mine', *Eastern Africa Social Science Research Review*, 17, 1-42 .

Biot, Y. 1990. *How Long Can High Stocking Densities be Sustained?* Paper to the Technical Meeting to the Savannah Development and Pasture Production, Woburn, United Kingdom, 19-21 November, 1990.

Boserup, E. 1981. *Population and Technology*, Basil Blackwell, Oxford.

Botswana Society. 1984. The *Management of Botswana Environment*, Proceedings of a Workshop, Gaborone, 1984.

Botswana. 1986. *North East District Development Plan 1983-1989*. Ministry of Local Government and Lands, Gaborone, Botswana.

Central Statistics Office. 1994. *Population Report 1989-1991*, Gaborone.

Clarke, J. I. 1995. 'Introduction', in J. I. Clarke and L. Tabah (Eds), *Population-Environment-Development Interactions*. Committee for International Cooperation in National Research in Demography (CICRED), Laxenburg, Austria, pp. 1-10.

Cooke, H.J. 1991. 'Environmental conservation: the road to survival', *KCS Newsletter*, 33: 4-16.

Cooke, H.J. and R.M.K. Silitshena. 1986. *Botswana – An Environmental Profile*, UNEP Desertification Branch, Nairobi.

Cooke, J. 1981. 'Environmental change at Oliphant's Drift in Kgatleng District, South East Botswana', *Botswana Notes and Records*, 13, 123-129.

Dahlberg, A. 1994. *Contesting Views and Changing Paradigms. The Land Degradation Debate in Southern Africa*, Nordiska Africa Institute, Discussion Paper 6, Uppsala.

Darkoh, M.B.K. 1999. 'Desertification and environmental management in Botswana', In *Environmental Planning, Policies and Politics in Eastern and Southern Africa*, Salih, M.A. and S, Tedla (Eds), Macmillan, London.

Darkoh, M.B.K. 1998. 'The nature, causes and consequences of desertification in the drylands of Africa', *Land Degradation and Development* 9: 1-20.

Davis, K. and M.S. Bernstan (Eds), 1991. 'Resources, Environment, and Population: Present Knowledge, Future Options', *Population and Development Review: A Supplement* 16, pp 19-29.

Ditlhong, M. 1997. *Poverty Assessment and Poverty Alleviation in Botswana.* BIDPA Working Paper 12.

Ehrlich, P.R., Ehrlich, A.H. and Holdren, J.P. 1977. *Ecoscience: Population, Resources, Environment,* W.H. Freeman and Co, San Franscisco.

Field, D.I. 1978. *A Handbook of Basic Ecology for Range Management in Botswana,* Land Utilisation Division. Ministry of Agriculture, Gaborone, Botswana.

Fortmann, L. 1989. 'Peasant and official view on rangeland use in Botswana – fifty years of devastation?' *Land Use Policy* 6, 197-202.

Government of Botswana. 1973. *Report on Rural Development.* Government Printer, Gaborone.

Government of Botswana. 1977. *Botswana's Accelerated Rural Development Programme 1973-76: Experiences and Lessons,* Government Printer, Gaborone.

Government of Botswana. 1990. *National Settlement Policy,* Government Printer, Gaborone.

Government of Botswana. 1992. *The National Conservation Strategy: National Policy on Natural Resources Conservation and Development,* Government Printer, Gaborone.

Government of Botswana. 1996. *National Population Policy (Draft) Ministry of Finance and Development Planning,* Government Printer, Gaborone.

Government of Botswana. 1997. *Population IEC Document (Draft) Ministry of Finance and Development Planning,* Government Printer, Gaborone.

Hansen, S. 1994. 'Population and Environment', in *Population Growth and Sustainable Development,* African Development Bank, Abidjan, Cote d'Ivore, pp 229-280.

Kalipeni, E. 1996. 'Demographic responses to environmental pressure in Malawi', *Population and Environment: A Journal of Interdisciplinary Studies,* 17, 4, 1-16.

Kgathi, D.L. 1997. 'Introduction', in Kgathi, D. L., Hall, D.O., Hategeka, A., and Sekhwela, M.B.M. (Eds), *Biomass Energy Policy in Africa: Selected Case Studies,* ZED Books Ltd, London.

Khupe, B.B. 1994. *Integrated water resource management in Botswana,* in Gieske, A. and Gould, J. (Eds), *Intergrated Water Resources* Workshop, Workshop proceedings, Gaborone, Botswana.

Kinlund, P. 1996. *Does Land Degradation Matter?: Perspectives on Environmental Change in North-eastern Botswana,* Almqvist and Wiksell International, Stockholm.

Kwerepe, R. 1997. 'Land Degradation and Desertification Processes', Paper Presented at the National Forum, National Action Programme on Convention to Combat Desertification and Drought, October 20-22, Gaborone, Botswana.

Lowe, M.S. and Bowlby, S.R. 1992. 'Population and Environment,' in A.M. Mannion and S.R. Bowlby (Eds) *Environmental Issues in the 1990s,* John Wiley and Sons, Chichester.

Martine, G. 1995. 'The Relations Between Population and Environment in the Context of Economic Globilization, Preliminary Considerations', in J. I. Clarke and L. Tabah (Eds), *Population-Environment-Development Interactions* CICRED, Luxenburg, Austria, pp 95-116.

Milas, S. 1987. 'Population and environment: the gathering crisis,' *The Courier,* 103, 35-47, May-June.

Miller, Jr., G.T. 1996. *Living in the Environment: Principles, Connections and Solutions,* 9th Edition, Wadsworth, New York.

Perkins, J.S. 1991. 'The Impact of Borehole Dependent Cattle Grazing on the Environment and Society of Eastern Kalahari Sandveld, Central District Botswana', Unpublished PhD Thesis, Sheffield: University of Sheffield.

Perkins, J.S. and D.S.G. Thomas. 1993. 'Spreading deserts or spatially confined environmental impacts?', *Land Degradation and Rehabilitation* 4: 179-194.

Poel, P. Vander. 1980. *Rainfall Erosivity and its Use for Soil Loss Estimation,* Division of Land Utilisation, Ministry of Agriculture, Gaborone, Botswana.

Rakgoasi, S. D. and Gaisie, S. K. 1999. Population growth and composition, in Gaisie, S. K. and Majelantle, R. G. (Eds), *Demography of Botswana: Change in Population Size and Structure,* Mmegi Publishing House, Gaborone, Botswana.

Rensburg, H.J. Van. 1971. *Range Ecology in Botswana,* FAO Technical Document No. 2 UNDP/SF/359 (BOT1), Rome, Italy.

Repetto, R. and Holmes, T. 1983. 'The Role of Population in Resource Depletion in Developing Countries', in UNFPA Readings in Population Research Methodology, Vol. 8. *Environment and Economy,* pp. 29–50.

Repetto, R. 1993. 'Population, resources, environment: an uncertain future', *Population Bulletin* 42, 2, 23-40.

Ringrose, S. 1987. 'The development of land use/land cover classification system and an application of its usage in Botswana', *Botswana Notes and Records*, 19, 75-82.

Segosebe, E.M. and C. VanderPost. 1991. *Urban Industrial Waste Pollution in Botswana: Practices, Attitudes and Policy Recommendations*, African Caribbean Institute.

Sekhwela, M.B.M. 1997. 'Environmental impact of woody biomass use in Botswana – the case of fuelwood,' in Kgathi, D. L., Hall, D.O., Hategeka, A. and Sekhwela, M.B.M. (Eds) *Biomass Energy Policy in Africa: Selected Case Studies*. ZED Books Ltd, London.

Thirwal, A.P. 1983. *Growth and Development with Special Reference to Developing Countries*, McMillan Ltd, London.

Toure, M. 1994. 'Population, urbanisation and migration in Africa', in *Population Growth and Sustainable Development*, African Development Bank, Abidjan, Cote d'Ivoire, pp 119-162.

UN 1984. *Population, Resources, Environment and Development, Proceedings of the Expert Group on Population, Resources, Environment and Development*, Geneva, 25-29 April, 1983.

UN 1991. *Population, Resources and the Environment: The Critical Challenges*, UNFPA. New York.

VanderPost, C. 1995. Population change and environmental problems in the Mid-Boteti region of Botswana, *GeoJournal*, 35,4, 521-529.

Vegten, J.A. Van. 1981. *Man-Made Vegetation Changes. An Example from Botswana's Savanna*, NIR Working Paper No. 40, Gaborone, Botswana.

Weeks, J.R. 1981. *Population: An Introduction to Concepts and Issues.* Wadsworth Publishing Company, Belmont.

White, R. 1992. 'Livestock Development and Pastoral Production in Communal Rangeland in Botswana', *Workshop on New Directions in African Range Management Policy*, Matopos, Bulawayo, Zimbabwe.

Williamson, D. 1989. *An Environmental Analysis of Botswana*, The IUCN Conservation Monitoring Centre, London.

Acknowledgement

This is a revised version of a chapter which first appeared in *The Journal of Eastern African Research and Development*, Vol. 29, 1997/98. It is reproduced here with the permission of the authors and publisher.

Chapter 7

Environment and Forced Migration: A Review

W. F. G. Cardy

Introduction

We are all migrants or descendants of migrants; our ancestors appear to have come from East Africa. A glance at the history of the development of mankind shows a consistent spreading of our homelands, a widening of our influence and a consolidation of our impact into every corner of the globe. Only in the Antarctic are we still trying to establish our hold.

We have not always been successful; we have not everywhere maintained sustainable settlements; we have suffered reversals as a result of climatic fluctuations and our over-exploitation of the environment. When these reversals occurred, and when the local conditions became intolerable, we either overcame the obstacles or we migrated to somewhere else.

For us as a species there always has been somewhere else to go. There is still room on earth for a larger population but never before have so many people shared this planet. There are more people alive now than the total of all the people who have lived and died in the two million or so years since modern man appeared. The resources we need for life are still available, but not always where they are needed, and they are being depleted rapidly. The pressures for migration have never been so great; the pressures that result from migration have never been so challenging.

The interaction of the environment and forced migration has received increasing attention since publication of a book entitled *Environmental Refugees* by Essam El-Hinnawi in 1985. This chapter addresses the environmental causes of migration and the environmental effects of migration as well as the research and actions that are required to address the issues.

The Issues

Our first priorities are to find and preserve an adequate environment for survival. If we cannot live in one location we move and try elsewhere. If others are already there, we may threaten them and their environment and compete for resources. Stress, unrest, conflict and even war may develop, frequently resulting in further migration. It has become clear that there are increasing numbers of migrants resulting from environment related causes, but there is a lack of rigorous data on their numbers.

People that migrate as a result of purely environmental factors often remain within their own country and are not technically refugees. The original UN convention definition of refugee is very restrictive and does not encompass environmental migrants and especially not the 'internally displaced'.[1] Officially, to become a refugee, a migrant must cross national borders and should be escaping persecution. There has been much discussion recently on definitions but the concept is not easy to define or quantify. There is a need for national and international attention to be drawn to the needs of the internally displaced, especially those suffering from environmental pressures. However, there is a proper concern about using the term refugee too loosely and so degrading its significance. Nevertheless, the term 'environmental refugee' has caught the attention of the media and the public.

The term 'environmental refugee' was given prominence by the United Nations Environment Programme's publication of the same name (El-Hinnawi, 1985). Since then the subject has received increasing attention and recognition. In 1987, there were estimated to be as many as 10 million environmental refugees in the world (World Resources Institute, 1988). Subsequently in a carefully prepared study, Myers and Kent (1995) estimated that there were some 25 million environmental refugees, and projected that there could be as many 50 million by 2010. These estimates have focussed attention on the precise definition of 'environmental refugee'.

The definition suggested by El-Hinnawi applies to 'those people who have been forced to leave their traditional habitat, temporarily or permanently, because of a marked environmental disruption

---

1. The term 'refugee' shall apply to any person who . . . owing to a well founded fear of being persecuted for reasons of race, religion, nationality, membership in a particular social group or political opinion, is outside the country, and is unable or, owing to such fear, is unwilling to avail himself of the protection of that country.

(natural and/or triggered by people) that jeopardised their existence and/or seriously affected the quality of their life. In this definition 'environmental disruptions' means any physical, chemical and/or biological changes in the ecosystem (or the resource base) that render it, temporarily or permanently, unsuitable to support human life. According to this definition people displaced for political reasons or by civil strife, and migrants seeking better jobs purely on economic grounds, are not considered environmental refugees.' Note that this definition does not qualify for recognition as a refugee under the Convention.

El-Hinnawi proposed that 'There are three broad categories of environmental refugees. First there are those who have been temporarily displaced because of an environmental stress. Once the environmental disruption is over and the environment rehabilitated, they return to their original habitat. This is usually the situation with populations displaced by natural hazards such as earthquakes or cyclones or an environmental accident' (for example an industrial accident that created temporary environmental disruption such as the chemical plant explosion at Bophal, India).

The second category of environmental refugees comprises those who have to be permanently displaced and re-settled in a new area. They are displaced because of permanent changes, generally man-made, that affect their original habitat – in the case of the establishment of large reservoirs for example, the relocation effects of nuclear accidents such as at Chernobyl, or of sea level rise expected as a result of global warming.

The third category of environmental refugees consists of individuals or groups of people who migrate from their original habitats, temporarily or permanently, in search of better quality life. Generally they only do this if the resource base in their original habitat has deteriorated to such an extent that it can no longer support them. Share croppers and smallholders whose lands are being waterlogged and salinised and who cannot afford capital investment necessary to reclaim them, often give up their holdings and migrate to nearby urban centres in search of other jobs.

All these causes result in migration within the home country, but the third in particular is likely to give rise to migrations beyond the home country borders and thus create those who may be officially defined as 'refugees' if they can establish 'persecution' at home (El-Hinnawi, 1985).

In addition to these specific causes other factors have an

influence. Migration in areas such as West Africa has long followed traditional routes and flowed back and forth as long as trade was possible. However Bimal Ghosh (1993) suggested that when 'trade flows freely, migration slows down' referring to involuntary migrations rather than those driven by economics. He concluded: Trade is not a complete answer to migration. But trade, aid and foreign direct investment can be woven into a new and dynamic fabric of economic co-operation between labour-surplus and capital-rich countries to make international migration more manageable and productive.

If nations are serious about reducing the threat of disruptive mass migration, generated or exacerbated by poverty and deprivation, they need to develop and implement effective trade policies, evolved in a co-operative framework. In 1992, some $500 billion a year was lost to the poor countries because of restrictions on, and unequal access to, international trade, financial and labour markets (this amounted to nearly ten times the aid provided to these countries) (UNDP, 1992).

Most of the demand for natural resources in the developing and less developed countries comes from the industrialised countries (Tolba, 1992). According to the United Nations Development Programme (UNDP, 1992) the richest 20 per cent of the population of the world in 1960 had incomes 30 times greater than the poorest 20 per cent. By 1990 the richest 20 per cent had incomes that were 60 times greater than the poorest 20 per cent. The gap between the rich and the poor within countries continues to expand in most countries of the world. Faced with an over-populated and degrading environment and visions of wealth in towns or abroad, it is not surprising that migrations are on the increase.

The Numbers Involved

It is not easy to determine the extent of the problem. There are fairly good statistics on the official refugee situation. Since 1991 the total of official refugees and asylum seekers worldwide climbed from 16.6 million to 17.6 million in 1992 and then gradually declined to 13.5 million in 1998 before increasing again to 14.1 million in 1999 (UNHCR, 2000). By definition, these figures do not include purely environmental migrants, although environmental factors may well have played a role in the displacement of many.

In addition to 'official' refugees, in many countries there are internally displaced people that do not have formal refugee status and consequently have no access to international help. In Sudan there are up to 4 million internally displaced persons and more than 400,000 Sudanese refugees in other countries. That is ten internally displaced for every official transboundary refugee. Close to 2 million live in shantytowns around the capital Khartoum which, they prefer to the miserable life in the camps. According to the UNHCR (*ibid.*) there are at least 21 million internally displaced people in the world. Many of these may well be environmentally induced migrants, but there is no way at present of determining how many.

In 1988 a Worldwatch report estimated that about 10 million people could be considered environmental refugees (Jacobson, 1988). In 1992 Westing (1992) suggested:

> The net numbers of both 'recognised' and 'unrecognised' refugees have been increasing steadily in recent years. Indeed the world-wide net increase in 'recognised' cross border refugees has been about 1 million per year; that of unrecognised cross-border refugees about half a million per year and that of unrecognised internal refugees about 1.5 million a year. What is being suggested here is that the ranks of all three of these categories are being swelled by environmental refugees rather than by political or social refugees.

Norman Myers (1993) reviewed the possible impact of global warming and climate change with a view to projecting migration pressures. Conservative estimates suggested that 30 million in China, 30 million in India, 15 million in Bangladesh and 14 million in Egypt may be displaced by sea level rise as a result of global warming. In China, India and Bangladesh twice as many may actually be affected in due course. An additional 50 million people may be displaced by famine induced by climate-change.

Further work by Myers and Kent (1995) resulted in detailed estimates of the likely numbers of existing environmental refugees. Using reasonable arguments based on a variety of available estimates they concluded that there were probably as many as 25 million environmental refugees at that time – people that had been displaced from their homeland as a result of environmental factors or events.

On the basis of this work they made projections that environmental refugees from all causes would reach 50 million by 2020 and around 150 million by 2050. This would represent an increase of environmental refugees to 1.5 per cent of the then predicted world's population of 10 billion, compared with 0.25 per cent of some 6 billion at the time of writing.

One billion people are dependent on agriculture for a living in the productive drylands of the world and are therefore at risk from drought and desertification. Another one billion people are living in urban shantytowns. Urban areas in Africa are still growing at rates of 6 per cent per annum as people leave their rural homelands. The Minister of Environment in India stated in 1994 that 30 000 new environmental refugees arrive in Delhi every month.

Lonergan (1995) divided examples of environmental events that may result in population displacement into five categories. He listed natural disasters, cumulative changes, accidental disruptions or industrial accidents, development projects and conflict and warfare. He noted that the severity of natural disasters in terms of their human impact has increased over the years rising from 28 million people affected in the 1960s to his estimate of 64 million in the 1980s.

Valuable recent data relating to this matter is offered by the statistical work of the Red Cross and Red Crescent Societies. Latest estimates from the World Disasters Report 1999, list those affected by environmental disasters at about 70 million annually in the 1970s; 163 million in the early 1980s; about 179 million annually between 1988 and 1992 and about 163 million between 1993 and 1997. The overall annual average of those affected for the period between 1973 and 1997 is estimated at 144 million. Of course, as the report notes, defining those affected is extremely arduous and different standards are used in forming the estimates. It is likely that reporting and quantification has improved over this period, so it may well be that a figure of around 160 million could be taken as a reasonable annual average expectation for those that will be 'affected' by environmental events. Most of those affected, however, are unlikely to be displaced, even temporarily.

A finer focus is provided by another table from this report, which shows the annual average number of people, reported made homeless by region and period (1973-1997). Homeless is defined as the number of people reported needing immediate assistance with shelter. These people were clearly displaced at least temporarily. Most will have returned to their home base, but some will undoubtedly have become migrants. The resulting estimate of the annual average homeless worldwide is some 4.87 million.

During the same period, the Report indicates that the average annual number of people reported affected by floods was some 66 million and by drought and famine some 61 million. However, the average number of people reported made homeless by floods was only 3.4 million and by drought and famine only 21,920. It is very likely that these last became environmentally induced migrants, but it is unlikely that many of the 3.4 million rendered homeless by floods, became actual migrants, as most will have returned to their home land once the waters receded.

The Red Cross work includes separate data on earthquakes, drought and famine, flood, high wind, landslide, volcano and others with a 'natural trigger'. Disasters with a 'non-natural trigger' are also identified, including accidents, technological accident, and urban fire. The tables could form a useful basis on which to build a more reliable assessment of the numbers of environmental refugees. The Red Cross says that natural disasters in 1999 were the worst on record, creating 58 per cent of the world's refugees. 'Everyone is aware of the environmental problems of global warming and deforestation on the one hand and the social problems of increasing poverty and growing shanty towns on the other', said Astrid Heiberg, the President of the International Red Cross, 'but when these two factors collide, you have a new scale of catastrophe.'

Security Issues

Environment-based conflicts are already happening and these frequently originate with conflicts over access to resources. The recent war in the Gulf was at least partly a reflection of this. Inequalities in access to natural resources or to an adequate environment lead to social instabilities that tend to occur first in or around the least developed countries. Tensions over shared resources range from individual killings over disputed land and water rights on the Tana River estuary (*Daily Nation*, Nairobi, 9 March, 2001) through those that arise periodically between Mauritania and Senegal over the shared Senegal River, to those between Turkey and Syria and Iraq and between India and Bangladesh.

Naigzy Gebremedhin (1991) illustrated the connection between environment and security in describing the flow of refugees between and within the States of Ethiopia and Somalia. 'Thousands of refugees have fled from the effects of environmental deterioration as well as the ravages of war. The two factors are difficult to separate. War has undoubtedly contributed to environmental degradation, but ecological crisis is itself a source of political conflict and instability in this region'.

As the Brundtland Report (*Our Common Future*, 1987) found, 'Poverty, injustice, environmental degradation and conflict interact in complex and potent ways. The immediate cause of any mass movement of refugees may appear to be political upheaval and military violence. But the underlying causes often include the deterioration of the natural resources base and its capacity to support the population'.

A report commissioned by the Ethiopian Relief and Rehabilitation Commission on the drought and famine in the early 1970s found that: 'The primary cause of the famine was not drought of unprecedented severity, but a combination of long-continued bad land use and steadily increased human and stock populations over decades'.

Once war breaks out the effects are exacerbated. In what is now Eritrea, the dusty highlands are deeply scarred by erosion; the Ethiopian army used to uproot trees to destroy the cover provided for Eritrean soldiers (*The Economist*, 1993). The effects of the widespread aerial defoliation during the war in Vietnam are well known. The massive effort in Iraq to drain the swamplands and displace the descendants of the ancient Sumerians is also well documented.

The movement of refugees in turn adds to political tensions between the neighbouring countries and ecological degradation in the host countries. The influx of a large number of environmental refugees places additional pressure and demand on water resources and grazing land. The arid region in Ethiopia and Somalia which hosts thousands of refugees suffers from a dearth of water (Gebremedhin, 1991). Iran also has recently been high on the list of countries accepting refugees, although not itself provided with abundant resources.

Most governments, including those of Ethiopia and Somalia, have adopted policies which regard refugee settlements as temporary phenomena. Such policies do not encourage the implementation of long-term development activities such as re-afforestation. As a result the deforestation around refugee settlements may be more pernicious and irreversible than deforestation elsewhere.

Environmental Impacts

The environmental impacts of migration include the following: over-exploitation of natural vegetation which may lead to increased soil erosion and gully formation, increased water run-off and siltation of rivers and reservoirs downstream; pollution of water and soils; depletion of water resources; changes in land use from sustainable to unsustainable; reduction of loss of 'reserve lands' for local population in drought periods; spreading of and decreased resistance to various kinds of diseases between hosts and refugees.

Refugees often live in camps with a population of several hundred thousand people. The causes of environmental degradation in these situations are similar to those existing in normal communities of a similar size but the effects are more pronounced. This is due to: disproportionately high population densities; the fact that refugee camps are often located in environmentally fragile areas; the frequently inadequate infrastructure; the fact that the incentives for refugees to maintain their environment may be low due to the trauma of resettlement or war, and because the land where they are living is not theirs.

As Richard Black pointed out, 'the size of the refugee and other displaced populations is a key factor determining the extent of the area's resources that must be drawn on to sustain them. The length of time over which the refugees have been displaced, the condition they are in, the opportunities for them to be gainfully employed and establish roots, and the length of time they must remain in the camp are all key factors' (Black, 1993).

Of critical importance also are the conditions under which they are settled. 'So often camps and settlements are established late, in a hurry, with inadequate planning and support and infrastructure. In such a situation protection of the environment is not a prime consideration.' Indeed Black found that only about half of the NGOs that he polled had yet developed firm guidelines for environmental protection (*ibid.*).

The impacts on the host environment may be extended to include those on the host people. Frequently they at first welcome the newcomers but then they find they have to share their resources and perhaps their reserves with the new arrivals. This may totally undermine the sustainability of their own livelihood patterns developed over years, which may well be needed to protect them against cyclical or unexpected environmental disasters. They may also find that much attention is being given to the refugees but none to their predicament. What appears to be unused land to those seeking a site for a camp, may have been the local communities' reserve, held as insurance against drought. Worse, the competition may reach the point where they themselves find they can no longer sustain themselves and they may themselves be forced to migrate, as a consequence of the impact of the refugee camp placed in their midst.

Susan and William Tamondong-Helin (1992), give a graphic account of these impacts in their paper 'Migration and the Environment: Interrelationships in Sub-Saharan Africa' including the migrations from Mozambique to Malawi. They described the impact of refugees on those that welcome them; the allocation of traditional reserve lands to refugee camps; the competition with refugees for wood and water and the long-term continuing impacts. The refugees may receive external support but the locals do not, although the locals may benefit somewhat from improved infrastructure built primarily for the refugees. The authors recommended very forcefully that 'donors and supporting organisations should look specifically at the Malawian (host) villagers, who are seriously affected by, but taken for granted in the midst of the refugee movement'.

As Black (1993) points out: 'problems may also arise because of the refugees' unfamiliarity with local environmental circumstances, they lack knowledge of local systems of sustainable management; ignorance of local regulations due to poverty or political insecurity; and an inherent short-term outlook deriving from their temporary status.'

Black emphasises that environmental assessment should not become a mechanism through which refugees are denied protection. Environmental assessment should focus on how the unavoidable environmental impacts from concentrations of migrants can be minimised.

Where there are environmental impacts, care should be taken to avoid rushing to resolve them too simplistically. Restoration efforts without the full involvement of local people and the use of locally adapted plant species are very likely to fail. The impact of the refugees on the host community makes it even more critical than normal that the community should be fully involved in the design and preparation of any local development programme. This takes time, which is often not available, and necessitates advance planning, but research could show the way to improved anticipation of migrant flows and improved planning for their handling.

At the national level there is a significant impact because many of the poorest and smallest countries are receiving large quantities of refugees. According to the USCR (2000) the ratio of refugees to host country populations is as high as 1:3 in Jordan, 1:11 in Lebanon, 1:17 in Guinea, 1:32 in Liberia and 1:36 in Iran. Given that these are figures for only the official refugees the unrecorded migrants probably tip the balance much further. Although many of these are not environmental refugees, the extra pressure on the host resources they create will only increase the likelihood of environmentally induced migration occurring in the future.

So far, there has been insufficient recognition of the fact that environmental migration may become an increasingly self-perpetuating problem; migration from one stressed environment creating dramatically increased pressure on another one and so on and on. This is not a sustainable pattern. Attention has to be given to the causes and more effort spent on preventing the problem as well as humanitarian but short-term assistance at times of crisis.

The Research and Action Programme

This review of the environmental causes and impacts of migration explains the growing concern on this topic, but it also illustrates the scarcity of solid information based on rigorous scientific data or research. Keeping track of migrants and refugees and enumerating them has never been an easy task (Harrell-Bond *et al.*, 1993). Identifying those that might be classified as environmental migrants as opposed to other causes is even harder.

If it can be shown that a specific number of migrants and refugees are the result of land degradation in their home territories, then that information should strengthen the case for reducing land degradation in those territories. The knowledge that some 2500 people were killed at Bophal, that several thousand families were moved from the toxic waste dumps at Love Canal and that 135,000 people were moved from a 750 sq. km area at Chernobyl may have helped support arguments for addressing the root causes; but Bophal and Love Canal are once again occupied. Where, as in India, some 20 million people are said to have been displaced by official projects, with irrigation projects alone displacing some 6.5 million (RPG, 1992), this knowledge has helped provoke and inform revisions in the approach of governments and developers. The Red Cross and Red Crescent work is particularly valuable here.

Defining, identifying, categorising and distinguishing environmental migrants and refugees would clearly improve the information base available and throw more light on their needs and problems. But would it do anything practical for the refugees themselves? It could even be counter-productive, in that some environmental refugees may be considered more responsible for their own plight than others (those resulting from land degradation more than those from floods and more than those from industrial accidents for instance). This could (it already does) lead to inequalities of treatment. Definitions and categorisation have been a tool for avoiding action as well as encouraging it. Considerations of national sovereignty, mandates of organisations and legal definitions have already affected and could rapidly increase the complexity of addressing the fundamental humanitarian issue.

The Root Causes

In 1990, Vaughan Robinson pointed out that it is essential to study the factors that encourage flight, and how and why different individuals respond to them in different ways. There is a strong case that most of the research and actions should be concentrated on the root causes of the problem, rather than on quantifying the symptoms (the migrants). The complexities of the causes shows where the focus of research should be.

Much has already been done in this area but there is need for a much more humanistic and less agro-economic, cash crop, production oriented approach. On the basis of this fresh approach, economic arguments could then be developed to show that money would be better spent on solving the problem before, rather than after, the event. But this is a necessary part of the much broader research agenda designed to improve the sustainable management of the productive drylands.

With regard to the longer-term, less dramatic, but even more difficult causes such as land degradation and the effects of drought, the Tamondong-Helins say 'The causes of land degradation within the semi-arid zones are multiple and often inseparable. Environmental problems do not occur in isolation and are often intimately linked with security problems and social stability. The causes of migration in this environment are influenced by many socio-political, economic and other environmental factors'.

The Special Needs of the Migrants

With respect to the migrants themselves, it seems appropriate to determine better their needs and learn how better to meet these needs within the constraints of the host environments. As Stuart Leiderman points out (1998), 'few genuine environmental refugees have told their story to the world'. This needs to be corrected. The humanitarian approach to dealing with the problem seems more appropriate than trying to narrow and categorise it. An essential element to recognise is the extent of local variations and the need for all research and proposed solutions to be based firmly on what the refugees and the host populations themselves perceive to be the problem. To the maximum extent possible the research agenda should be set by those involved, faced with the real life difficulties in the field.

Practical Approaches for Action

The need for practical approaches, for concrete solutions to the hard-to-express or define needs of the people involved, cannot be over-emphasised. This practical approach is emphasised by those in the field such as the Tamondong-Helins (1992): 'There is no easy way to fix the problems of environmental degradation in the semi-arid regions of Africa, but there are processes that contribute to desertification which are poorly understood. To wait until its complexity is fully grasped may be logical but it is not practical. There is a need for immediate action. There are no panaceas but constant, steady progress is imperative'.

Influencing Decision Makers

An area where research and action is needed, is on ways to increase public awareness and the pressure on politicians and industrial and commercial leaders, to address the causes of problems rather than only the results. The drylands appear to be rapidly approaching their limits of growth. In the humid areas in developing countries the toll from environmental disasters increases in proportion to the increasing population. The idea that continued economic growth, population growth, and increasing socio-economic development are all together possible forever is regarded by increasing numbers as wishful thinking. As Robert Stranks (1997) concluded 'there is a need for international and national institutions to address refugees and displaced persons in an integrated manner. The World Summit for Social Development stressed the interdependence of issues. The question of environmental migrants, with its environmental, social, political and economic context, is a manifestation of such interdependence'.

Planning

The fact is that we have to find a sustainable means of living in a sustainable environment. That means planning for the cyclical crises, managing the risks and results of natural and man made disasters, mitigating the effects, allowing for migration and coping with it, but striving all along to reduce the need for it. The question for research is how can this actually be achieved, in the real world where politicians have to make decisions?

Apart from nomads, of whom there are few left, and semi-nomadic pastoralists, most people prefer to remain stable. Forced migration is nobody's wish. Despite this, in 1998 for the first time, according to the International Federation of Red Cross and Red Crescent Societies, more people were forced to leave their homes because of environmental disaster than because of war. Many were able to return, but the less fortunate became migrants.

It is possible to identify 'hot-spots' and high-risk areas and to prepare contingency plans which will help avoid, or mitigate, the death and hardship that may follow from predictable environmental events. At a workshop held during the preparation of the Convention to Combat Desertification, the 'Almeria Declaration' was issued. This proposed that it may be time to recognise a 'Right to Remain' on established family lands, and there is a right to expect support from their national governments to withstand human or environmental pressures to migrate.

The Search for Security

As Arthur Westing (1992) wrote, it has become clear 'that a long-term solution to the problem of environmental refugees can only be reached through the achievement of overall environmental security. And environmental security, in turn, appears to be attainable only within a framework of comprehensive human security, which inter alia takes into account notions of carrying capacity for humans and of sustainable development. It is clear that in the long term the poignant dilemma of the ever-increasing numbers of refugees can be addressed only in terms of achieving environmental security within a framework of comprehensive human security'. This is grand argument but the politicians faced with making the decisions and deciding on priorities need to know what to do to bring it about in concrete terms. Researchers need to be specific as to how to get there from here.

Minimise the Environmental Impacts

With regard to the environmental impacts of refugees, there are again questions to be answered. The immediate problem is apparent – the poor or even critical state of the newly arrived migrants and of the environment in and around their camps or new settlement areas. Don't we know already the steps that need to be taken? Shouldn't the money be better spent on practical on-site actions? These would be developed and evaluated constantly in the light of experience. The need is for applied research, based on the needs in the field, and the application of solutions for those needs on a real time basis. This could be in the form of on-going pilot programmes designed to improve the understanding of the causes and relieve the impacts of the results of environmental migration.

Clarify Environmental Impacts

Black has pointed out that there is need for a clearer understanding of what is meant by environmental impact and environmental degradation. Environmental impacts such as vegetation changes can be temporary, long-term or irreversible. There is a distinction to be made between these and actual land or soil degradation of a permanent nature. Forms of environmental change other than soil degradation and deforestation are also important, for instance health and disease-related impacts (for crops, cattle and humans).

Opportunities for Post Impact Studies

There is also a need to take every opportunity to study the longer-term impact of refugees on the environment and especially to determine the extent of environmental recovery after the refugees return home. This would also be important input to the assessment of the wider issues of dryland development sustainability. Note that it is important to distinguish the environmental impacts of refugees from those occurring already, or that would have occurred anyway.

The Search for 'New Paradigms' of Development

Priorities for research and policy in refugee-affected areas coincide with other development concerns such as local resource management strategies and the way that these withstand or adapt to pressure brought by increasing populations in refugee affected areas (Black, 1992).

There is a close relationship here to the search for new paradigms of rural development in the drylands. There is a realisation that the top-down approach does not work, that 'trickle-down' benefits are hard to achieve and that radically new approaches have to be taken to achieving sustainable dryland management. There is now widespread agreement that these must include the full and willing participation of the people on the land. Ways are being considered and tested now so that this can be encouraged and supported by international and national agencies.

A key factor in the implementation of the UN Convention to Combat Desertification, is how the significant change of approach that it advocates can be brought about. The ability to link national ratification of an international treaty to effective action at the village level is the key to its success. In the long-term, the key to successful handling of refugees will also depend on improved involvement of the affected people in the planning and support of migrant movements. This will be even harder to achieve but efforts should be made to explore ways to do it effectively.

Conclusion

The conclusions of the above discussions on the causes of environmental migration, and on the environmental effects of migrations, are that they both need to be considered and addressed within the broader context of the development and management of sustainable environments. Access to resources and resource use are keys to the consideration of both cause and consequence. Ensuring sustainable management is essential in the long run. To achieve this, comprehensive, but flexible, planning and socially rational management are required.

Many likely causes of environmentally induced migration can be predicted. Active volcanoes and earthquakes on the Pacific Rim; cyclones and flooding in Bangladesh; prolonged drought, locusts, land degradation in the Sahel; accidents at nuclear installations and industrial plants, tornadoes and hurricanes in the United States; increasing numbers of landless through sub-division of smallholdings at death; salinisation of irrigated lands; all of these and many others can be more or less predicted.

Effective planning would enable prompt and effective action to be taken so as to minimise the social and environmental disruption caused by the events and by ensuing migration. Suitable locations could be identified in advance during national planning exercises as potential sites for temporary settlements. Preparation for the more predictable events such as floods could stretch to increased provision of flood-proof shelters and temporary accommodations. Models could be prepared which would help achieve a fast start to the establishment of necessary support and infrastructure.

The role of industry is of particular importance, in so far as many of the industry–related accidents or impacts are predictable (although usually regarded optimistically as unlikely). The responsibility of industry for environmental impacts resulting from their activities, and thus for the social consequences is becoming clearer to the public, whatever legal protections may have been established. It is in industry's long-term interest, as much as the public's, to ensure that adverse environmental consequences are minimised and to gain from the positive image of being able to show that they are accepting their responsibilities by taking all reasonable precautions and also planning for the accidents that always might happen.

The conclusions of the meeting held in 1992 by the Refugee Policy Group (RPG, 1992) drew many of these threads together. The meeting emphasised that ways must be found to enable policy makers and planners to design and implement programmes that promote more rational use of the environment. At the same time we must recognise that large numbers of people will be forced from their homes in the coming years because they will not be able to sustain themselves. The goal should be to minimise mass environmental movements while handling those that occur with rational, humane policies, prepared, in so far as possible, in advance. There should, in effect, be adequate national and international support for those seeking to exercise their Right to Remain.

The agenda for the future (based on the RPG's conclusions) should:

- Address the causes of environmental migration and develop strategies to mitigate their impact and deal with the complex and interlocking issues that prompt these movements.
- Anticipate possible environmental disruptions and consequent migration scenarios, review policies and plan appropriate responses in advance.
- Develop a better understanding of the development - environment - migration chain of events and devise alternative policy approaches.
- Develop better information on the incidence and causes of environmental displacement and on the quantities of people involved.
- Develop closer trans-national co-operation between countries and among international organisations.
- Develop more co-operative working relationships among the various involved groups.
- Improve assistance to victims of disruptions so as to reduce the need to migrate.
- Provide assistance to help returnees and assist others with their integration in host communities.
- Consider new international instruments to help victims of serious environment disruptions.
- Focus greater attention on the needs of host populations and seek comparable support for the host populations affected by the arrival of immigrants.

The UNFPA (1993) brought it all into focus when it said that 'Given secure rights to resources and reliable access to services, people will behave in ways that demonstrate their confidence in the future. Under such conditions, even poor families are likely to invest their time and scarce resources in improving the environments in which they live, giving thought among other things to birth spacing, and smaller families. In other words, when treated with respect, given support and conditions in which they can feel secure, individuals and families will confront their problems and develop their own solutions'.

The key question then is how to provide these conditions so that the populations of the world that are at a high risk to environmental factors should have a secure enabling environment in which to protect themselves and their families. In their book, *More People, Less Erosion'* Tiffen *et al.* (1993), examined the interactions between people and their environment over a period of 60 years in Machakos, not far from Nairobi, in Kenya. Between 1930 and 1990 the population of the District increased more than fivefold. Nevertheless, the environment in 1990 was considered to be in much better condition than in 1930. Soil erosion had been slowed or reversed; predictions of woodfuel crisis had not been fulfilled; agriculture production per head and per hectare was higher and new technologies had been adopted as appropriate.

The knowledge that success stories such as this do exist, and the conviction that they can be achieved elsewhere, is fundamental in the drive to turn around the atmosphere of failure and despair that so easily pervades discussion of dryland environmental development and of how to prevent the causes of much environmentally forced migration. It is clearly more important to focus research and action on ways of achieving results, on ways to avoid or alleviate the effects of environmental stress on people, and of people on the environment, rather than on prolonged and arcane discussions of how an environment refugee might properly be defined. If research can improve on and demonstrate the possibilities of success, and can help convince government leaders, public opinion leaders and the young, that success is possible when the root causes are addressed, then research will have played an outstanding role in the task of addressing the causes and results of environment related migrations.

References

Black, R. 1993, *Refugees and Environmental Change: Global Issues*, Report prepared for OAD Population and Environment Research Programme, University of Bradford.

El-Hinnawi, E. 1985, *Environmental Refugees,* UNEP. Nairobi.

Gebremedhin, N. 1991, 'The Environmental Dimension of Security in the Horn of Africa: The Case of Somalia', *The Life and Peace Review,* 5, 1, 11-19.

Harrell-Bond, B., Voutira, E. and Leopold, M. 1993, *Counting the Refugees: Gifts, Givers, Patrons and Clients,* Refugee Studies Centre, Oxford.

International Federation of Red Cross and Red Crescent Societies, *World Disasters* Report 1999.

Jacobson, J. 1988, '*Environmental Refugees: A Yardstick of Habitability?*', Worldwatch, Washington.

Leiderman, Stuart M. 'Environmental Refugees Require Restoration, Not Development', Paper Presented at the 24th Annual Third World Conference Chicago, March 1998.

Lonergan, Steve. *Population Movements and the Environment,* RPN 18 Jan 1995 www.fmreview.org/rpn181.htm.

Myers, N. 1993, 'Environmental Refugees in a Globally Warmed World', *Bioscience,* 43, 11, 23-41.

Myers, N. and Kent, J., 1995. *Environmental Exodus: An Emergent Crisis in the Global Arena,* Climate Institute, Washington DC.

Robinson, V. 1990, 'Into the Next Millennium: An Agenda for Refugee Studies', A Report on the First Annual Meeting of the International Advisory Panel, *Journal of Refugee Studies,* 3, 1, 1-46.

Royal Geographical Society, 1993, *Geography and Refugee Policy,* Geographical Intelligence Paper No. 1, London.

Stranks. Robert T. 1997. 'Policy Commentary: Environmental Refugees?', Department of Foreign Affairs and International Trade, Canada.

Tamondong-Helin, S. and Helin, W. 1992, *Migration and the Environment: Interrelationships in Sub-Saharan Africa,* Field Staff Reports, 1990/91, No. 22, Universities Field Staff International and the Natural Heritage Institute.

The Refugee Policy Group (RPG), 1992, *Migration and the Environment,* The International Organisation for Migration, Geneva.

The Economist, 4 December, 1993, 'Eritrea, Hungry Again'.

The Economist, 3 March, 2001. 'When is a refugee not a refugee?'

Tiffen, M., Mortimore, M. and Gichuki, F. 1993, *More People, Less Erosion: Environmental Recovery in Kenya,* John Wiley and Sons, Chichester.

Tolba, M., El-Khouly, O. 1992, *The World Environment 1972-1992: Two Decades of Achievement,* UNEP, Chapman Hall.

UNDP, 1992, *Human Development Report,* Oxford University Press.

UNFPA, 1992, *The State of World Population,* United Nations Population Fund.

UNFPA, 1993, *The State of World Population,* United Nations Population Fund.

UNHCR, 1991, *Some Environmental Considerations in Refugee Camps and Settlements,* Report No. 10.

UNHCR. *The State of the World's Refugees 2000,* United States Committee for Refugees, World Refugee Survey 2000, Oxford University Press.

Westing, A. 1992, 'Environmental Refugees: A Growing Category of Displaced Persons,' *Environmental Conservation,* 19, 3, 23-61.

World Commission on Environment and Development, 1987, *Our Common Future,* Oxford University Press.

Chapter 8

International Migration in Southern Africa: A Retrospective Benefit-Cost Analysis

A. Rwomire

Introduction

International migration is one of the most enduring and controversial issues that countries in Southern Africa have to grapple with. To a great extent, the political economy of the region revolves around the pervasive system of migrant labour operated by the Republic of South Africa (RSA) as the centre of attraction and the Southern Africa Development Community (SADC) countries as the peripheral providers of relatively cheap labour.

This chapter examines the theory and practice of international migration with special reference to Southern Africa, especially during the apartheid era. Reviewing the migration process from economic, political and social perspectives, the chapter investigates the linkages between immigration and development and underdevelopment in region. It reviews the characteristics, determinants, dynamics and consequences of migration. It concludes that due to uneven development and South Africa's dominant position in the region, migration could adversely affect the development process in the supplier countries. The chapter is of the view that despite recent political changes in the region, SADC countries are unlikely, at least in the short run, to extricate themselves from South African based economic dependence and migratory labour entanglements.

Conceptual and Contextual Dimensions

Human migration is deeply rooted in both the past and present functioning of modern society. In Africa, migration is partially traceable to the arbitrary demarcation of national boundaries by the colonial masters. With reference to Southern Africa, the most dominant type of migration, besides that of refugees, is labour migration. Leistener (1988: 112) argues that next to rail and road transport, the oscillating movement of workers across political boundaries is the most visible and critical feature of regional interdependence in Southern Africa. As such, migration has had a profound impact on the lives of millions of people throughout the region.

Migration can be defined as 'the movement of individuals or groups which involves a permanent or semi-permanent change of usual residence' (Wilson, 1985: 144). Although the exact definition of a migratory move depends on the issue at hand, it is worth noting that there are three dimensions that are usually taken into account. Firstly, there is the permanence of a move. Thus, tourist trips (regardless of what distance), commuting and nomadic movements, are strictly not part of migration. The second dimension involves the distance of the move, or precisely the crossing to some administrative boundary, internal or international. The third variable concerns the time within which moves have taken place, thus leading to the concept (either explicit or implicit) of a migration defining period (Wilson, 1985: 145). This includes movements such as those of labourers in harvest times.

Migrants are often classified for research purposes according to whether they crossed political boundaries, and if so, what kind of boundary (country line, state line, international border), and also according to the points of origin and destination (Weeks, 1989:186). The major difference, however, is usually between internal and international boundaries.

According to Adepoju (1984:441), internal migration, ' in principle implies movement of people within a geographically defined territory unrestricted by legal constraints'. It involves a permanent or semi-permanent change of residence within a specified country or territory. An example of internal migration is that of rural-urban migration. It should be noted, however, that the movement does not only involve the mere change of residence but also a change in terms of the country and perhaps the region or state within a country. When one moves from the area of his/her origin he/she is referred to as an out-migrant and he/she is termed an in-migrant in the area of his/her destination. With reference to internal migration both the areas of origin and destination are parts of the same country.

International migration, on the other hand, refers to movement across national boundaries or even between continents. International migration, unlike internal migration, is more complex and difficult to accomplish in that, 'an international migrant is invariably confronted with a series of sometimes complex regulations relating, at first, to exit from the country of origin, and later, entry into, residence within, and exit from the receiving country' (Adepoju, 1984: 411). Furthermore, international migration is far more likely to include a cultural change such as a change of language, customs and politics – in general, a change of lifestyle and worldview than a move internally. International migration is termed immigration from the point of view of the nation from which the move takes place and immigration in the receiving nation.

The Southern Africa region, the focus of the present discussion, consists of thirteen countries that constitute the Southern African Development Community (SADC): Angola, Botswana, the Democratic Republic of Congo (DRC), Lesotho, Malawi, Mauritius, Mozambique, Namibia, South Africa, Swaziland, Tanzania, Zambia and Zimbabwe.

Southern Africa is one of the world's major regions in which international migration takes place. A significant proportion of the migrants in the region is composed of labour migrants who provide much-needed human power. It also tends to be of temporary nature with a contract lasting up to a maximum of two years. Further, it also involves illegal migration across SADC countries, especially countries like South Africa, Botswana, and Swaziland.

The major international flow of migrants in Southern Africa is the movement of workers into South Africa from neighbouring countries. As an economic giant in the region, South Africa plays a prominent role as the country of immigration. This is largely attributed to the country's well-developed economy and infrastructure, which serve to attract people fleeing famine, drought, poverty, environmental degradation, conflict, and human rights violation (Carim, 1995).

Virtually all SADC countries continue to export labour to South Africa. Moreover, most of them also depend on South Africa's railways, harbours, airports and other services for their exports and imports (Thahane, 1991: 69; Adepoju and Oucho, 1995). The majority of the migrants work on the farms and in the mines. It is noteworthy, however, that South Africa as well as other SADC countries do import professionals and technicians from other African countries and overseas. These professionals include engineers, doctors, business people, university professors, teachers and so on.

For those countries that export labour, migration has become an important feature of their economies. However, the importation of manual labour by South Africa seems to be decreasing and this is expected to have negative effects on the socio-economic conditions of labour exporting countries. The cause of decreasing labour importation would appear to be the increase in unemployment in South Africa (Appleyard, 1998:161). In due course, the unemployment rate is bound to rise in these countries and this could lead to social ills such as poverty, armed robbery, prostitution, drug trafficking, and other crimes.

Characteristics of the Migrants

Migrants in Southern Africa, as in the rest of the developing world, share certain characteristics. In most instances, maturity tends to increase one's chances of migrating. Young adults between the ages of 20-34 are much more mobile than other age groups. Previously, migration was male-dominated, but as the educational opportunities of women have expanded, the relative proportion of female migrants has increased. Fundamentally, women migrants can be classified into two categories: the 'associational' migrants, that is, wives and daughters accompanying the 'primary' male migrant, and the migration of unattached females (Todaro, 1987: 257). The latter type has increased in recent years. Also, smaller families with pre-school going age children and those with no families are more likely to move than those with older children.

In Southern Africa, as elsewhere, there seems to be a positive relationship between the level of educational attainment and migration. People who have college or university degrees are more likely to migrate than those with less education. Notwithstanding, most migrants in Southern Africa tend to be poor, landless and unskilled individuals.

Generally, labour migrants who work in the mines and farms in South Africa are school dropouts. It would appear that migration for such people is not a matter of choice.

Causes of Migration

Migration is a response of human beings to economic, social, demographic and physical forces in the environment. In other words, most population movements are purposeful. Such movements are conditioned by the desires to satisfy certain human needs or by the threat of discomfort. Thus, there are positive and negative aspects of the migration-provoking situation.

Migration may occur as a search for an opportunity to improve one's lot in life. In this case the area of destination exerts a pull on the migrant. However, migration also occurs as an escape from unfavourable social or economic conditions which constitute an expulsive push. In short, the pull factors are associated with the destination area whereas the push factors are associated with the area of origin.

During the past century, certain economic, political and social pressures have shaped population movements in Southern Africa. By and large, as a growing body of the relevant literature indicates, economic and political factors have played a key role in stimulating migration throughout the region. To a great extent, migration is a reflection of the imbalances between and within sectors of the economy and between countries and regions. As Adepoju (1984: 444) has observed, 'international migration signifies inequalities in development, employment opportunities and especially income and living conditions between countries, particularly between developed and developing countries'. In other words, a majority of people migrates with the ultimate intention to improve their economic well being.

The most commonly cited theory of migration has, as indicated previously, been the so-called push-pull theory. The argument behind this theory is that people move because they have been pulled or attracted to some place elsewhere (Weeks, 1989: 189). It can be observed, however, that overall pull factors are of greater significance than the push factors. No doubt, the conditions in the place of origin may be so bad that they may actually force one to migrate. Nevertheless, none of such currents can compare in volume with those arising from the desire inherent in humans to improve themselves in material respects (Ogden, 1984: 24). People who are either unemployed or dissatisfied with their present jobs are more likely to move than those with contrasting characteristics. Indeed, unemployment and underemployment are some of the major sources of migration in Southern Africa.

Other theories that have dominated migration studies include the Lewis two-sector model and the Harris-Todaro model. These theories stress the fact that economic variables dominate most explanations of why people migrate. They argue that an individual's drive to migrate is determined, to a large extent, by wage differentials between the modern industrial sector or economy and the traditional subsistence agricultural sector or economy in another (Thahane, 1991: 66).

This implies that 'the high rates of rural-urban migration are, in part, a consequence of official neglect of rural economic conditions, despite rhetorical commitments to rural development' (Bilsborrow, 1998). As long as the gap between the two sectors of an economy continues to widen, people will migrate from the rural sector to the urban sector. People will be inclined to move to urban areas in order to improve their welfare and that of their families. At any rate, investments tend to be concentrated in the urban areas, especially in developing countries.

The Harris-Todaro model further notes that the decision to move can also be affected by the chances one has of finding a job in the host country. For example, a person who has relatives in the host country is more likely to migrate than someone who does not. These contacts provide information on employment, wages, working conditions, social life, and shelter to potential or actual migrants (Thahane, 1991: 66). Because of limited job opportunities in their home country, people may be forced to migrate to where jobs are in abundance. Furthermore, some professionals, after training in developed countries, may be forced to migrate because their skills are so specialised that they cannot be easily used in their home country.

According to Marxists and Neo-Marxists, migration is inextricably intertwined with the development of capitalism. Thus, migration is viewed as one aspect of the divisions which exist between the core and the periphery, between cities and the country-side, and between metropolitan and satellite areas (Ogden, 1984: 25). Owing to technological underdevelopment and the lack of capital in the periphery, labour tends to be alienated from the means of production in the subsistence sector, consequently creating a reserve 'army' of the unemployed. The lesser bargaining power of labour against that of capital guarantees that labour is exploited in order for capital to flourish. The process of capital accumulation makes the periphery dependent on the centre for employment, with incomes in the form of remittances and modern goods consumed in the periphery. The Marxist analysis has been used to explain labour migration in Southern Africa and the dependency relationship between South Africa and surrounding labour-supplying economies (Thahane, 1991: 67).

The reasons associated with migration are not solely economic. In some instances, political variables are of greater importance than economic ones (Adepoju, 1984: 445). In countries like Angola, DRC, Liberia, Rwanda, Sierra Leone, Somalia, Sudan, and Somalia, governments have for several years lost their legitimacy. This has created a situation of high incidence of political violence and instability. One effect of such an unsettled state of affairs is the rise in the number of refugees. It has been estimated that sub-Saharan Africa contains more than 6 million refugees, more than half of the world's total (Sybella, 1994). In Southern Africa, countries such as Angola, Mozambique, South Africa, and Zimbabwe have at different times faced political violence and instability. Namibia could also be added to the list if the recent granting of refugee status to more than one thousand Namibian citizens by Botswana is anything to go by. This followed their illegal entry into Botswana under the leadership of Muyango who had escaped allegedly after committing political crimes in Namibia. Owing to civil strife in Angola and Mozambique during the 1990s, political instability and war displaced millions of people who were then forced into the neighbouring countries. In Mozambique alone, the civil war caused about 1 million deaths while 3 million people were uprooted and rendered homeless (Solomon, 1994: 65).

Furthermore, new migration streams have emerged in response to the favourable situations created by the formation of economic unions such as SADC, South African Customs Union (SACU) and South African Labour Commission (SALC). These arrangements promote free movement of labour between nations.

Apartheid did not only kill people, it also destroyed their land and environment. Prior to the 1990s, apartheid legislation, the industrial colour-bar, migratory labour and the demarcation of South Africa between 'white' and 'Bantu' territories, did combine to confine Blacks on limited, unfavourable land area and forced them into a reservoir of low-skilled labour (Rogerson, 1982: 51). The 'bantustans' or 'homelands' were located in the former Native Reserves and mainly comprised impoverished, eroded, overgrazed and mineral-deficient land that accounted for less than one-seventh of the total land area of South Africa (Hurd *et al.*, 1989: 124). Blaikie and Brookfield (1987: 109) have observed that environmental degradation in Southern Africa could be the worst in the continent. For one thing, enforced migration and land annexation by colonial settlers led to 'instant overpopulation'. Migratory movements compounded the situation across international boundaries, from countries of low economic development to those of precocious economic development.

Thus, international migration is partly responsible for land degradation in Southern Africa. As Darkoh (1987) has pointed out, land degradation can be attributed to socio-economic and institutional factors, especially the marginalization of some racial and ethnic groupings. Although natural disasters have contributed to land degradation in Southern Africa, sight must not be lost of man-made causes, including racial and ethnic competition, apartheid, war and skewed distribution of resources.

Benefits and Costs of International Migration

Migration has dynamic consequences for the migrants themselves, for the areas from which they come, and the areas to which they go. Its effects can be divided into three categories: economic, political and social.

There are several major economic benefits that accrue to the sending country. Firstly, unemployment and underemployment are some of the main problems facing countries in Southern Africa. By exporting part of their labour force, domestic unemployment in these countries is reduced. Thus, international migration can be seen as a form of indirect aid, for the host country employs the sending country's unemployed and/or underemployed labour force. For example, in 1982, the 13 659 Swazi immigrants employed in South Africa were equal to 17.7 per cent of the 77 253 workers in paid employment at home (Leistener and Easterhuysen, 1988: 119).

Undoubtedly, the migration of unskilled and unemployed workers acts as a safety value for employment opportunities in the labour exporting country. For example, about two-thirds of the working class in Lesotho are migrant labourers. If this free movement were stopped, the lives of these people and their families would be put at risk. Moreover, the out-flow of redundant labour from the rural sector raises the land labour ratio, thus improving productivity, raising rural incomes and stimulating the rural sector growth (Todaro, 1987: 268).

Moreover, by allowing disadvantaged sections of the population to migrate and earn higher wages, international migration is presumed to help improve income distribution in the sending country.

Secondly, labour migrants provide the sending country with much needed income. The remittances sent to families by migrants lead to an increase in local consumer demand and savings, thereby creating an incentive and a source of much needed foreign exchange for physical capital formation which would spur long-run economic growth (Todaro, 1987: 267). Moreover, the remitted incomes help to uplift the standard of living of the migrants' families and also raise their countries' GNP. For example, remittances by Basotho migrants rose from 38.7 per cent of GNP in 1978 to 52.4 per cent in 1982 (Leistener and Esterhuysen, 1988: 117). Thus, migrants' remittances do stimulate economic growth in the sending country. In many cases, before migrating a potential migrant's marginal product at home may be lower than the cash income he gets as a migrant labourer. Increases in per capita income is felt when the additional income is invested or consumed in the home country, leading to improvements in social welfare, labour productivity and output, and modernisation of the economy.

Thirdly, the sending country saves money with regard to the training of migrants. On their return, migrants will bring back new skills and experiences which will aid in the development of a primitive industrial economy, thereby contributing to increased human capital formation without the need for expensive domestic training programmes (Todaro, 1987: 268). Furthermore, productivity in the home country is likely to increase as migrants return with new, appropriate techniques of production which they have acquired abroad.

Looking at the benefits of migration to the host country, the first point to make is that the country (for example RSA) solves its problem of labour shortages. For the recipient country, migrant workers are useful in that they usually take the worst jobs that native workers are not interested in. For instance, the majority of the labour migrants in South Africa work in the mines and in the farms. However, it is noteworthy that employers exploit many of the migrants. For example, many Mozambican men earn as little as R30 (about US$3) per month and some women have been sold for as little as R200 (Solomon, 1994: 67).

However, it is not always the unskilled workers who migrate. It would appear that many developing countries are experiencing the contemporary problem of 'brain drain', that is, ' a phenomenon which involves the free or forced, permanent or temporary movement of highly skilled persons such as human doctors, university professors, teachers, engineers, technicians, etc., from one occupation/profession/trade or organisation or country to another' (Kabumba, 1988:4). Since the socially ambitious and educationally and occupationally talented people are among the most likely to migrate, their departure contributes to the potential for further economic stagnation and decline. Indeed, in some Southern African countries, the mass exodus of teachers, doctors, nurses and other professionals to South Africa and overseas countries has caused a serious concern among government circles. During the apartheid era graduates from the 'Homelands' tended to leave for the 'White areas' in RSA where job prospects were better than at home. Also, professionals such as teachers and engineers from Lesotho and Swaziland have migrated to South Africa. Thus, South Africa 'milks the cow which it never fed' in so far as it utilises already well groomed human resources without incurring the cost of education and training.

The migration of highly skilled persons can also be attributed to local 'market saturation'. For instance, unemployment among South African teachers is reported to be high while the demand for teachers in Botswana is high also. Consequently, many unemployed teachers in South Africa have migrated to Botswana.

The fact that migrants spend some of their money in the host country such as RSA implies that more jobs are created therein due to the expanding markets for various goods and services. However, this is debatable as money from taxes may instead be used to buy the kind of machinery that cuts down on manual labour. Furthermore, social dividends like company taxes help to build South Africa's infrastructure. Again, immigration provides a large labour supply, which serves to maintain a low wage scale for the migrants. If the migrants are skilled, the host country gains from their skills and knowledge.

The cost of migration to the receiving country includes

expenditures on building more health and social facilities including housing, to cater for the additional population.

Furthermore, migration may increase socio-cultural tensions, especially when migrants come in large numbers and introduce values and norms that may clash with those prevailing in the receiving society. This is particularly applicable to the multi-racial and multi-ethnic societies of Southern Africa. In some instances, tensions and conflicts have often risen between the national labour force and the migrant labour force due to competition over jobs.

Migration is also considered by some to be the second biggest problem facing Botswana and South Africa after HIV/AIDS. Media reports show that in 1999 more than one thousand illegal immigrants were repatriated from Botswana. Of these, 90 per cent were Zimbabweans while others were from countries such as Britain, Pakistan, China and many others. As such, illegal migration can have negative social and economic consequences for the host countries, as there is a possible link between migration and social problems including crime and AIDS.

The costs of migration to the supplying country are many and multidimensional. On the economic front, 'output and income per head in the supplying area will fall on the migrants' departure if his marginal product was greater than his wage or if his labour generated substantial externalities in the economy' (Nattrass, 1976:44). Labour migration is intimately associated with male depopulation in countries like Lesotho and Swaziland, and this adversely affects these countries' long term development prospects. Moreover, the sending country incurs the cost of decreasing taxation due to the departure of the taxable population and subsequent reduction of output. Migration also constraints the country's long term development policies in that the local remuneration systems have to continually compete with South Africa's relatively high wage rates. Indeed, the sending countries continue to lose their skilled manpower largely because their salary scales do not compare with South Africa's.

There is considerable controversy surrounding the role of remittance in promoting long-term economic growth through capital formation. The argument is that the immigrants' households spend a large chunk of these remittances on private consumption, with very little going to savings (Todaro, 1987: 267). Without doubt, the earnings of the migrant workers increase the buying power of the population, thus benefiting the sending country in terms of sales tax. However, the remittances may be spent on consumer goods and cattle with very little going to the improvement of the agricultural sector. It should also be borne in mind that the wages of the migrants are just bachelor wages. As more people rely on remittances, a country may be forced to increase its imports. This works to the detriment of the sending country in that the foreign exchange, which could have been used for long-term investment in the country, is used to satisfy the growing appetites for consumer goods. Locally produced goods often suffer in that imported goods are thought to be superior and cheaper.

According to the Lewis-model, the rural sector in developing countries suffers from both underdevelopment and unemployment as a result of its large labour-land ratio, which leads to low productivity and low average income (Todaro, 1987: 268). Thus, the exodus of redundant labour from this sector is viewed as a blessing in disguise. As the migrant workers leave and find well paying jobs elsewhere, pressure on scarce land is reduced for those remaining. Consequently, the higher the land-labour ratio the higher the incomes for those left behind due to greater productivity per worker. Higher incomes may be an incentive for the remaining workers to make improvements to their land, thereby raising productivity and incomes even further.

Notwithstanding this labour migration has had negative effects on the sending country in that if the immigrants send enough money to their families, the need to increase agricultural output may decline as leisure would be substituted for labour (Todaro, 1987: 267). Migrants include the young and able-bodied men so that farming is left in the hands of women, children and the older adults. As a result, rural production lags behind and a country is forced to import foodstuff. This is particularly so because the older adults tend to be conservative and will not readily accept new methods of production. Labour migration in the states bordering RSA has, to some extent contributed to the failure of the Rural Development Programmes (RDPs). In the case of Lesotho, the absence of decision-makers, opinion leaders and innovators has led to adverse consequences. Insufficient male labour, for example, has contributed to low crop yields, especially where the furrows produced at ploughing time might be too shallow for maximum plant survival and growth (Smith, 1982:81). In Swaziland, due to the withdrawal of labour from the agricultural sector, labour shortage has at times constrained cotton and tobacco production.

As noted earlier, migrant workers acquire skills while working abroad and later use these on returning home. Firstly, it is assumed that the jobs in question are of greater skill than those available at home. The second assumption is that on their return, workers will be able to apply their new skills locally. However, this assumption is questionable. If a person is unskilled, he or she stands little chance of getting a skilled job in the host country. Failing to find a job, immigrants, especially the skilled ones, may opt to take a job that is inferior to their qualifications. As a result, far from obtaining new skills workers may be de-skilled. Secondly, as for those who are lucky to find the opportunity to climb up the ladder and obtain better skills, chances of returning home are very slim, as prospects for advancement there may be non-existent. Those who ultimately return home may discover that their skills are not applicable to the prevailing level of technology (Todaro, 1987: 268). They therefore become 'over-skilled' with regard to domestic production. Moreover, when the migrants finally return because of old age or disease contracted in the mines, there are no pensions or insurance schemes to pay for their care (Thahane, 1991: 83). They consequently become a burden to an economy which has not benefited from their productive years and has, in fact, suffered from the loss of their services.

In Southern Africa, so it would appear, international migration does not promote income redistribution. The ever-growing economic disparities between South Africa and adjacent peripheral economies can partially be blamed on labour migration. Firstly, labour-supplying countries incur the cost of raising the potential migrants. They provide for various services from the time they are born to the time that they are ready to enter the labour force. Eventually, when they begin working they are lost from the economy, for their governments do not recover these costs, as remittances are not taxed. Secondly, during their working life, migrants generate surplus product, which is appropriated by the mining companies (and farmers) in the form of higher profits, which are reinvested in South Africa. As capital is accumulated, especially in RSA, the development gap between the labour-importing and labour-exporting countries widens. Understandably, investors establish their enterprises where the able-bodied persons are situated. During the apartheid era, as Lemon (1985) has pointed out, immigrants spent four fifths of their wages in the 'white' areas of South Africa.

There are also political consequences of international migration. Reports indicate that refugees have accounted for more than half of all international migration since the end of World War II and constitute an increasing share of current flows (Todaro, 1987: 269). It has been estimated that one in every 200 Africans is a refugee (Turok, 1987: 10). Refugees pose serious problems since they tend to move suddenly, travel in large groups, and cannot be planned for. A sudden influx of refugees may seriously overburden the economy of a developing country or upset a delicate racial, ethnic or religious balance within that country (Todaro, 1987: 269). During the past three decades, political instability and violence in Africa have caused millions of people to flee from their home countries to neighbouring states. With regard to RSA thousands of Basotho, Angolans, Mozambicans, Zambians, and Zimbabweans have illegally crossed borders into South Africa which plays host to approximately four million illegal migrants (Chigodo, 1997: 1). These migrants have been accused of distorting economic performance, threatening border security, and indulging in organised crime. South Africa has approached the issue quite pragmatically by constructing an electric barbed wire along its borders. Hundreds of illegal migrants who tried to cross the fence have died in the process. This can create and/or exacerbate the political tensions existing between the countries involved. Furthermore, refugees take away jobs from nationals, thus adding to a country's social pressures. In all, refugees pose a serious threat to the stability and peace of the receiving country. As the saying goes, a hungry man is an angry man: poverty breeds instability.

In the case of South Africa, the existence of migration labour has not only allowed the country to grow and become an economic giant in the region, but it has also enabled her to exercise some great political leverage in its relationships with SADC and other African countries. South Africa often threatened to expel large numbers of miners in its fight against economic sanctions which until the early 1990s were imposed by the international community. Likewise, South Africa has sometimes used migrants as weapon in its disagreements with its neighbours. For example, in September 1986 the recruitment of Mozambican miners was suddenly halted in order to pressure Mozambique to reconsider its disagreements with South Africa (Thahane, 1991: 80). In some instances, South Africa has turned neighbouring countries into labour reserves by entering into coercive and clandestine agreements with some political leaders in the region. For instance, the then government of Leabua Jonathan was able to survive the aftermath of the 1970 general elections in Lesotho due to the intervention of the South African government. Although the opposition had won the elections, the results were nullified and a state of emergency declared. However, consequent upon Jonathan's move to the ideological left, the South African government supported dissident elements within the Lesotho government against Jonathan, closed the border and manipulated the political events in Lesotho towards a bloodless coup. Following the coup, the border was reopened and a large number of Basotho were recruited to work in the South African mines.

The social effects of international migration are as overwhelming as the economic ones. As Houghton (1960:189) has observed, migration is 'an evil canker at the heart of our whole society, wasteful of labour, destructive of ambition, a wrecker of homes'. For the individuals concerned, migration is an occasion of immense grief, an impediment to normal social relationships. Some of the negative social effects of migration include the following: illegitimacy, bigamy, prostitution, homosexuality, drunkenness, violence, and the breakdown of parental authority as direct effects of migration. The indirect results of the lifestyle of the migrants include venereal disease, tuberculosis, malnutrition and beriberi (Lemon, 1987: 200). Other effects include destitution and vagrancy, the spread of diseases, an increase in the crime rate and a reduction in the normal birth rate (Ndongko, 1991: 242).

The social consequences of international migration in Southern Africa include general social disorganisation stemming from the prolonged absence of men from their families. Men are deprived of the role of leadership in their families, while women are left feeling lonely and helpless, eagerly waiting for letters and money from their husbands. For the women it is difficult to live with 'ghost' husbands. This situation creates numerous psychological and social problems, which are difficult for many women to handle. No matter how hard the woman may try she cannot effectively play the man's role. The children are likely to grow up undisciplined. With no father figure to emulate the boys develop identity problems. Needless to say, a majority of juvenile delinquents comes from broken or disorganised families (Coleman and Cressey, 1999; Eitzen and Zinn, 1992).

Apart from disrupting family and community life, international migration exerts a negative effect on the demographic composition and social structure of both donor and host areas. The demographic composition is influenced by the selective nature of migration, particularly the selectivity by age (Weeks, 1989: 196). The sending country surely loses people from its young population. Since a majority of migrants fall under the childbearing age, this will have a negative effect on the population growth, particularly of the receiving area.

Negative effects of migration can be immense, especially when people migrate in large numbers, for it alters patterns of social relationships and social organisation. The extended kinship relations are weakened and local economic, political and social institutions have to adjust to the increasing number of people serviced (Weeks, 1989: 197). Moreover, the mixing of different racial and ethnic groups results in cultural transfusion and social disorganisation.

The Future of International Migration

Given the winds of change blowing in Southern Africa and more especially the significant changes that have been initiated in South Africa in the past decade, one wonders what the future holds for the region, particularly with reference to international migration. It is evident that international migration, especially labour migration, will persist. It can be argued that as long as the discrepancy exists between the economic levels of nations and regions, it would be difficult to stops the flow of migration movements (Tabbarah, 1984: 437). Migration will not only continue to take place between the developing countries in Southern Africa, but a substantial number of people are expected to move from the region to the more developed countries. It is predicted that a large percentage of whites in South Africa will migrate to the West in the next few years. In fact, the exodus has already begun.

Several countries in Southern Africa are currently facing recession and high unemployment rates. With high unemployment rates, migrants cannot expect to secure high-paying jobs immediately. Unable to find jobs, they are more likely to turn to the informal sector for survival. The heavy influx of immigrants obviously creates problems for urban authorities. For example, the acute shortage of housing in urban areas has resulted in the appearance of shantytowns that skirt most primate cities in the less developed and semi-developed countries, including those in Southern Africa.

International migration in Southern Africa is likely to continue for many years to come. However, some predict that over time the demand for foreign labour in South Africa will decline and that the government of South Africa and its business community have the ability to accelerate the process (Thahane, 1991: 84). This could have extremely damaging effects on the labour supplying countries. By being so tightly integrated into the regional economy dominated by South Africa such that no aspect of their life has been left untouched, these countries may in future find themselves encumbered by South Africa's decision to expel foreign workers. With South Africa now facing its own problems such as recession and high unemployment, dependence on foreign labour reserves has to be cut down. As previously noted, remittances have economically become important for supplying countries. Lesotho, in particular, depends on migration more than any other labour supplier. For example, mine wage remittances constitute about 50 per cent of the country's Gross National Income. Thus, if labour imports were to be cut down, this would result in immediate term government cash crisis, drastic cuts in purchasing power and a lowering of living standards in Lesotho and other labour exporting countries.

Since the abolition of apartheid, people's perception of South Africa has changed, positively. Owing to the country's improved political image, many multinational corporations have invested in South Africa. The international community's attitudinal change has enhanced the dominance of South Africa, which some people regard as the only hope for Africa's development. Indeed, it is widely recognised as a development bridge for the transfer of skills and capital to other parts of the sub-Saharan Africa. Internal migration is also expected to rise since discriminatory laws, for example the Land Act of 1913, have been repealed. Those who will be flocking permanently to urban areas with their families will dominate migration movements. International migration is likely to continue unabated as long as the abovementioned socio-economic problems persist. At any rate, cutting down on foreign labour would probably increase the flow of clandestine migrants as the labour sending countries are also facing high unemployment rates.

Understandably, some Southern African countries seem to be worried about the adverse effects of extreme dependence on the export of (migrant) labour to South Africa. This trend makes it highly important for these major labour-exporting countries to pursue vigorously strategies to stimulate employment opportunities in order to reduce migration of their nationals. Moreover, there is a need for them to formulate policies to combat brain drain. They should be prepared for the unpredictable future.

References

Adepoju, A. 1994. *Brothers at War, Dissidence in Southern Africa*. British Academy, London.

Adepoju, A. 1984. 'Linkages between internal and international migration: the African situation', *International Social Science Journal*. 36, 3, 441-452.

Adepoju, A. and Oucho J. O. 1995. *International Migration, Human Resources Utilisation and Regional Development in Sub-Saharan Africa,* Union for African Population Studies, Dakar.

Appleyard, R. 1998. *Emigration Dynamics in Developing Countries*: Vol 1, Sub-Saharan Africa, Ashgate Publishing Company, Aldershot.

Bilsborrow, R. E. 1998. *Migration, Urbanisation and Development: New Directions and Issues,* UNFPA, New York.

Blaikie, P. and Brookfield H 1987. *Land Degradation and Society,* Methuen, London.

Carim, X. 1995. 'Illegal migration to South Africa', *Africa Insight,* 25, 4, 221-222.

Chigodo, T. 1997. 'The plight of SADC labour migrants', In *Africa News*. The all African Press Service, Nairobi, pp 1-3.

Coleman, J. W. and Cressey, D. R. 1999. *Social Problems,* Longman, New York.

Darkoh, M. B. K. 1987. 'Socio-economic and institutional factors behind desertification in Southern Africa', *AREA,* 19, 1, 25-33.

Eitzen, D. S. and Zinn, M. B. 1992. *Social Problems,* Allyn and Bacon, Boston.

Houghton, H. D. 1969. 'Man of two Worlds', *The South African Journal of Economics,* September.

Hurd, G., Ashton, D. N., Brown, R. K., Dunning, E., Jewson, N.D., Johnson, T. J., McIntosh, M. and Zubaida, S. 1989. *Human Societies: An Introduction to Sociology,* Routledge, London.

International Labour Organisation, 1981. *Black Migration to South Africa,* ILO, Geneva.

Kabumba, I. 1987. 'Human resources development, employment and brain drain', Paper presented at the AAPAM Conference, Kampala, Uganda.

Karen, A. F. Kenneth, H. H. and Linda, G.M. 1993. *Demographic Change in Sub-Saharan Africa,* National Academy Press, Washington DC.

Ken, C. 1994. *Sustainable Development for a Democratic South Africa,* Eathscan Publications Ltd, London.

Kowet, D. K. 1987. *Land, Labour Migration and Politics in Southern Africa: Botswana, Lesotho and Swaziland,* Offcenter, Uppsala.

Leistener, E. and Easterhuysen, P. 1988. *South Africa in Southern Africa: Economic Interaction,* Africa Institute, Pretoria.

Lemon, A. 1985. *Apartheid in Transition,* Gower, London.

Nattrass, J. 1976. The *Migrant Labour system and South African Economic Development, 1936–1970,* University of Natal, Natal.

Ndongko, W. A. 1991. 'Labour migration and regional economic co-operation and integration in Africa,' *Labour and Society,* 16, 3, 231-249.

Ogden, P. 1984. *Migration and Geographic Change,* Cambridge University Press.

Ransford P 1990. *In Search of a Better Life: Perspectives in Migration from the Carribean,* Praeger, London.

Rogerson, C. M. 1982. 'Apartheid, decentralisation and spatial industrial change', In Smith, D. M. (Eds), *Living Under Apartheid,* George Allen and Unwin, London.

Smith, D. M. 1982. *Living Under Apartheid,* George Allen and Unwin, London.

Solomon, H. 1994. 'Migration in Southern Africa: a comparative perspective', *Africa Insight,* 24, 1, 23-36.

Stoneman, C. and Thompson, C. B. 1991. 'Southern Africa after apartheid', *Africa Recovery.* 4, 106-123.

Sybella, W. 1994. *One Day We Had to Run,* Evans Brothers, London.

Tabbarah, R. 1984. ' Prospects of international migration', *International Social Science Journal.* UNESCO, 36, 3, 425-440.

Thahane, T. T. 1991. 'International labour migration in Southern Africa', in Papademetriou, D. G. and Martin, P. L. (Eds), *The Unsettled relationship: migration and Economic development.* Greenwood Press, New York.

Timberlake, L. 1985. *Africa in Crisis,* Earthscan Publication Ltd. London.

Todaro, M. P. 1987. *Economic Development in the Third World,* Longman, New York.

Turok, B. 1987. *What Can Be Done?,* Zed Books Ltd, London.

United Nations (ECA) 1985. *Migratory Labour in Southern Africa,* New York.

Weeks, J.R. 1989. *Population: An Introduction to Conceptual Issues,* Wordsworth Inc, Belmont.

Wilson, C. 1985. *A Dictionary of Demography,* Blackwell Ltd, New York.

World Bank, 1978. *Migration from Botswana, Lesotho and Swaziland,* World Bank Washington DC.

Chapter 9

Desertification: Myths and Realities

O. B. Smith and S. Koala

Introduction

An intense debate over the desertification phenomenon has been raging for well over two decades, since the advent of the Sahelian drought in the early 1970s. The debate has generated some new knowledge, raised interest in, and stimulated commitment to controlling the phenomenon, but has also generated some myths around it, to the extent that some have begun to question the existence of desertification. The issues around which certain myths have been built and which tend to detract attention from the pressing realities of combating it and alleviating its effects include:

- The historical perspective (is it a recent or an old and time tested phenomenon?);
- the definition and its extent (what is desertification, and how common or serious is it?);
- the causes and process (what are the responsible factors and how does it develop?);
- the consequences (what are its visible and not so visible effects?); and
- the remedies and control (how can we combat it?).

This chapter presents an overview of these issues, with the goal of identifying and contrasting the myths and realities, because some of the former have the potential to confuse issues and negatively impact on policy decisions, actions and approaches required to effectively tackle the problem.

Historical perspective

Desertification is not a new problem, and archaeological and historical evidence suggest it began several centuries ago (Olsson, 1981; Grainger, 1990), and could be traced back to the medieval, and even the neolithic period (Spooner, 1989). Recent events following the 1968-73 African Sahel drought that thrust the issue unto the global agenda, have created the impression or the myth, that it is a recent and pressing issue.

In the distant past, inappropriate traditional agricultural methods were apparently responsible for the degradation of the Negev several centuries ago. According to Evanari *et al.*, (1971), soil evidence indicated that a system of narrow terraced wadis and farm units with small watersheds which fitted well into natural soils landscapes without damaging the environment was abandoned for larger catchments systems that eventually led to soil erosion, silting up and destruction because yield demands were greater than the soil and water resources the landscape could bear. Thomas and Middleton (1994) suggested, however, that climate change might have also contributed a fair bit, since they indicated that geological and geographical evidence suggest that the Negev enjoyed a humid climate from about 100 BC to about AD 300 after which it became drier, reaching a critical threshold at about AD 500. Another lesson of land abuse that led to degradation from the distant past comes from the Maya areas of Central America. According to Olsson (1981), soil evidence suggest that the Maya farmed the Valle de Naco in Honduras more than 1000 years ago. With time, the population expanded into the hills around the valley site, and cleared the upland forests for cultivation. Eventually, the upland was degraded, and its yield declined, the valley was flooded because of excessive soil erosion in the upland, the population declined and the civilisation collapsed.

More recently, reports of land degradation leading to desert like conditions were quite frequent in the early part of this century. Incompatibility between the farming systems, climate and soil status, as well as a premium for political and economic considerations over sustainable land management approaches were responsible for the unprecedented wind erosion in the American Great plains with the resultant ecological catastrophe and human misery associated with the Dust Bowl of the 1930s 1950s and 1970s. At about the same time the American Great Plains Dust Bowl was raging, Stebbing (1938) called attention to the ongoing forest degradation in West African

Sudano Sahelian region. A deforestation process, which according to his observations, led to erosion and was helped along by sand blown from the Sahara being deposited on the denuded land. A process he referred to as the encroaching Sahara.

Finally the term desertification coined by the French scientist and explorer Louis Lavauden, in 1927, and popularised by Andre Aubreville in 1949 was in popular language several decades before the Sahelian drought of the 1970s.

In reality therefore, desertification is not a new post-1970 Sahelian drought phenomenon, but an age-old phenomenon dating back several centuries. What is new is its conceptualisation and popularisation as one of the creeping environmental phenomenon of devastating local consequences, but with equally alarming global impact, and to which the global community should pay commensurate attention.

Definition

What is desertification? This question has been answered in many different ways by the many scientists and institutions who have attempted to define and describe what desertification is all about. In a review of the literature on desertification, Glantz and Orlovsky (1983) reported encountering more than one hundred definitions of desertification. Since its popularisation by Aubreville (1949), the term has apparently presented a conceptual and descriptive dilemma to researchers and institutions, each of which have emphasised aspects and perspectives related to their disciplines or institutional interests, and have defined desertification differently with respect to a number of important characteristics. The result is a great diversity of definitions, with some contradiction, misunderstanding, confusion, institutional and disciplinary biases and myths about the nature of the phenomenon.

In its original form as presented by Aubreville (1949), who is considered to be the father of the term, desertification was referred to as a human induced process of degradation that transformed tropical forest into savanna, and savanna into desert-like regions. Since then, others have come up with definitions that consider the phenomenon as a process of change (Rozanov, 1990), as an event or the final phase of the process of change (Mainguet, 1991), or a combination of both. Still others have included in the definitions, a number of factors to indicate the form of change (destruction of the biological potential of land - UN, 1977), the location of the change (usually drylands - UNCED, 1992), the reversibility of the change (Nelson, 1990), and the cause of the change (people related activities - Thomas and Middleton, 1994; climate variability - Tucker *et al.* 1991; or both climate and people related activities - UNCED 1992). We reproduce below, some of the more prominent definitions to capture the variability hinted above.

- Desertification is the notion that the extent of deserts - dry areas with few plants - is increasing, usually into the semi-arid lands (Warren and Agnew, 1988).
- Desertification is the impoverishment of terrestrial ecosystems under the impact of man. It is the process of deterioration in these ecosystems that can be measured by reduced productivity of desirable plants, undesirable alterations in the biomass and the diversity of the micro and macro fauna and flora, accelerated soil deterioration, and increased hazards for human occupancy (Dregne, 1986).
- Desertification is the diminution or destruction of the biological potential of the land that can lead ultimately to desert-like conditions. It is an aspect of the widespread deterioration of ecosystems and has diminished or destroyed the biological potential, i.e. the plant and animal production, for multiple use purposes at a time when increased productivity is needed to support growing populations in quest of development (UN, 1977)
- Desertification is a process of sustained land (soil and vegetation) degradation in arid, semi-arid and sub-humid areas, caused at least partly by man. It reduces both resilience and productive potential to an extent which can neither be readily reversed by removing the cause nor easily reclaimed without substantial investment. (Nelson, 1990).
- Desertification is land degradation in arid, semi-arid and dry sub-humid areas resulting from various factors, including climatic variations and human activities (UN, 1992). This working definition was agreed upon recently at the 1992 United Nations Conference on Environment and Development.

This plethora of definitions might give the impression that the phenomenon is well understood, but that is the myth. In reality it hides a lack of knowledge of what it is. This apparent uncertainty has led to a misunderstanding as to how to tackle it effectively. This scientific and/or institutional based paradigm diversity is unfortunate, because policies are often guided by scientific knowledge and outputs, and probably accounts for the perceived lack of progress in combating desertification. As a final reality check on the issue of definitions, we call attention to the fact that current definitions which focus exclusively on drylands, would exclude Aubreville's observation as to where it could occur. We believe the process could occur anywhere except perhaps in the desert itself, and make a plea to broaden the contemporary view as to where the process might occur.

Extent and Magnitude

The literature on desertification suggests that about 33 per cent of earth's surface is susceptible to desertification. This proportion represents all of the global dryland (arid, semi-arid and dry sub-humid) totalling about 51,720,000 square kilometres (5.2 billion hectares). Apparently 70 per cent of this land mass is already subjected to varying degrees of degradation. According to Dregne and Chou (1992), the proportion is that high because of the dominance of rangelands which have a high percentage of degraded land. These figures, are often quoted, but have also equally been challenged as to their accuracy. Thus, Warren and Agnew (1988) queried the claim of desertification threat to 33 per cent of earth's surface in the following terms 'An example of a misleading statistic is the claim by the UNDC/PAC (1987) that 33 per cent of Earth's land surface is threatened with desertification . . .' On the other hand, Dregne and Chou (1992) admitted that the information base upon which the estimates in their report were made was poor, because anecdotal accounts, travellers description, personal opinions, local experience and some research reports provided most of the data. UNEP has also confirmed the problem of inadequate data bases. Franklin Cardy, the former Director of the Desertification Control Programme at UNEP, in a recent article (Drylands, people, and desertification control - UNEP looks ahead), indicated that there is no question that the scientific basis for determining the extent of desertification is too subjective, and that there are clear inadequacies in the type of data base used to construct the Atlas of Desertification published by UNEP in 1992. He concluded that more detailed information of better quality is urgently needed, although the shortcomings of available information do not necessarily negate the increasing scope, severity and worldwide impact of desertification.

In reality therefore, the extent and magnitude of the phenomenon is not yet accurately known, for two reasons. On the one hand, the poor data base from which the estimates are calculated, and on the other hand, the exclusion from the 'affected areas' of areas where the aridity index – the ratio of average annual precipitation to potential evapotranspiration is greater than or equal to 0.05, and less than or equal to 0.65 (Table 9.1).

Table 9.1: Ecosystems Affected or Threatened by Desertification According to the Aridity Index

Ecosystem	Aridity Index	Desertification Status
Hyper-Arid	<0.03	Not affected
Arid	0.03-0.2	Affected
Semi-Arid	0.21-0.50	Affected
Dry-Sub-Humid	0.51-0.65	Affected
Sub-Humid and Humid	>0.65	Not affected

Process and Causes

Process

The process of desertification is not one of advancing deserts but usually starts in small areas or pockets of land where due to a number of factors, the vegetative cover is reduced to the extent of exposing the soil to erosion. If neglected, these small pockets of land will continue to grow, and finally coalesce to form larger seriously degraded areas. This is the type of observation made by Aubreville and the message he intended to convey when he used the term desertification to describe 'how deserts are being born today under our eyes'. The myth of the advancing desert has, however, gained ground and the Sahara is sometimes pictured as a sea of sand dunes invading the Sahel and swallowing large chunks of farm land.
Mobile sand dunes are portrayed as creeping forward onto cultivated land, damaging crop, burying houses. As indicated by

Nelson (1990), the advancing sand dune is a popular myth and/or misconception of desertification. Although it could be found in localised cases as in some parts of North Africa, Mauritania and Sudan.

This myth was popularised by Hugh Lamprey an ecologist who compared the position of the edge of the desert in Western Sudan at two different times, in 1958 and 1975 and concluded that 'it is evident that the desert's southern boundary has shifted south by an average of about 90 -100 km in the last 7 years'. The Sudanese Government readily accepted this statement, and added that it is currently advancing at the rate of 5-6 km per year (DECARP, 1976). Others have since disproved Lamprey's findings which were not based on a sound scientific and methodological approach. Dregne and Tucker (1988) noted that a permanent vegetational shift of 5-6 km per annum could only be established with perhaps 30 to 40 years of observation by meteorological satellites and ground studies, a far cry from Lamprey's approach of a one time aerial survey comparison to a vegetation map drawn by others. Evidently, as has been indicated by others, the process is not one of advancing deserts emanating from desert margins, but rather a complex process of invasive degradation radiating from centres of unsustainable land use. Unfortunately, although sufficient evidence has been provided to dismiss the myth of the spreading desert, the idea appears firmly implanted in many minds, including those of decision- makers who make crucial policies for combating desertification.

Causes

There is no shortage of causes, contributory and predisposing factors for land degradation in the drylands, and the raging debate in the literature identifies three schools of thought: those who identify natural climatic variations as the major cause, those who lay the blame squarely on human activities and those who attribute equal responsibilities to climatic variations and human activities. Climatic variations and human activities related to land use are so intricately linked that to try to separate them into primary and secondary causes would be foolhardy, as they often pose the chicken and egg problem. We group them therefore under direct causes, in contrast to socio-economic and political factors, which constitute the indirect causes of desertification.

Direct Causes

Climate

The debate on the link between climate change or climatic variation and desertification is inconclusive. Nelson (1990) tipped the balance in favour of a direct linkage when he concluded that about 70 per cent of the desertification problem can be attributed to natural events such as uncontrollable climatic events and population growth, and advocated climate research and population work to tackle the problem. A similar view was held by the Swedish team that worked in Sudan for over 30 years, when they concluded among other things that no widespread changes in vegetation cover took place that could not be explained by climatic variations, and that crop yield variations could mainly be explained by climatic variations, rather than a secular trend (Olsson 1983). Others (WMO, 1982; Hellden, 1992), however, have concluded that there is no adequate long term data sets to support such causal relations between climate change and desertification. Nevertheless, there is some evidence of increasing aridity in the Sahel over the last 2-3 decades (Hulme *et al.*, 1992), and a two way interaction between drought and desertification is evident. According to Warren and Agnew (1988), drought can precipitate land degradation by reducing water supplies in a system that is already imbalanced through overexploitation. Land degradation in turn may contribute to drought by the feedback mechanisms involving surface albedo, soil moisture and possibly dust.

People Related Activities

These are a complex mix causes all linked to land use systems that can be grouped into four categories: deforestation, cultivation, rangeland grazing and inappropriate use of irrigation technology which could result in salinization – a form of land degradation. There is some consensus that all of these activities could overcome the resilience of the dryland and cause degradation. Deforestation for fuel wood, agriculture, pasture land and industrial uses goes on unabated even in the drylands where vegetation is sparse, to meet increasing demand for land and food. The removal of trees exposes the soil to the sun, wind and water, and eventually to erosion by these elements. According to Glantz and Orlovsky (1983), cultivation practices that can lead to desertification are: land clearing practices such as slash and burn, cultivation of marginal land, cultivation of poor soils, reduced fallow periods, and the indiscriminate use of mechanized farming. Uncontrolled use of these practices may lead to overcultivation, which has been identified as the principal cause of dryland degradation by a number of people (Warren and Agnew 1988).

Indirect Causes

If we accept the suggestion that deforestation, overcultivation, over grazing and poor irrigation management are direct causes of desertification, it is pertinent to ask why people would knowingly damage their environment. The answer lies in the explanation given by Warren and Orlovsky (1988) that land degradation takes place in societies under stress. They used the example of the stress of drought and the effect of the stock-market crash of 1929 on the price of wheat that led or contributed to the Dust Bowl of the 1930s, to illustrate the point. We identified the following social, economic and political factors as stress factors that contribute to and exacerbate the above direct causes.

The Population Pressure

A number of attempts have been made to suggest that population increases have no direct bearing on land degradation, and some studies have found no correlations between population data and land degradation (Olsson, 1985). It has even been suggested that a selective decrease in population may be the operative mechanism (Kates *et al.* 1977), as in West Africa where degradation is occurring in the face of a decrease in active male population, or in developed drylands, where degradation is occurring as the human population decreases Nevertheless, it is clear that population growth puts pressure on natural resources in drylands in at least two ways.

Firstly, increased population means an increased demand for food, which has to be produced by increasing productivity (yield per ha) or production (area under cultivation) In most cases, the additional food comes from the latter, including an expansion into marginal and fragile land that are highly susceptible to degradation.

Secondly, increased family size means a parcelling of land resources with each subsequent beneficiary owning an increasingly smaller plot, which is then over cropped, with all of the known consequences.

The Stress of Poverty

With a shortage of capital resources, poor people would exploit their limited resources to meet immediate and pressing needs even if such short term exploitation would compromise the long term stability and viability of such resources. Hence an indisputable impact of poverty on desertification.

The Stress of Land Ownership

Changes in land tenure systems with the advent of political independence in many of the West African countries for example, have led to a cohabitation of modern private and traditional communal land tenure systems. New political realities have, however, sufficiently weakened the latter, to make the tragedy of the commons a common phenomenon. As indicated by Hardin (1968), it would therefore be in the interest of the pastoralist who owns his livestock individually, but his land communally, to overgraze for a personal short-term gain, and a societal long term loss.

The Stress of the New Global Economic Order

Declining commodity prices and unfavourable trade terms, could and do encourage or compel dryland farmers in developing countries to degrade their land resources, in the race to produce more to increase dwindling returns from increasingly poorly priced primary production.

The Stress of Inequitable Distribution of Resources

Land resources are usually inequitably distributed between men and women, small-scale producers and large commercial concerns, pastoralists and crop farmers. It is the poor fragile land that pastoralists who need good land for pastoral strategy have access to. Women, who carry a large part of the burden of farming, have no access to and cannot own land in most part of Africa. Such marginalized groups that have no access to or control over land resources can not be expected to exploit such resources in a sustainable manner, and often do not.

These are some of the social, economic and political factors that enable, encourage or even compel people to adopt destructive land use practises. If the stress from these underlying factors are not removed, efforts directed at the direct causes will only yield short-lived results, and the consequences of desertification will be intensified.

Effects and Consequences

The effects and consequences of desertification are well known, and include loss of vegetation, wind and water erosion, loss and/or impoverishment of land resources, reduced food production and productivity, food insecurity, loss of biodiversity including medicinal plants, climate related disasters, including health hazards from dust particle movements, environmental refugee problems, loss of livelihood system, disruption and break up of social structures etc. What is not well known and which have generated some debate, controversies and myths are the costs associated with these effects (it has been suggested that too much money has already been spent on desertification control), and the nature and extent of off-site effect of desertification (it has been suggested that desertification is exclusively a southern problem, where it is most severe, and its effects are most acute).

Several attempts have been made to calculate the economic impact of desertification, made up of income forgone as a result of land degradation and the cost of controlling and repairing damaged land (Aveyard, 1988. Dregne and Chou, 1992). According to Dregne and Chou 1992, not all desertified land can be reclaimed economically, and they estimated that 52 per cent of all desertified land can pay back the cost of rehabilitation. The income forgone by not rehabilitating this land they estimated at $28 billion annually, while the annual rehabilitation cost was estimated at $11 billion, and they concluded that the benefits are 2.5 times higher then the costs. Estimates from UNEP are of the same order, with rehabilitative costs of about $10 - $22.4 billion estimated at half the cost of income lost ($45 billion) due to desertification. It should be noted that none of the estimates included the difficulty of assessing the social cost of human misery suffered by the affected people. In other words, the reality belies the myth that too much money has been spent on mitigating the effect of desert-ification.

The claim that desertification is a southern problem with only on-site and no significant off-site impact needs to be re-examined in the light of recent evidence of transcontinental dust travel. Desertification is not considered a global environmental problem, because apparently it does not pass the 'transboundary test', i.e. no cause-effect relationship has been established where its action in one territory causes environmental damage in another territory or to the global commons. Environmental events that currently enjoy this status are: ozone depletion, biodiversity (loss to humanity), Amazonian deforestation (reduced carbon sink). Recent evidence, however, suggest that remote off-site effect of desertification due to transcontinental dust travel, qualifies desertification as a global environmental concern.

Recently in April 1998, American scientist and climatologists tracked a dust storm from Spokane across the country to Washington. The dust was determined to have come from China. On the day it reached Washington, the temperature was apparently 10 degrees F higher than normal and the air was stagnant. According to McNew (1998) meteorologists confirmed that such dust particles aside from lowering visibility, could cause adverse health effects, block sunlight and affect climatological patterns, and that for these reasons it was important to understand how and where dust storms form, how long they last and what can be done to prevent them or hinder their travel.

This was not the first documentation of transcontinental dust travel. Apparently, more than100 million tons of dust per annum is blown westward over the Atlantic from Africa (Middleton *et al.* 1986), dust coming from desertified land. More recently, Macleod (1998) indicated that evidence exist for a strong correlation between the summer climate of the Sahel and the frequency and intensity of hurricanes formed in the Eastern Atlantic Ocean. He argued on the basis of a model being tested that dusty Sahelian airs are a large source of energy used for hurricane formation and maintenance as it blows across the Atlantic. He concluded that if this model is proven, then the dust in the Sahelian airs is a factor in causing billions of dollars of material and economic damage to the US and the Caribbean regions, in addition to social and human costs of livelihoods and lives. These are strong evidence of the transboundary nature of desertification, and bely the myth that it is a purely a southern problem.

Solutions and Remedies

There can be no single set of solutions, for such a complex problem. An integrated approach that considers as many of the causative factors as possible has to be adopted. We believe that the Convention to combat desertification provides such a framework, and has all of the required elements. Some of these elements are:

- Develop a knowledge base through appropriate research to unravel some of the obscure issues surrounding the phenomenon.
- Transfer, share, and disseminate such knowledge and required technology from north to south and vice versa.
- Establish partnerships at the local level where the problem is most felt. The battle will be won at the local level, so establish partnerships and cooperation among all stakeholders - government, NGOs, communities and so on.
- Adopt the bottom-up approach, to ensure that decisions on the design and implementation of programmes are taken with the participation of populations and local communities who best understand their environment.
- Approach the complex problem with an integrated approach that addresses the physical, biological and socio-economic aspects of the processes of desertification within a National Program of Action.
- Pay a special attention to the economic environment, both internationally and within nations. Poverty, population growth, land ownership, local and political organizational controls, trade, and marketing arrangements - the underlying driving forces or indirect causes must be tackled, to relieve the stress they impose, in order to have long term and permanent solutions.

Concluding Remarks

As indicated earlier, the objective of this chapter is to draw attention to some of the myths surrounding the desertification debate, and confront them with the realities so that the former do not distract attention from the pressing task at hand - that of controlling and combating desertification. Unfortunately, such myths tend to attract attention, and even when sufficient evidence is provided to dismiss them, tend to linger in many minds, including those of decision makers who formulate and implement crucial policies. For example, although there is general consensus that the desertification process is not about advancing deserts, the concept is still to be put to rest, perhaps because it catches and attracts attention. The other side of the coin, however, is that it may lead to the utilization of inappropriate control techniques, and wastage of resources. Efforts must therefore continue to separate myths from realities, and all concerned must devote their resources, energy and time to finding long lasting solutions that will benefit not only drylands population abut also the global community at large.

References

Aubreville, A. 1949. *Climats, Forets et Desertification de l'Afrique Tropicale*. Paris, Societe d'Editions Geographiques, Maritimes et Coloniales.

Aveyard, J. 1988. 'Land degradation: Changing attitudes – why?', *Journal of Soil Conservation*, 44, 46-51, New South Wales.

DECARP, 1976. *Sudan's Desert Encroachment Control Rehabilitation Program* (DECARP), Khartoum, Sudan.

Dregne, H. E. and Tucker, C. J. 1988. Desert encroachment. Desertification Control Bulletin 16:16-19.

Dregne, H. E. and Chou Nan-Ting 1992. 'Global desertification and costs', in *Degradation and Restoration of Arid Lands,* Texas Tech. University, Lubbock.

Evanari, M., Shanon, L. and Tadmor, N. (1971). *The Negev: The Challenge of a Desert*. Harvard University Press, Cambridge, Massachusetts.

Glantz, M. H. and Orlovsky, N. S. (1983). 'Desertification: a review of concept'. *Desertification Control Bulletin* 9, 15-22.

Grainger, A. 1990. *TheThreatening Desert,* Earthscan, London.

Hardin, G. 1968. 'The tragedy of the commons', *Science*, 162, 1243-1248.

Hellden, U. 1992. 'Desertification – Time for an assessment?'. *Ambio* 20, 372-383.

Hulme, M., Marash, R. and Jones, P. D. 1092. 'Global changes in humidity index between 1931 and 1990.' *Climate Research,* 1-22.

Kates, R. W., Johnson, D.L. and Johnson Haring, K. 1977. 'Population, society and desertification', United Nations Environmental Programme, World Desertification Conference Document A/Conf. 74/8.

Macleod, N. H., 1998. 'The role of airborne Sahelian dust in the frequency and intensity of hurricanes formed in the Eastern Atlantic Ocean.' Third World Foundation paper, Washington DC.

Mainguet, M. 1991. 'Desertification'. *Natural Background and Human Mismanagement*. Springer-Verlag, Berlin.

McNew, J. G. 1998. 'Specks of dust make worldwide impact', On-line, *http://www.disasterrelief.org/disasters/980722wind.*

Nelson, R. 1990. *Dryland Management: The Desertification Problem*, World Bank Technical Paper Number 116, Washington D.C.

Olsson, G.W. 1981. *Soils and the Environment: A Guide to Soil Surveys and their Applications*. Chapman and Hall, New York, N.Y.

Olsson, L. 1983. 'Desertification or climate ? Investigation regarding the relationship between land degradation and climate in the central Sudan.' *Lund Series in Geography Series A*, 60.

Olsson, L. 1985. 'An integrated study of desertification: applications of remote sensing, GIS and spatial models in semiarid Sudan', Meddelanden fran Lunds Univesitets Geogrfika Institution, Avhadlingar, 98.

Rozanov, B. G. 1990. 'Global assessment of desertification: status and methodologies', in *Desertification revisited: Proceedings of an ad hoc Consultative Meeting on the Assessment of Desertification*, 45-122, UNP-DC/PAC, Nairobi.

Spooner, B. 1989. 'Desertification: the historical significance', in Hess Ashmore and S.O. Katz (Eds). *African Foods Systems in Crisis. Part One, Microperspectives*, pp 111-162, Gordon and Breach, New York.

Stebbing (1938). 'The advance of the desert', *Geographical Journal*. 91, 356-359.

Thomas S. G. and Middleton, N. J. 1994. *Desertification: Exploding the Myth*, John Wiley and Sons, Sussex.

Tucker, C. J., Dregne, H. E. and Newcomb, W. W. 1991. 'Expansion and contraction of the Sahara desert from 1980 to 1990.' *Science* 253, 299-301.

UN, 1977. 'Desertification: an overview', in *Desertification: Its Causes and Consequences,* Pergamon Press, Oxford.

UNCED, 1992. *Earth Summit '93.* The United Nations Conference on Environment and Development, Rio de Janeiro.

Warren, A. and Agnew, C. 1988. *An Assessment of Desertification and Land Degradation in Arid and Semi-Arid Areas,* International Institute for Environment and Development, Paper 2, London.

World Meteorological Organization (WMO), 1983. *Report on the ECA Discussion on the Climatic Situation and Drought in Africa,* Geneva.

Chapter 10

The Nature, Causes and Consequences of Desertification in the Drylands of Africa

M.B.K. Darkoh

Introduction

The term 'desertification', as pointed out in the previous chapter, was first coined in 1927 by the French scientist and explorer Louis Lavauden and first popularised in 1949 by the French forester Andre Aubreville. The latter, generally considered as the father of the term, used it to describe a general process of degradation starting with deforestation - not necessarily in drylands - and ending in land turned into desert (Aubreville, 1949). He was working in the sub - humid parts of West Africa. Unfortunately, the name includes the word 'desert'; so people assume it refers to productive land being covered by sand dunes. This popular image of sand dune encroachment is only a minor part of the problem. Sometimes thousands of kilometres away from the margins of the Sahara, Gobi, Atacama and other so-called 'true' deserts, desertification is taking place. The situation has been likened to a skin disease in which existing eruptions worsen and coalesce with new outbreaks of the disease (Tolba, 1984). According to UNEP (1992a), the idea of shifting sand dunes or desert encroachment has detracted scientific and public attention from the real issue. The process of desertification is not shifting sand dunes but rather 'patches of increasingly unproductive land breaking out and spreading over hundreds of square kilometres' (*ibid.*). Odingo (1990) has reviewed the different definitions of desertification and their programmatic consequences for UNEP and the international community. Until UNCED, UNEP itself had attempted to allay confusion by defining desertification as 'land degradation in drylands resulting from adverse human impact' (Dregne *et al.*, 1991).

Most current definitions include some mention of land degradation, the loss of productivity, the combined roles of climate and human activity, and the fact that desertification is restricted to drylands. But these definitions are too general for estimating the actual extent and degree of desertification in a particular area. Most in common use are criticized for their vagueness on key issues (is it a process or a state?; are processes like deforestation and salinization causes or symptoms?; what exactly is the role of drought?).

The latest and most widely accepted definition is the internationally negotiated one which defines desertification as land degradation in arid, semi-arid and dry sub-humid areas resulting from various factors including climatic variations and human activities (UNCED, 1992). In the context of this new definition, as in previous definitions proposed by UNEP and others (see Odingo, 1990), land is taken to mean the terrestrial bio-productive system that comprises soils, vegetation, other biota and the ecological and hydrological processes that operate within the system.

Degradation implies diminution or destruction of the biological potential (resource potential) by one or a combination of processes acting on the land. The concept of land degradation is broader than soil degradation since it deals with the whole ecosystem in which soil is just one of the components. The processes of land degradation include water erosion, wind erosion and sedimentation by these agents, long-term destruction of vegetation and diminution of many plants and animal populations, or decreases of crop yields where relevant, and salinization or sodication of soils. The 1994 UN International Convention to Combat Desertification described land degradation as the reduction or loss of the biological or economic productivity and complexity of the land (INCD, 1994). Human activities include overcultivation, overgrazing, deforestation, poor irrigation practices and any other inappropriate land use practices.

Desertification is often confused with drought or desiccation. Drought refers to the naturally occurring short-term (1-2 years) phenomenon when precipitation is significantly below normal recorded levels. Usually such temporary deficits in rainfall can generally be accommodated by existing ecological, technical and social strategies. Drought is generally a dry period from which an ecosystem recovers rapidly after the rains return. Desiccation refers to longer-term (decadal order) deficits in rainfall which seriously disrupt ecological and social patterns and require national and global responses. Recovery after desiccation is much slower, for trees may have died and vegetation may then take years to recover. Responses include management of population movements and the development of alternative livelihood systems.

However, it does not necessarily follow that drought or desiccation per se will give rise to or cause desertification in dryland areas. Much depends on the resource management in these areas. When human misuse or mismanagement of land weakens the natural system, drought and desiccation often lead to desertification. While desertification should not be confused with drought and desiccation, the definition provided by UNCED cites climate variation as a direct causal factor and implicitly links desertification with climatic variation or climate change. Climatic variation or climate change refers to short-term climate variability and longer-term climatic trends or shifts caused by natural mechanisms or by human activity (Kelly and Hulme, 1993). Climate change does cause global warming often through greenhouse gas emissions. Natural climate change, which typically operates at a slow pace is not a problem. Climate has been changing constantly for hundreds of millennia. As a result of the slow advance of natural processes, the planet has warmed and cooled, passing regularly from ice ages to warm, interglacial periods. These gradual transitions, often spanning thousands of years, have allowed life on earth to adjust relatively smoothly to each new climatic equilibrium. Nonetheless, during these transitions, the boundaries of ecological communities have shifted, the associated human cultures have flourished and, occasionally, disappeared.

However, something important has changed recently. During the last two centuries, the natural greenhouse effect (see Box 10.1) has become the 'greenhouse problem'. In the foreseeable future, rising concentrations of greenhouse gases (GHGs) threaten to induce rapid shifts in global and regional climate regimes, disrupting economic systems and inflicting significant economic damage on the affected societies.

The United Nations Global Environmental Facility (GEF, 1993, 1994) identifies two sources of greenhouse gases that affect the climate. First are the GHGs that have a direct effect on climate. Such gases are CO_2, methane (CH_4), nitrogen dioxide (NO_2), and the chlorofluorocarbons (CFCs), hydro-CFCs (HCFs), hydrofluorocarbons (HFCs), and other chlorine compounds. The first three of these compounds - CO_2, CH_4 and NO_2 have important natural as well as

Box 10.1: The Greenhouse Effect

The 'greenhouse effect' is the name given to a biogeographical process that has been essential to the evolution of life on earth. For two billion years, natural background concentrations of certain gaseous compounds have trapped heat close to the earth's surface, warming the planet. These gases, principally water vapour, ozone, and carbon dioxide are transparent to incoming sunlight in the short-wave portion of the electromagnetic spectrum. But they absorb and remit some of the outbound, long-wave,infra-red radiation from the earth's surface. Part of this infra-red radiation is re-emitted upward towards outer space; the remainder is re-emitted downward toward the planet's surface.

For millennia, this downward re-emission has been to raise the average temperature at the surface by about 33°C (59°F) above what it would otherwise have been. Without this natural greenhouse effect, the average temperature on earth would have been - 18°C, and water would have been present only as ice. The natural greenhouse effect allowed liquid water to remain stable over most of the earth's surface, providing the fundamental substrate for biological activity as we know it (GEF, 1993).

anthropogenic sources. Most of the chlorine compounds that have greenhouse effect are of anthropogenic origin and their sources are well established (GEF, 1994).

The second group comprises emissions of gases that have a negligible direct greenhouse effect, but are indirectly affecting the climate through their impact on chemical and physical processes in the atmosphere and thereby on the GHGs that are affected by chemistry. Source gases that belong to this group are nitrous oxide (NOx), carbon monoxide (CO), and hydrocarbons. GHGs that may be affected by emissions of these gases are methane and ozone. Ozone (O_3) is not emitted, but is formed in the atmosphere. It affects climate through interaction with both long-wave radiation and solar (short-wave) radiation. Emissions of sulfur dioxide (SO_2) may also affect climate through the formation of sulfate aerosols. These aerosols reflect solar radiation and may also affect the radiation budget through changes in cloud optical properties. The climate impact of this second group of anthropogenic gases cannot be assessed with satisfactory accuracy at present as too little is known about their impact on GHGs and the aerosols in the atmosphere (GEF, 1994). Nevertheless, it is believed that the indirect climatic impact of such gases can be important.

Since the beginning of the Industrial Revolution, GHG emissions from human activities have grown steadily. Because the rates of emissions of these gases have exceeded the ability of natural processes to remove them from the atmosphere, their atmospheric concentrations have increased, enhancing the natural greenhouse effect. The building up of the gases creates the greenhouse problem and, if current trends continue, may warm the planet at an unprecedented rate (GEF, 1994).

In tropical lands, land use practices that might have impacts on regional or even global climate include large scale clearing of rainforests, the burning of biomass, paddy field cultivation, cattle breeding and overgrazing of semi-arid pasture grounds (Jacobeit, 1991). These land use types modify climatic parameters such as land surface evapotranspiration, concentration of atmospheric trace gases and aerosol particles, albedo and roughness of the earth's surface. Many effects appear to be small-scale effects which sometimes compensate each other. However, a variety of greenhouse gases released from tropical areas under land use seem to be increasingly enhancing the earth's greenhouse warming.

In arid and semi-arid lands, human-managed activities like overgazing, coupled with degradation of the natural vegetation can lead to changes in particular climatic parameters. Initially, surface albedo (coefficient of reflectivity) could rise by roughly 5 per cent increasing up to 10-15 per cent under conditions of desertification (Jacobeit, 1991). Reduced vegetation cover and increased surface albedo may however also result from naturally occurring periods of drought. Reduced vegetation cover also increases dust emission in arid and semi-arid lands due to intensified deflation and thus contribute to the anthropogenic particle emission, one third of which are due to land-use-caused deflation. (Jacobeit, 1991). However, since arid and semi-arid lands are climatically determined, any changes in climate which result in an expansion or contraction will alter the extent of the area in which desertification can be considered to occur. Deciding the precise contribution of climate variation to the problem of desertification is, however, problematic.

Causes and Linkages of Desertification

Climatic variation apart, desertification is, first and foremost, the outcome of resource management failure (Kelly and Hulme, 1993). Its anthropogenic processes include overcultivation, overgrazing, deforestation, and poor irrigation practices. (UNEP, 1992 a, b, and c). These processes are the result of excessive human pressures on resource ecosystems. These pressures are fueled by local forces (Fig. 10.1) such as increases in human numbers and escalation of human

Fig: 10.1 Desertification and its Linkages

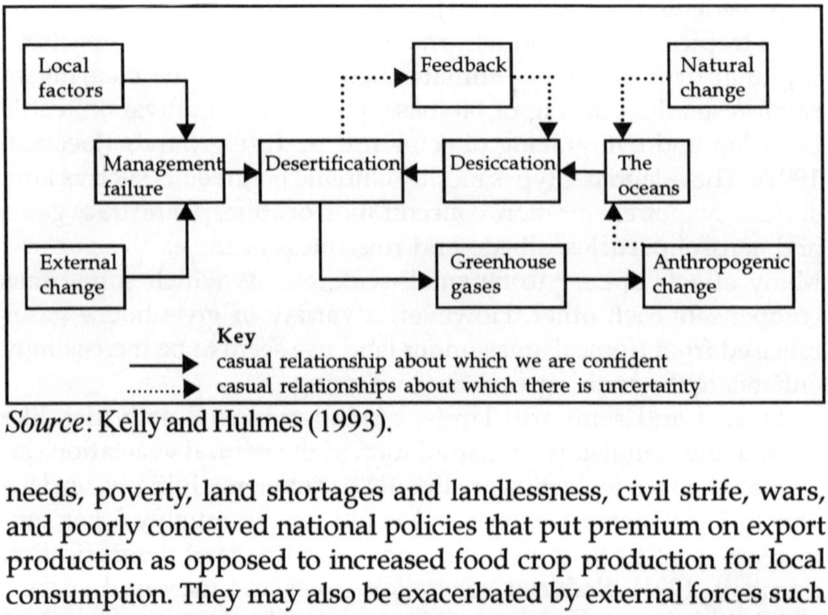

Source: Kelly and Hulmes (1993).

needs, poverty, land shortages and landlessness, civil strife, wars, and poorly conceived national policies that put premium on export production as opposed to increased food crop production for local consumption. They may also be exacerbated by external forces such as the state of the global economy, commodity prices, the debt burden, brain drain, terms of trade, protectionism and import barriers in developed countries which make poverty alleviation more difficult for exporting countries and which may cause them to accelerate rates of natural resource exploitation by preventing diversification.

The process may be aggravated by climate change (Kelly and Hulme, 1993): in particular, by prolonged drought and desiccation (Figure10.1). Desiccation itself could be the result of natural mechanisms with the climate system such as the influence of ocean temperature patterns (Kelly and Hulme, *ibid.*). Desiccation may also be caused by desertification, through surface-atmosphere interaction, or it may possibly be the result of global warming, which can be defined as climate change resulting from greenhouse gas emissions discussed already. Finally, dryland degradation can contribute to global warming through its effect on the sources and sinks of greenhouse gases (Kelly and Hulme, *ibid.*). The relative importance of all the above factors may change with sustained degradation or decreased rainfall. Balling (1992), Hulme and Kelly (1993), and Williams and Balling (1994) have provided an extensive discussion on the interactions of desertification and climate, with a fairly comprehensive review of the existing literature on climate and human impact on dryland environments.

That climate change does occur is now an established fact. In the African Sahel, as Nicholson (1978) Kelly and Hulme (1993), and Hulme and Kelly (1993) have amply demonstrated, annual rainfall during the most recent decades (1961-1990) has been between 20 and 40 per cent less than it was from 1931 to 1960. Within contiguous Africa, Kelly and Hulme (*ibid.*) have shown that there has been a net shift of land toward aridity, especially toward hyperaridity, and a consequent net loss of semi-arid and dry subhumid land.

However, determining the precise contribution of climate change to the problem of desertification, as we have noted, is not an easy matter. In general, one can generalise in the light of recent research findings that there is a causal link between climate change and desertification. While the relative contribution of climate change to desertification is difficult to determine, there is little doubt, however, that climate variation can aggravate the desertification problem, especially where resource management failure has occurred or where, as a result of natural or anthropogenic disturbances, prevailing management systems of land use in arid and semi-arid lands exceed their resilience thresholds.

As the case histories of four different regions which are undergoing or have undergone desertification – the Sahel, the Euphrates – Tigris basin, the Aral Sea region and the Dustbowl of the USA – show (Darkoh, 1995), a common structure underlies the desertification process. In arid and semi-arid areas, systems formed by man and natural resources evolve to cope with the maximum possible ranges of local climate variability (Puigdefabregas, 1995). Eventually such systems may be perturbed by a factor that produces a sudden increase in resource availability or resource consumption, as for example, a humid climate anomaly as in the Sahel or a technological change allowing access to new resources as in irrigation development in the

Basin of the Euphrates and Tigris (the 'Garden of Eden') and the Aral Sea Region of the former Soviet Union, or a change in the market condition as in the Dustbowl of the USA. The result of such disturbances is a burst of the exploitation rate, by the inflow of population from the outside, as in the case of the Sahel and the Dustbowl, or by increase in per capita ability to exploit resources as in the Garden of Eden and the Aral Sea Region. Figure 10.2 highlights the structure of the desertification process outlined above.

Once a disturbance like the humid rainfall anomaly in the Sahel, for example, ceases, resources continue to be rapidly depleted by the large population and the system reaches its resilience threshold. At this point, unless the pressure on resources decreases, or the management system changes to more intensive or sustainable forms capable of maintaining equilibrium between resource and human utilisation, desertification accelerates as the land continues to lose its biological and economic productivity.

However, desertification is still a controversial issue (Thomas and Middleton, 1994) because of the ambiguity about whether it is a process or a condition and also because of the difficulties of regarding processes like overcultivation, deforestation and overgrazing causes or symptoms.

The desertification concept is still evolving and a more rigorous definition is needed in terms of measureable 'indicators'. Furthermore, despite the general consensus, there are those (e.g. Arnalds, 2000) who feel that the current definition is too narrow and limits political and economic actions and programmes aimed at combating desertification or reversing land degradation. There has, therefore, been a plea for a broader perspective (see, for example, Smith and Koala in in the previous chapter) as severe land degradation leading to nearly total devastation of ecosystems does occur outside the regions specified in the UN definition, as exemplified by the Icelandic case history (Arnalds, 2000). Scientifically, the term still remains vague despite the fact that this has been pointed out more than ten years ago (Forse, 1989; Hellden, (1991). These weaknesses hinder efforts to solve real problems in threatened areas.

The scientific and social status of desertification as an environmental issue has recently become a bone of contention (Thomas 1997). Science has been viewed as responsible for some of the difficulties that have clouded debates about, and solutions, to desertification. Thomas *(ibid.)* notes that scientific contributions to

Fig. 10.2 Outline of the Desertification Process

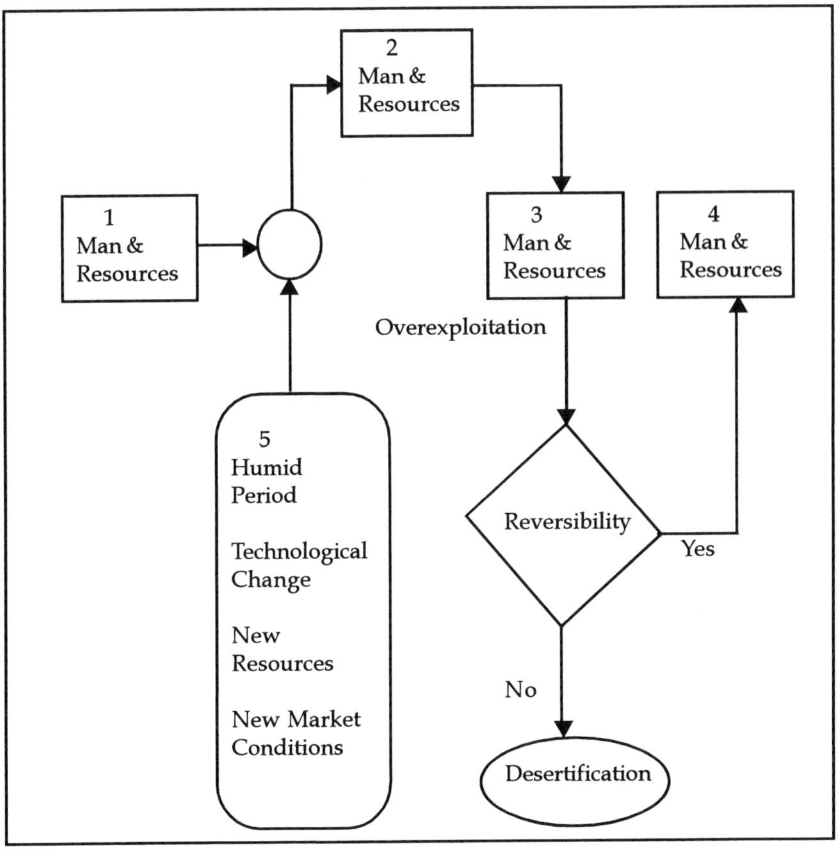

Source: Puigdefabregas (1995).

Square boxes represent the system formed by man and local resources. Boxes 1 and 4 stand for systems in equilibrium with the normal long-term variability of resources in the area. Box 2 represents the system stressed in box 5. Box 3 stands for the system in overexploitation regime, once the disturbance has ceased.

desertification have not always been viewed in a positive light and science has been blamed in some quarters for failure to resolve the desertification problems. Such pessimistic views of science do not augur well for progress at the anti-desertification front. As Thomas *(ibid.)* has rightly pointed out, science does not necessarily provide instant and simple, easily applied solutions to environmental problems. Scientific research findings are iterative and not always definitive. Paradigm shifts occur and the way science tackles problems does not necessarily correlate with the way decision makers visualize or portray problems at the society-environment interface. Despite doubts raised about the role of science, scientific research has helped to enhance our understanding of the functioning of dryland environmental systems. Since UNCOD, science has been involved in research that has attempted to identify the onset and impacts of drought and desertification at different spatial scales and monitoring the operation of systems in relation to degradation and recovery processes. Science has been helping to clarify what desertification actually is and how it relates to various environmental conditions and changes. Most recently, with the coming into force of the Desertification Convention, scientific research explicitly being promoted by the Convention's Committee for Science and Technology has been grappling with the problem of reliable data and information sources which are an essential precondition to forecast and combat future desertification and to mitigate the effects of on-going processes (European Commission, 2001). From the findings of science we now know that desertification is a human as well as a physical problem. Scientific research on desertification in Africa should continue to make its debut in enhancing our understanding of desertification processes and impact as well as finding solutions to pertinent problems (see Chapter 4).

Desertification in Africa

Drylands in Africa, including hyper-arid deserts, comprise 1,959 million hectares (ha) or 66 per cent of the continent and about one third of the world's drylands (UNEP, 1992a). One third of this area is hyper-arid deserts (672 million ha) that are uninhabited, with the exception of sparse tiny oases, while the remaining two thirds or 1,287 million ha are composed of arid, semi-arid and dry sub-humid areas with a population of about 400 million (two thirds of all Africans).

According to UNEP's latest assessment (1991), 1.9 million ha of irrigated croplands (or 18 per cent of their total area), 48.86 million ha of rainfed croplands (or 61 per cent of their total area), and 995.08 million ha of rangelands (or 74 per cent of their total area) in Africa are affected by desertification at a moderate or higher degree, as shown in Table 10.1.

Table 10.1: Status of Desertification / Land Degradation in the Drylands of Africa

Irrigated Lands			Rainfed croplands			Rangelands			Total agriculturally used drylands		
Total (M ha)	Degr (M ha)	(%)	Total (M ha)	Degr (M ha)	(%)	Total (M ha)	Degr (M ha)	(%)	Total (M ha)	Degr (M ha)	(%)
10.42	1.90	18	79.82	48.86	61	1342.35	995.08	74	1432.59	1045.84	73

Source: UNEP (1992a: 81).

Three distinct areas of the continent are at most risk: Mediterranean Africa, Sudano-Sahelian region and the Kalahari-Namib Region in Southern Africa (Figure 10.3). One third of Africa is affected by desertification and 73 per cent of the total agriculturally used drylands are degraded (see Table 10.1).

Recurrent droughts constitute a permanent factor of life throughout the drylands of Africa; virtually every year there is a drought in some part of the continent or another. Major droughts, however, regularly affect large portions of the drylands. Such disastrous events occurred in 1968-1973, 1982-1985 and 1990-1991, when many countries of Africa experienced substantial food shortages. With each drought cycle, desertification increases.

Data provided by UNEP (1992a) reveals that all major factors of desertification in Africa remain unabated, leading to increasing land degradation despite modest efforts to arrest it. Although satellite data show rather large fluctuations in the rainfall-dependent northern and southern boundaries of green biomass production zones, both seasonal and annual, the overall trend is negative. There are clear manifestations of continued ecological degradation.

Nevertheless, there seems to be little evidence to link desertification with the notion of desert advance in Africa. Earlier observations and assertions have been made that the desert was advancing at approximately 5 km a year (Lamprey, 1975).

This has now been disproved by Hellden (1991) whose work in the Sudan shows no such advance and Tucker *et al.*, (1991) who assert that patterns of vegetation cover in these areas are dependent on rainfall. Hellden (*ibid.*) has further asserted that contrary to

Figure 10.3: Africa: Areas at Risk of Desertification

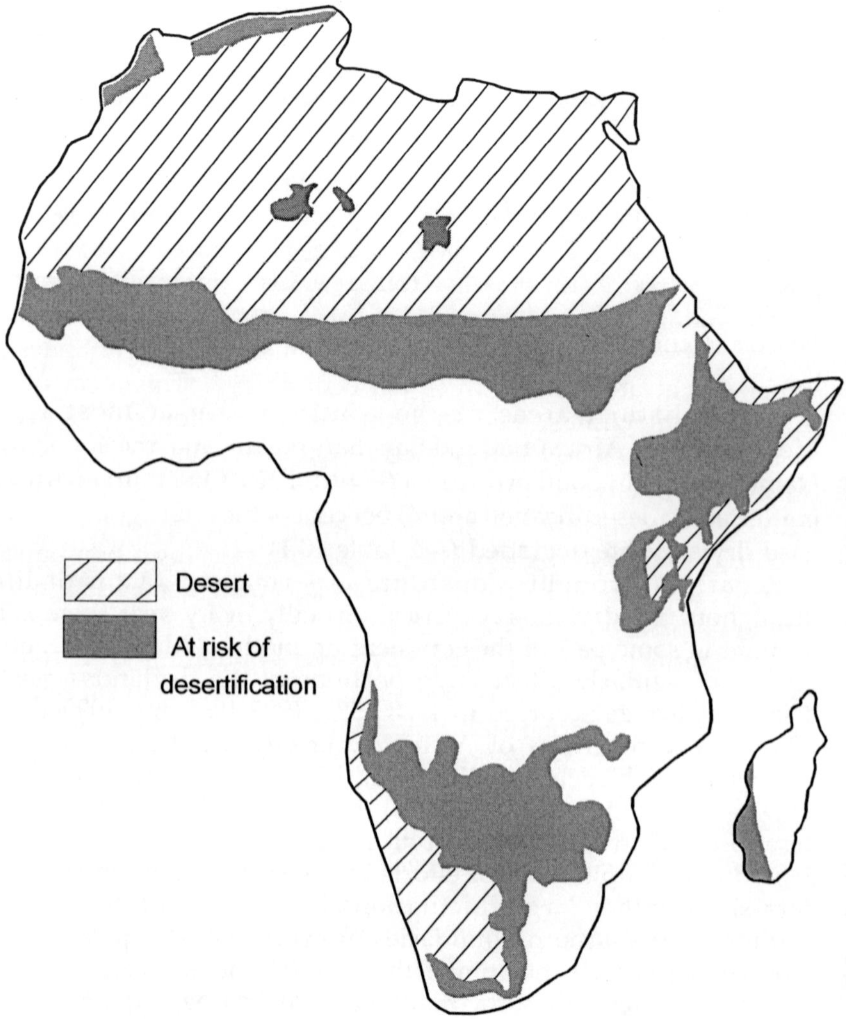

arguments advanced, there is no evidence that patches of deserts outward from villages and water holes within the drylands of the Sahel area, for instance. More sustained research work at micro-scales in other parts of Africa are needed to determine the impact of human settlements on the phenomenon of desertification to corroborate or disapprove the latter findings of Hellden.

The Human Causes of Desertification in Africa

In Africa, a leading cause of desertification is human population pressure which leads to over-exploitation and intensified stresses on the natural environment. Africa's population has doubled in the past three decades to about 708 million (1994) and continues to expand at a rate of some three per cent a year. This means that Africa's farmers must feed an additional 21 million people every year, whether the weather is good or bad. By comparison, North America adds fewer than three million people to its population each year.

Rapid population growth with concomitant high population densities and rapid urbanisation implies growing demand for more land for crop production as well as wood for fuel, because there are more mouths to feed. Continuous rapid population growth also implies increased use of land in economies dependent on rainfed cultivation. Where suitable land for agriculture is plentiful, increasing population growth and high densities do not immediately pose a problem; but in areas of land scarcity or where population/ land resource ratios are high, further increases in population growth can imply shortening of fallow cycles as land comes under increased pressure, leading to consequent declines in productivity, unless, of course, more intensive forms of land use such as the application of manures and fertilisers can be adopted.

Increased population pressure on the fragile and vulnerable soils of Africa's dryland regions, leads to over-exploitation of water, land, forest and pasture resources through over-cultivation, overgrazing and deforestation. Over-cultivation, overgrazing and deforestation therefore constitute the principal threats to the livelihood of millions of people in this continent. These are the foremost causes of soil erosion and the rates of soil erosion in Africa are among the highest in the world. Already, soil erosion caused by overgrazing, over-cultivation and deforestation has robbed the susceptible drylands of Africa of more than a quarter of their top soils (Table 10.2 and 10.3). Loss of topsoil is the greatest problem, accounting for over ten times as much water erosion as terrain deformation (Middleton and Thomas, 1997: 58). The resulting degradation of productive lands has led to declining production and intensified food insecurity (Figures 10.4 and 10.5).

Many of the susceptible and highly populated upland areas of the Sahel such as the northern Ethiopian Highlands, the Ennedi Highlands of Chad and adjoining Darfur in Sudan, have been highly degraded. So also are the Atlas and Rif Mountains in Africa north of the Sahara, where 'badlands' have developed in places. Other susceptible dryland areas include Western Madagascar, Central Tanzania and the pastoral and arable savanna lands of Southern Africa.

Table 10.2: Main Causes of Soil Degradation in the Susceptible Drylands of Africa

Causes	Area Affected (M ha)	Area Affected (%)
Overgrazing	194.4	15.1
Overcultivation (agricultural activities and over exploitation)	115.9	9.0
Deforestation	22.0	1.7
Total	332.3	25.8

Source: UNEP (1992 b).

Table 10.3: African Susceptible Drylands by Degrees of Soil Degradation (%)

Degree	Area (M ha)	Area (%)
Light	144.2	11.2
Moderate	112.2	8.7
Strong	72.8	5.7
Extreme	3.1	2.0
Total degraded	332.3	27.6

Source: UNEP (1992b).

Ambiguities, however, surround the impact of population growth and density on desertification (UNSO, 1992; Darkoh, 1994). Caution, therefore, must be exercised in evaluating the effects of rapid population growth, since Africa has also been shown to offer good examples in recent times of cases of 'more people, less erosion' (Tiffen *et al.*, 1993; Darkoh, 1994). According to UNSO (1992), the connection between land degradation and population is seldom direct. Furthermore, while, population may be one cause of land degradation, it is rarely the primary one and never alone. It would appear that in the drylands of Africa, population as a factor in desertification or land degradation commonly acts in conjunction with other factors such as drought, desiccation, rural poverty (which translates into inability to invest in land management), land shortages and landlessness, land ownership and tenure patterns, and national policies which lead to excessive development of cash crops.

However, while the linkages between desertification and population growth are seldom direct or clear cut, there seems to be evidence of a clear link between desertification and population movement or migration in the arid and semi-arid lands of Africa. The migration generally takes three forms. These are: rural-urban migration, i.e. that caused by the urban pull; rural-to-rural flows; and emigration or the steady loss of labour especially of able-bodied males which is keenly felt in rural areas. All three types are common in the drylands of Africa. They are particularly pronounced in the Sudano-Sahelian countries. In Kenya for example, this author's research reveals three migratory trends: permanent immigration, temporary or seasonal emigration and return migration (Darkoh, 1994). The first two are of particular importance to environmental issues in the drylands. Several factors influence these migration trends. These include climate, security, tradition, availability of central services, famine relief, gainful employment in towns, irrigation schemes and seasonal labour in high potential districts.

With high population densities in the high potential areas, more and more people are moving from these areas in search of farming land and employment opportunities (Darkoh, 1982a, and b; 1991b, 1994). Environmentally, permanent immigration into the arid and semi-arid lands is the most important form of migration in Kenya as it is these immigrants who often cause land pressure and import inappropriate technologies that lead to land degradation. They also disrupt the indigenous management systems which are based on appropriate and locally adapted technologies.

Fig.10. 4: Per Capita Agricultural and Food Production in Sub-Saharan Africa, 1962-88
(index 1980=100)

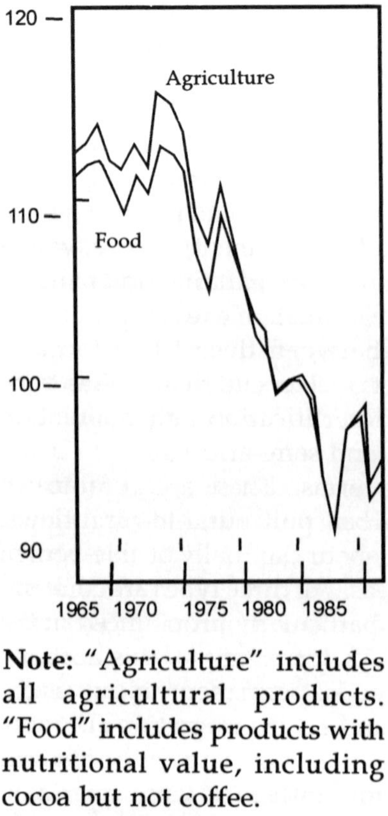

Note: "Agriculture" includes all agricultural products. "Food" includes products with nutritional value, including cocoa but not coffee.
Source: FAO data.

Fig. 10.5: Sub-Saharan Africa's Food Trade, 1961-87
(millions of tons)

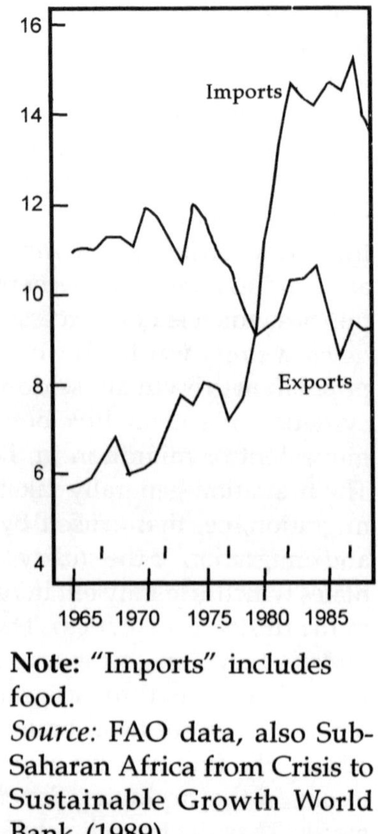

Note: "Imports" includes food.
Source: FAO data, also Sub-Saharan Africa from Crisis to Sustainable Growth World Bank (1989).

Sometimes this permanent movement of population into the drylands is part of a deliberate government policy of solving the population-resource problem by encouraging planned settlement and cultivation in the wetter margins of the arid and semi-arid lands, or irrigation schemes (Darkoh, 1992; Campbell, 1981, 1986). From evidence gathered from fieldwork in both Kenya and Tanzania (Darkoh, 1982, 1990, 1996a), permanent and seasonal migration within the drylands, especially movement of farmers from the over populated high potential areas (where arable land is in short supply) to the arid and semi-arid lands appears as one of the primary factors posing a threat to the environment in the drylands. It has given rise to landuse conflicts. These conflicts arise as a result of intrusion of agriculture into lands traditionally used for domestic stock. Partly, because there is a lack of coordinated land use policies for the drylands, there is competition for resource use between the various production sectors. The major contenders are agriculture, livestock, wildlife and human settlements. Often, the weakest sector – pastoralism – bears the brunt of this fierce competition for resource use, consequently becoming increasingly marginalized.

The conflicts are especially intense in the key production areas within the drylands. These include the riverine forests along the main water courses, the natural forests, swamps and hilly areas. The dynamics of the conflicts in these key production areas, which are currently some of the most threatened productive lands within the drylands need to be studied and remedial policies and strategies devised to protect them from further degradation.

Civil strife is one of the complicating factors influencing resource systems and availability of food in many parts of the continent. Many parts of Sub-Saharan Africa have suffered from breakdown of law and order repeatedly during the last decades. The problems in Ethiopia, Somalia, Sudan, Rwanda, Burundi, Chad, Angola, Mozambique, among others, in these drylands are well known. Being short-term in itself, this factor contributes greatly to the long-term process of land degradation in many ways, partly by displacing people and leaving land unattended – which is not always good for the natural recovery of land, partly by diverting resources that might otherwise be used for environmental and development purposes and partly by deterring people from their normal way of life and forcing them to untried methods and new areas. Civil strife also gives rise to the mass exodus of people, who are often quartered in refugee camps in neighbouring countries. Land degradation around such refugee camps is a common phenomenon in several countries e.g. Tanzania and Zaire.

The Office of the UN High Commission for Refugees (UNHCR) found that, based on 1989 estimates, roughly 11 million trees were cut for shelter needs during the initial period of refugee influxes in Africa (Cardy, 1994). This represents the deforestation of over 12,000 hectares. In addition, about 4 million tons of woodfuel were consumed by refugees in Africa.

In North Africa overgrazing has led to moderate to severe desertification of rangelands in arid and semi-arid zones of Algeria,

Libya, Morocco and Tunisia. There has also been widespread destruction of woodlands as a result of clearing for fuelwood, agriculture, or grazing and fires. Salinization of irrigated lands is reported from all these countries as well as Egypt where it is linked with inadequate drainage. Many irrigated lands are threatened by sand encroachment resulting from heavy grazing or the abandonment of irrigated lands.

If we look at the remaining Sub-Saharan African countries in the drylands, especially the Sudano-Sahel and Southern Africa today, we find basically similar problems: fragile soils being degraded through improper cultivation practices, fuelwood cutting leading to deforestation, and overgrazing destroying the ground cover over large areas, all of which combine to leave the land more vulnerable to drought and soil erosion (Darkoh, 1988, 1989, 1993). As a result, vast tracts of land are being transformed into 'dust bowls', losing their productivity and impoverishing their populations. As Whitlow (1980) has shown for Zimbabwe, a major cause of soil loss in southern Africa is vegetation destruction. The forces and dynamics involved are illustrated in Figs. 10.6 a and b.

Caution, however, must be exercised in assessing the effects of these macro-and micro-factors or processes. For example, in Africa as a whole, while the incidence of deforestation resulting from fuelwood requirements can have some serious effects because fuelwood and charcoal are critical resources for the poor, recent research has revealed that these effects tend to be generally localised in the dryland areas, especially around settlement nodes. Most local people only collect dead wood of selected species. UNSO (1992) has noted that in a purely rural macrocosm, with dispersed settlementsfuelwood seldom has been a great problem. Except in some highly localised or highly populated rural areas such as the Ethiopian Highlands and parts of Central Tanzania, little evidence exists to suggest that rural household energy consumption is responsible for large-scale deforestation. Mounting evidence notably from Zambia, Kenya, and Sudan would seem to point out that it is rather the urban demand, usually for charcoal, that leads to the extensive cutting down of forest. The commodity status of charcoal makes it an attractive choice for entrepreneurs who can derive incomes from its production and distribution (Darkoh, 1990, 1991 a and b, 1994).

Furthermore, while the causes of overgrazing are sometimes mechanically attributed to large growing numbers of the animal

Figure 10.6a: Some Reasons for the Destruction of Woody Vegetation in Zimbabwe

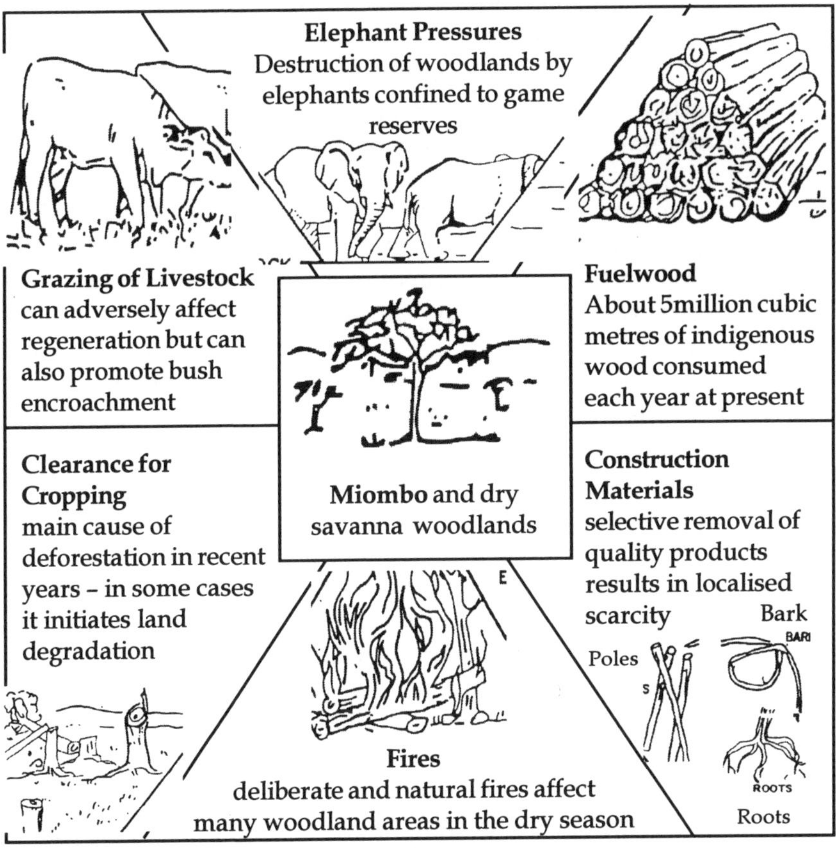

Figure 10.6b: Dynamics of Vegetation Changes in Relation to Agricultural Activities

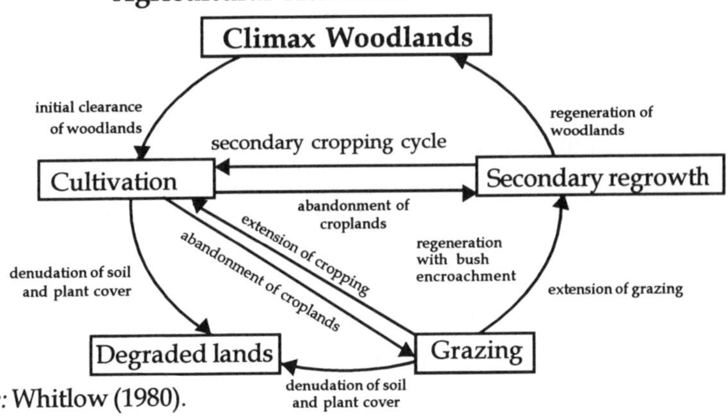

Source: Whitlow (1980).

population, recent evidence seems to point out that, in areas where overgrazing occurs in the drylands, there are other crucial factors which exacerbate the situation. These include population pressures in rainfed croplands and increasing encroachment of cultivators on adjacent rangelands which tend to diminish the areas of available grazing lands and intensify overstocking and overgrazing.

An additional factor is the historical one, where, because of colonialism, traditional systems of pastoral resource management have been undermined. The imposition of political boundaries, restrictions on movement of nomads, resettlement schemes or sedentarisation and concentration of men and animals around new high capacity watering places, the development of towns and villages, commercial agriculture and new land tenure systems have disrupted the mobility and seasonal grazing patterns of nomads and their herds and reduced the area they formerly occupied. Denied access to traditional grazing reserves, these herdsmen have been pushed to marginal lands. Overuse of such restricted areas has eroded the basis for their livelihood and initiated a downward spiral of poverty, degradation and marginalisation.

Like the population factor, however the connection between land degradation and pastoralism is not that straight forward. Except in limited areas where populations are high, according to UNSO (1992), grazing has probably not inflicted more damage than drought and desiccation in the rangelands. UNSO further notes that the problem is not just one of too many animals relative to available grazing areas, but that the state of the rangeland ecosystems in arid and semi-arid areas has more to do with the highly irregular supply of rainfall than anything else.

While the literature may disagree on the issues of overgrazing, population and rural household fuelwood consumption, there is general agreement that a major cause of desertification in the drylands of Africa is inappropriate agricultural practices. Among these are poor agricultural management, inappropriate crops, overirrigation and inadequate drainage, and inappropriate use of chemical fertilisers. Other factors include the expansion of cash cropping into marginal areas, inappropriate agricultural machinery, pricing policies that reduce the value of local product, and land tenure policies that promote short-term exploitation over long-term sustainable development. Expansion of agriculture is also a prime cause of deforestation (agrodeforestation). Sedentarization or villagization of nomads without improved lifestyles and management practices have also done little to prevent degradation in the drylands.

In summary, the human causes of desertification in Africa are still not fully understood, although they are generally better known than the physical causes. Changing paradigms and varying views seem to be working against a consensus of opinion. While some researchers have focused on the more immediate causes others have concentrated on the deeper root causes. The range of solutions has depended on whether one or the other has been emphasised.

While the causal factors of desertification have been studied extensively, one dimension that has escaped attention is the spatial scales (Lambin, 1993). Much of the controversey surrounding the subject relates to the discrepancy between cause and effect - the different spatial scales at which the effects are perceived and the causes (responsible for the threats) are unfolding. Desertification is best detected and probably only conceived on a continental and sub-continental scale -- the macro-management level. Yet, it is a product of innumerable landuse decisions at the local scale - the micro-management level.

Impact of Desertification and Drought

While the causes of desertification in Africa are not fully understood, there is still a much greater problem with assessing its impact. Part of the problem is due to the lack of reliable data: few African nations have accurate statistics on it. Existing data, even at the national level are not based on quantitative measures, but are best-guess estimates at a high level of generality, which means that they cannot be used as a baseline against which to measure change. Another aspect of the problem is the little attention given by researchers to the socio-economic indicators of desertification. To date, for example, there are no significant studies that have assessed and monitored human reactions to desertification. A third aspect of the problem relates to the difficulties of separating the effect of changes caused by desertification from those of other events and policies.

Desertification in Africa, as elsewhere, reduces the productivity of land and deprives people of biological resources that are f the same economy. Prolonged periods of drought under these circumstances lead to hunger, malnutrition and starvation, high infant mortality and accelerated rural migration. Loss of biodiversity in cultivated plants and domesticated animals, and in wild foods so important when agriculture fails at times of drought, is a direct threat to food security (IPED, 1994). Desertification is therefore a major scourge in the drylands of Africa (Darkoh, 1993).

In the Sahel, desertification has brought an alarming drop in agricultural production: millet, sorghum and groundnut harvests have been critically low in Mali since the 1970 drought (Lo, 1994). Production has dropped by 50-80 per cent compared to the situation in the 1930s and the loss per year in income is estimated at US$5.7 million. In Senegal, groundnut production has fallen to 800 kg per hectare since 1991, having reached 1,100kg per hectare a quarter of a century ago; and the 'Groundnut basin' has moved southwards to less degraded soils (Figure 10.7).

Figure 10.7: Movement of the Groundnut Basin, Senegal

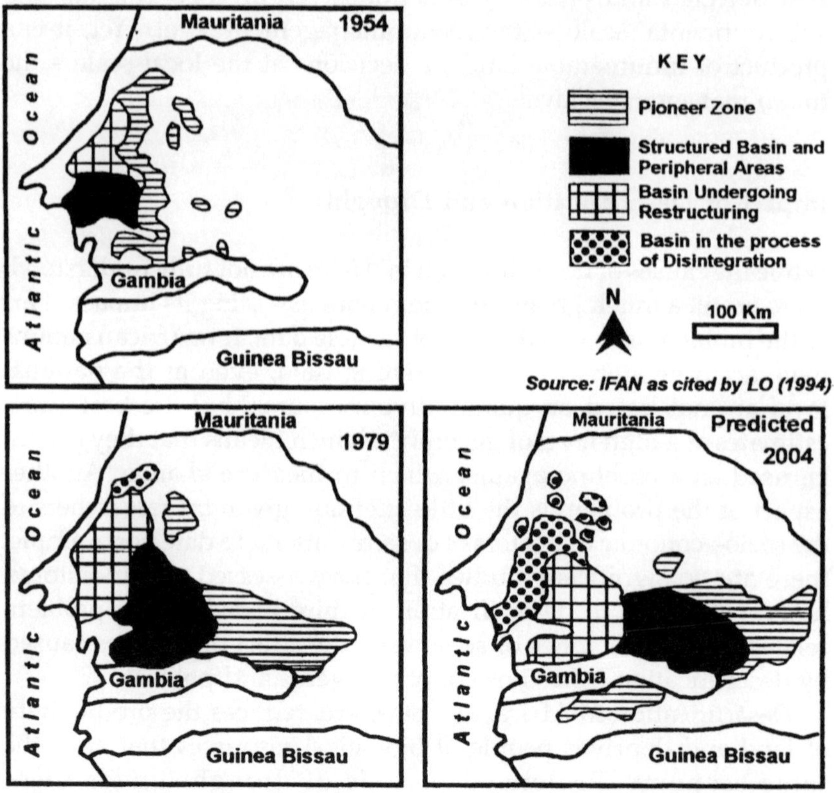

Source: IFAN as cited by Lo (1994).

The great drought in the Sudano-Sahel region of the early 1970s, claimed about 250,000 lives. Millions more were reduced to destitution, provoking mass migrations to urban areas in search of work and relief. The 1982-85 drought affected the entire Sub-Saharan region. The worst affected country was Ethiopia where an estimated one million people starved to death from the combined effects of drought and civil war.

Drought has accelerated the migration of farmers from the countryside to the cities, putting additional pressure on basic city services such as water and sanitation. Water-dependent factories and mines shut down, throwing thousands out of work. Many schools and hospitals also had to close, unable to operate without water. At the height of the 1992 drought, Bulawayo almost ran out of water completely.

The agroeconomic effects of drought have included lower and variable per hectare yields, reduction in acreage of cropped lands, less high-yield flood cropping, diminished rangeland productivity, changes in pastures and in the composition and size of herds, and lower prices as herdsmen flood the market with sickly cattle seeking to sell them before they die.The effects of desiccation on croplands and rangelands have been much more serious than those of droughts. As noted in Chapter 4 (p.57) many Sahel peasant and pastoral communities have been decimated by the desiccation of the last 20 years.

Unfortunately, despite growing concern, there are no exact and reliable figures available to quantify the economic costs of desertification either for individual farmers or society as a whole in Africa. However, some partial estimates have been made. Cardy (1993) has cited one unpublished World Bank Study which estimates that the equivalent of 20 per cent of the annual GDP of one Sahelian country could be lost through capital depletion of natural resources. In Madagascar, average loss from soil erosion is estimated at US$100-150 per hectare each year, the equivalent of 70 to 100 per cent of the average Malagasy's annual income (Randrianarijaoa, 1983). In Ethiopia, soil erosion is estimated to result in annual loss in grain production of 1 million tons (Milas and Asrat, 1985). In Zimbabwe, the financial cost of erosion on a per hectare per year basis varied from US$20 to US$50 on arable land to US$80 on grazing land (Ponzi, 1993). In Namibia, the monetised costs of land degradation at household level aggregated over a standard period for a communal area (Uukwalundi) were estimated at US$2,065 -

US$2565 (Quan *et al.*, 1994).

One effect of soil erosion as we have noted has been a declining or stagnant production of food and cash crops for millions of small farmers. The resulting shortages of food and cash has led to a search for supplementary income earning opportunities through part-time jobs and migration to urban areas. Sometimes farmers have had to contend with the problem of reduced yields by growing less nutritious substitutes such as cassava.

Deforestation, apart from precipitating and accelerating soil erosion, has had tremendous socio-economic impact in Africa. Fuelwood availability has become a major problem in parts of the Sahel and limited parts of the eastern and southern African region. Where dry dung is used in place of scarce fuelwood as in Lesotho, Ethiopia and other parts of the Sudano-Sahel region, the soil is robbed of its natural replenishment. The resulting loss in soil fertility reduces harvests, although the magnitude of this loss is not easily quantifiable.

More readily documented is the social cost in terms of scarcity and hardship – especially, for women and children who do most of the wood collecting. In central Tanzania, firewood has become so scarce that the average household spends well over 250 working days per year gathering its fuelwood supply (Mnzava, 1979). Similar desperate situations are found in Ethiopia, Lesotho, Somalia and northeast Kenya, among others.

Ecological and social mismanagement of irrigation schemes accounts for a good share of the productivity losses the UN attributes to desertification. Through their role in the spread of schistosomiasis, poorly planned irrigation projects can involve health costs that, for the infected individuals at least, surely outweigh the benefits.

Desertification translates into a spiral of declining production, increasing poverty and diminished potential productivity. It exacerbates poverty which, in turn, exacerbates desertification because, as the pressure increases, people are forced to exploit their land to survive. In doing so, they further diminish its productivity - and the cycle continues. The result is seen today in the Ethiopian Highlands and all across the Sudan-Sahel: starvation, death, disease and the exodus of millions of environmental refugees moving in desperate search for survival to the urban areas or to other less degraded lands elsewhere.

Directly or indirectly, desertification slowly erodes the genetic base for human staple food and undermines the whole production system. Entire societies and cultures are now threatened. The pastoralists of the Sahel are a case in point. For most, the loss of their livelihood means a life in relief camps or in shanty towns mushrooming around Sahelian cities and those of the countries to the south.

Finally, desertification and resource-scarcity can provoke social unrest, political and armed conflict. Several governments, most notably the Haile Selassie regime of Ethiopia, have been swept from power by the suffering and unrest associated with drought and famine. With continuing degradation and increasing scarcity of natural resources, the struggle and competition for the remaining resources are likely to become a potent source of conflict among communities and countries in the African drylands (Ghai, 1992).

Failed Plans and Actions

Over the past three decades, there have been major efforts to combat desertification in Africa. In 1968, African governments under the auspices of the Organisation of African Unity (OAU) adopted the 'African Convention on the Conservation of Nature and Natural Resources', in Algiers, Algeria. Contracting parties agreed to adopt the necessary measures to ensure conservation and sustainable utilization of natural resources in accordance with scientific principles. The Lome Conventions also took up the issue of desertification seriously. For example, Lome III (signed in 1984) dealt exhaustively with the safeguarding of natural resources, and especially desertification control. The UN General Assembly addressed the problem of desertification in Resolution 3202 (VI) of 1 May 1974. This resolution recommended that the international community undertake concrete and speedy measures to arrest desertification and assist the economic development of affected areas. The UN Economic and Social Council's Resolution 1978 (LVII) of 16 July 1974 requested all concerned organisations of the UN system to pursue a broad attack on the drought problem. The United Nations Conference on Desertification (UNCOD), convened in September 1977 in Nairobi Kenya, is the most important global conference organised to initiate concerted international action to combat desertification (United Nations, 1978).

Organised under the auspices of UNEP, UNCOD drew up a Plan

of Action to Combat Desertification (PACD). The immediate goal of the PACD was to prevent and arrest the advance of desertification, and where possible to reclaim desertified land for productive use. The ultimate aim was to sustain and promote, within ecological limits, the productivity of arid, semi-arid and dry sub-humid areas that are vulnerable to desertification. The PACD recommended 28 measures for improving productivity and sustainability of the major types of land use, coupled with attempts to prevent soil erosion and reclaim land already desertified (United Nations, 1978).

As part of the attempts to arrest desertification and mitigate the effects of drought in Africa, a confusing array of organisations have been set up during the last three decades (Impact Team, 1992). These include: the Permanent Inter-State Committee on Drought Control in the Sahel (CILLS) established in 1973; the United Nations Sudano-Sahelian Office (UNSO) established in 1973 as a mechanism for the co-ordination of UN efforts to assist Sahelian countries in combating drought; the Committee of Ministers on Desertification (COMIDES) with its headquarters in Dakar; the Intergovernmental Authority for Drought and Development (IGAAD); the Centre d'Agrometeorogie et d'Hydrologie Operationnelle (AGRHYMET) in Niamey; the Drought Monitoring Centre (DMC) in Nairobi; the Sahara and Sahel Observatory (OSS) and African Desert and Arid Lands Committee (ADALCO) established in 1987 to combat desertification in the Kalahari-Namib region.

Despite the plethora of plans, activities and organisations to combat desertification at national, regional and international levels, the efforts have met with limited success. The limited success in Africa has been attributed to many factors: lack of political will; lack of funds; lack of people's participation; lack of co-ordination; destructive land tenure policies; war; misdirected research priorities; failure to include desertification control activities in national development plans; and failure of African governments to make desertification a priority (Darkoh, 1989, 1993). An external evaluation of the PACD in 1990 identified several of the factors mentioned above as causes of the failure of the PACD even though it concluded that the PACD should remain the global strategy for desertification control, with slight revisions that should give greater emphasis to socio-economic factors associated with desertification (UNEP Governing Council, 1990).

Furthermore, because of tremendous debt burdens (currently

standing at over 90 per cent of GNP for Sub-Saharan Africa) and the Structural Adjustment Programmes imposed by the IMF and the World Bank, national governments in Africa find it difficult to finance the expenses of environmental programmes designed to protect natural resources. Governments and national policies in Africa consequently still promote inappropriate agricultural policies that place a premium on higher priced non-food cash crops that generate foreign exchange often through unsustainable practices. Increasing protectionism of First World countries have caused international markets for the products, and thus the earning capacities of these poor countries, to shrink further. The direct environmental effects of all these factors is often to force these countries to over-intensively exploit their natural resources and eat away at their capital stocks.

Also, much of the financial and technical assistance many of these African countries received from the developed countries seemed to have gone largely to showpiece projects and into measures aimed at appeasing the more politically advantaged urban populations. By comparison, rural populations which tend to lack political clout - especially in the more remote arid and semi-arid regions - are all but ignored.

The environmental and social problems facing the drylands of Africa are of such magnitude that governments alone cannot handle them. The people themselves whose survival is at stake must be given the rights and the power to determine their own future. Despite all the intervening efforts and plans of action, including activities and pressures from NGOs and international conventions (see below), African governments still suffer from failures or refusals to devolve power to the people and to link environmental degradation to economic policy (Impact Team, 1992). Partly because of this, many programmes to combat desertification and drought have stymied for lack of local support, or have been undermined by conflicting trade and agricultural policies. Like donors, African governments prefer to invest in projects that would yield quick visible results that can be easily quantified in economic terms, and environmental programmes do not yield such immediate returns. It has also been estimated that 90 per cent of the activities related to desertification in Africa are not directed to field projects, but are channeled into planning and co-ordination, infrastructure provision and public awareness programmes (Stiles and Brennan, 1986).

Nowhere did the problems associated with desertification combat become so clear as the Earth Summit or the United Nations Conference on Environment and Development (UNCED) convened in Rio de Janeiro in June 1992 and its concurrent or subsequent negotiations on conventions on climate, biodiversity and desertification. UNCED (1992) presented to the international community concrete evidence demonstrating the complexity of the problem of desertification and the difficulties which arose when efforts to combat it were implemented at national and regional levels. The main conclusions reached at UNCED showed that, apart from a few countries which have a national strategy for combating desertification based on and integrated into social and economic development plan, desertification is only tackled at national level within the framework of cooperation projects financed completely by outside sources. Plans to combat desertification reached at national and regional levels have often encouraged curative intervention by the authorities while neglecting to develop subsistence policies for the local population with their age-old skills and know-how. The destruction of traditional production systems further aggravates social problems and poverty in the affected regions.

Agenda 21, the blueprint for action adopted at UNCED, appealed to the poor nations to give priority in combating desertification to the implementation of preventive measures, and mapped out national strategies for combating desertification and drought. Chapter 12 of *Agenda 21*, which specifically deals with desertification, outlines several measures that can promote sustainable development and counter the threat of desertification and drought. These include: information and monitoring systems, soil conservation, integrated programmes for eradication of poverty and promotion of alternative livelihood systems, integration of anti-desertification programmes into national development plans, drought preparedness and drought relief schemes, and popular participation.

Apart from the adoption of the special chapter in *Agenda 21* on desertification and drought, UNCED also achieved some measure of success on international negotiations on climate change and biodiversity. The resulting international Framework on Convention on Climate Change (FCCC) signed by 153 nations during the Earth Summit, has since been ratified by over 50 nations and subsequently has come into force. The FCCC has the objective of stabilizing GHG concentrations at a level that would prevent dangerous anthropogenic interference with the climate system. The Biodiversity Convention, which has also come into force, has the objective of equitable and sustainable use of biodiversity resources. It adds a new impetus to conservation, particularly in the African countries, where the genetic erosion of agricultural crops, loss of species and natural habitats and their long-term effects are significant. The International Convention to Combat Desertification signed in Paris in June 1994 had as its main objective the combating of desertification and the mitigation of the effects of drought, particularly in Africa through effective actions at all levels, supported by international co-operation and partnership arrangements, in the framework of an integrated approach which is consistent with *Agenda 21* (INCD, 1994). It recognizes that achieving this objective will involve long-term integrated strategies which focus simultaneously, on affected and threatened areas on management of land and water resources, leading to improved living conditions, in particular at the community level. The major message emanating from the Convention is that effective programmes for combating desertification must involve the very people who are affected by it.

Much general research has been carried out in many parts of Africa on desertification or land degradation. Now, more than ever, research is needed at the local and community level in order to understand better the multi-faceted causes likely to lead to environmental stress and human reactions to desertification and assess realistically what can be achieved through local and community-based actions and policies. It is time to identify and address the root causes of desertification at the field level in the drylands of Africa (Dowdeswell, 1993).

Conclusion

The conditions that were enunciated by the Brundtland Commission (WCED, 1987) and UNCED's *Agenda 21* with regard to the poor, remain unchanged in the developing African countries. For instance, levels of poverty are on the increase and population growth in Sub-Saharan Africa where the ecological base is most fragile and deteriorating, is projected to be the most rapid (Titi and Singh, 1994). External and internal central urban institutions and individuals such as governments and commercial interests continue to extend their power, ownership and exploitation of rural areas; they compete for and appropriate resources, i.e. forests, fisheries, and grazing and ranching lands resulting in the decline of biological productivity; rural communities are marginalised from the political process; and their latitude to migrate or fend for themselves is reduced, thus generating conditions for local population growth and pressure on the environment.

Development interventions targeted towards the poor in the arid and semi-arid regions of Africa have not yielded expected results and the countries in these regions remain among the most disadvantaged in the world. This failure has been attributed partly to the failure of external intervenors to familiarize themselves with the strategies employed by the poor in order to support and strengthen them (Titi and Singh, 1994). Where desertification control projects have been implemented, the programmes stem from insights and models of development tested elsewhere, rather than from those based on experience, local knowledge and adaptive strategies of the affected people. Many researchers have drawn attention to this failure and lack of appreciation or understanding of indigenous management systems by both national and international intervenors and the result has been a proliferation of research on indigenous knowledge, coping and adaptive strategies.

It is now recognized that the poor in these socio-ecological systems use diverse adaptive and coping strategies to deal with ecological, social, political and cultural risks (Chambers and Conway, 1992; Titi and Singh, 1994; Oba, 1994; Darkoh, 1996b). These strategies need to be strengthened by public policies as well as interventions by international and national institutions and organisations. The welfare of the people in the affected dryland areas needs to be put at the centre of the development agenda and those adaptive strategies of their livelihood and production systems that confer drought resistance or lessen their susceptibility to famine bolstered (Darkoh, 1996). Also, within the framework of desertification control at the local and community level, project activities must strengthen peoples' standard of living through, for example, the introduction of alternative livelihood systems in order that the population can attain an economic standard which makes it less vulnerable to stress and more resistant to drought and famines.

Presently many interventions from outside are weakening instead of strengthening local adaptive strategies, livelihood and production systems. For example, the adoption of Structural Adjustment Programmes (SAPs) and the subsequent policies of trade liberalisation to achieve equilibrium and economic growth in many of the African countries have worsened the standards of living and undermined efforts at achieving food security (Titi and Singh, 1994; Khor, 1994). The ability of the poor to sustain adaptive strategies towards sustainable livelihoods is constrained by their vulnerability to loss of access and assurances over resources, as well as the lack of relevant and timely information to transform access and assurances into capacities to evolve strategies to deal with change (Chambers and Conway, 1992; Singh and Titi, 1993). The need to facilitate some measures of empowerment to give the people of the drylands opportunities to shoulder the risk of change towards sustainable livelihoods has become critical (Singh and Titi, 1995).

At the national and international levels, wealth created by trade is one of the essential means of empowerment and poverty alleviation. Barriers to trade at the national and international levels can create impediments for the achievement of sustainable development, especially for the countries in the drylands of Africa. One of the great ironies of the current relations between the poor countries and the rich industrialised countries is that the beneficial effects of billions of dollars in development aid may be cancelled out by deleterious effects of trade barriers. Greater access to developed country markets and reduced subsidisation of developed country exports could do more for development and environment in Africa than a tripling of development assistance (Stevens, 1995). Because of the imperatives of the Structural Adjustment Programmes (SAPs) adopted by most of the affected African countries, these poor nations in the drylands of Africa understand the link between trade and sustainable development. The threats protectionism and environmental standards in developed countries and SAPs pose to poverty alleviation, environment and sustainable development, call for greater access by the poor African countries to developed country markets and a fundamental revision in the design and implementation of structural adjustment programmes.

Furthermore, whilst the peoples and governments of these poor African nations (and others in the rest of the developing world) are being persistently asked to be conscious of the importance of curbing destructive activities and achieving sustainable development of their resources, we must also caution the peoples and governments of the developed countries to curb the manner and rate of consumption of their own and other's resources (Sadik, 1993). Controlling desertification and land degradation in Africa has global implications. It can simultaneously contribute to minimization of global warming/climate change in two ways. First, by the effect on atmospheric concentrations of carbon dioxide through rehabilitation and/or conservation of biomass. Second, by minimizing the possible effects of devegetation on surface reflectivities and surface-water transfers. Two further issues of potential global climatic significance are the role played by changing surface character on local or perhaps global rainfall patterns and the role of aggravated dust emissions. On the latter, for example, dust emission levels in the Sudano-Sahel zone of Africa have risen considerably in recent years with potential global effects. Clearly desertification control in Africa is in the interest of global well-being, and thus requires funding from globally-oriented funds.

References

Arnalds, O. 2000. 'Desertification: an appeal for a broader perspective', pp. 5-15 in O. Analds and S. Archer (Eds) *Rangeland Desertification*, Kluwer Academic Publishers, Dordrecht, The Netherlands.

Aubreville, A. 1949. *Climats, forets et desertification de l'Afrique tropicale*, Societe d'Editions Geographiques et Coloniales, Paris.

Balling, B. Jr. 1992. 'Impact of desertification on regional and global warming', *Bulletin of the American Meteorological Society*, 72, 232-34.

Campbell, D.J. 1981. 'Land use competition at the margins of the rangelands: an issue in development strategies for semi-arid areas', pp 39-61 in G. Norcliffe and T. Pinfold (Eds), *Planning Africa Development*, Westview Press, Boulder.

Campbell, D.J. 1986. 'The prospect of desertification in Kajiado district, Kenya', *The Geographical Journal*, 152, 44-55.

Cardy, F. 1993. 'Desertification - a fresh approach', *Desertification Control Bulletin* 22, 4-8.

Cardy, F. 1994. *Environment and Forced Migration*, UNEP, Nairobi.

Chambers, R. and Conway, G.R. 1992. 'Sustainable Rural Livelihoods: Practical Concepts for the 21st Century', *IDS Discussion Paper* 296.

Darkoh, M.B.K. 1982a. 'Population expansion and desertification in Tanzania', *Desertification Control Bulletin* 6, 26-33.

Darkoh, M.B.K. 1982b. 'Desertification in Tanzania', *Geography 67*, 320-331.

Darkoh, M.B.K. 1988. 'Desertification in Africa', *Journal of Eastern African Research and Development* 19, 1-50.

Darkoh, M.B.K. 1989. *Combating Desertification in the Southern African Region,* UNEP, Nairobi and Moscow.

Darkoh, M.B.K. 1990. 'Trends in natural resource use and prospects for sustainable resource management in Kenya's arid and semi-arid lands', *Land Degradation and Rehabilitation* 2, 177-190.

Darkoh, M.B.K. 1991a. 'The fuelwood crisis in Kenya's ASAL', *Resources* 2 7- 13.

Darkoh, M.B.K. 1991b. 'Land degradation and resource management in Kenya', *Desertification Control Bulletin 19,* 61-72.

Darkoh, M.B.K. ed 1992. *African River Basins and Dryland Crises,* OSSREA, EPOS and The Department of Human Geography, Uppsala University, Sweden.

Darkoh, M.B.K. 1993. 'Desertification: the scourge of Africa', *Tiempo 8,* 1-6

Darkoh, M.B.K. 1994. 'Population, environment and sustainable development: Desertification in Africa', *Desertification Control Bulletin 25,* 20-26.

Darkoh, M.B.K. 1995. *Desertification: What it is and What Papua New Guinea Can Learn From It,* University of Papua New Guinea Department of Geography Monograph Series No. 1, Port Moresby.

Darkoh, M.B.K. 1996a. 'Environmental Problems in Kenya's arid and semi-arid lands', pp 26-143 in M.J. Eden and J.T. Parry (Eds), *Land Degradation in the Tropics,* The Commonwealth Foundation, London.

Darkoh, M.B.K. 1996b. 'Towards an adaptive and community-based approach to the management of natural resources in the drylands of Sub-Saharan Africa', pp 73-96 in A. Hjort af Ornäs (Ed), *Approaching Nature From Local Communities,* EPOS, Linkoping.

Dowdeswell, E. 1993. UNEP's approach to the convention to combat desertification, Remarks presented at the meeting of the Experts Group of the OAU, May, UNEP, Nairobi.

Dregne, H., Kassa, M. and Rosanov, B. 1991. 'A new assessment of the world status of desertification', *Desertification Control Bulletin,* 20, 6-18.

European Commission, 2001. *Desertification Convention: Data and Information Requirements for Interdisciplinary Research,* European Commission, Brussels, Belgium.

Forse, B. 1989. 'The myth of the marching desert', *New Scientist,* 1650, 31-32.

GEF 1993. *Implementing the Framework Convention on Climate Change Incremental Costs and the Role of the GEF,* Working Paper No. 4, The Global Environmental Facility, Washington DC.

GEF 1994. *A Review of Country Case Studies on Climate Change,* Working Paper No.7, The Global Environmental Facility, Washington DC.

Ghai, D. 1992. 'The social dynamics of environmental change in Africa', *Whydah, African Academy of Science Newsletter 21*(a) 1-8.

Hellden, U. 1991. 'Desertification: time for assessment', *Ambio 20,* 372-383.

Hellden, U. 1994. 'Forum: Desertification, fact or fiction?', *Our Planet,* 6, 33.

Hulme, M. and Kelly, M. 1993. 'Exploring the links between desertification and climate change', *Environment* 35, 1-11, 39-45.

Impact Team 1992. 'A Convention for Africans', *Impact: Newsletter of the Climate Network Africa,* 6, 3-10.

INCD 1994. *Elaboration of an International Convention to Combat Desertification in Countries Experiencing Serious Drought and/or Desertification and Particularly in Africa,* Final Negotiations Text of the Convention, United Nations, Geneva.

IPED 1994. *Effects of Desertification and Drought on the Biodiversity of the Drylands.* International Panel of Experts on Desertification, Preliminary Executive Summary Report, United Nations, Geneva.

Jacobeit, J. 1991. Climatic impacts of tropical landuse practices pp.69-89 in W. Erdelen, *et al.* (Eds), *Proceedings of International and Interdisciplinary symposium on Tropical Ecosystems,* Verlag J Margraf, Weikersheim.

Kelly, M. and Hulme, M. 1993. 'Desertification and climate change', *Tiempo 8,* 1-7.

Khor, M. 1995. 'Structural adjustment degrades', *Our Planet 7,* 1, 20-21.

Lambin, E. 1993. 'Spatial scales and desertification', *Desertification Control Bulletin,* 23, 20-23.

Lamprey, H. 1975. *Report on the Desert Encroachment Reconnaissance in Northern Sudan*, UNESCO/UNEP, Paris and Nairobi.

Lo, M. 1994. 'A global environmental phenomenon', *Our Planet, 6*, 34.

Middleton, N. and Thomas, D. 1997. *World Atlas of Desertification* (Second Edition), UNEP and Arnold, London.

Milas, S. and Asrat, M. 1985. 'Eastern Africa's spreading wastelands', *Desertification Control Bulletin 12*, 34-40.

Mnzava, E.M. 1979. *Village Afforestation: Lessons from Experiences in Tanzania*, Forestry Department of Tanzania, Dar Es Salaam.

Nicholson, S.E. 1978. 'Climatic variation in the Sahel and other African regions during the past five centuries', *Journal of Arid Environments 1*, 3-24.

Oba, G. 1994. *The Role of Indigenous Range Management Knowledge for Desertification Control in Northern Kenya*, Research Report No. 4, EPOS Environmental Policy and Society, Uppsala and Linkoping Universities, Sweden.

Odingo, R.S. 1990. 'The definition of desertification and its programmatic consequences for UNEP and the international community', *Desertification Control Bulletin 18*, 31-70.

Ponzi, D. 1993. 'Soil erosion and productivity: a brief review', *Desertification Control Bulletin, 22*, 31-70.

Puigdefabregas, J. 1995. 'Desertification: stress beyond resilience, exploring a unifying process structure', *Ambio* 24 (5), 311-313.

Quan, J., Barton, D. and Conroy, C. 1994. *A Preliminary Assessment of the Economic Impact of Desertification in Namibia*, Ministry of Environment and Tourism Windhoek: Randrianarijaoa, P. 1983, 'The erosion of Madagascar', *Ambio* 12 (6), 308-311.

Sadik, N. 1993. Statement by Dr. Nafis Sadik, Executive Director of the United Nations Population Fund at the Round Table on Population, Environment and Sustainable Development, 24 November 1993, Geneva.

Singh, N.C. and Titi, V. 1993. *Empowerment for Sustainable Development*, Working Paper, IISD, Winnipeg.

Singh, N.C and Titi, V. 1995. *Empowerment for Sustainable Development: Towards Operational Strategies*, IISD, Winnipeg.

Stevens, C. 1995. 'International trade and environmental standards', *Human Potential in Environment and Development*, 1 (1), 9-11, International Academy of the Environment, Geneva.

Stiles, D. and Brennan, R. 1986. 'The food crisis and environmental conservation in Africa', *Food Policy*, 6, 298-310.

Thomas, D.S.G. 1997, 'Science and the desertification debate', *Journal of Arid Environments, 37*, 599-608

Thomas, D.S.G. and Middleton, N.J. 1994 *Desertification: Exploding the Myth,* John Wiley and Sons, Chichester.

Tiffen, M., Mortimore, M. and Gichuki, F. 1993. *More People-Less Erosion: Environmental Recovery in Kenya,* Wiley and Sons, London.

Titi, V. and Singh, N.C. 1994. *Adaptive Strategies of the Poor in Arid and Semi-Arid Lands: In Search of Sustainable Livelihoods,* Working Paper, IISD, Winnipeg.

Tolba, M.K. 1984. 'Harvest of dust', *Desertification Control Bulletin* 10, 2-4.

Tucker, C.J, Dregne, H.E. and Newcomb, W.W. 1991. 'Expansion and contraction of the Sahara Desert from 1980 to 1990', *Science* 253, 299-301.

UNCED 1992. *Earth Summit Agenda 21: Programme of Action for Sustainable Development,* UNEP, New York.

UNEP Governing Council 1990. *Evaluation of the Plan of Action to Combat Desertification,* UNEP, Nairobi.

UNEP 1992a. *UN Feature – Desertification, the problem that Won't Go,* UNEP, Nairobi.

UNEP 1992b. *World Atlas of Desertification,* Edward Arnold, London.

UNEP 1992c. *Status of Desertification and Implementation of the United Nations Plan of Action to Combat Desertification,* United Nations Environment Programme, Nairobi.

United Nations 1978. *United Nations Conference on Desertification, 29 August - 9 September 1997, Round-up, Plan of Action and Resolution,* United Nations, New York.

UNSO, 1992, *Assessment of Desertification and Drought in the Sudano-sahel Region 1985-1991,* United Nations Sudano-sahelian office, New York.

WCED 1987. *Our Common Future,* Oxford University Press, New York.

Whitlow, J.R. 1980. *Deforestation in Zimbabwe, Some Problems and Prospects,* Government Printer, Harare.

Williams, A.J. and Balling Jr. R.C. 1994. *Interactions of Desertification and Climate,* WMO and UNEP Nairobi and Geneva.

World Bank, 1989. *Sub-Saharan Africa: From Crisis to Sustainable Development*, World Bank, Washington DC.

This chapter is a slightly modified version of an article first published in *Land Degradation and Development 9*, 1-20 (1998). Copyright John Wiley and Sons Ltd. The author and publisher wish to acknowledge the permission of Wiley and Sons for modification and reproduction of the text.

Chapter 11

The Sahel Dust Bowl: Processes and Control Strategies

O. B. Smith and S. Koala

Introduction

Aubreville (1949) was the first to popularise the term desertification to describe the process as well as the event of changing productive land into a desert. Since then, the term has presented a conceptual and descriptive dilemma to scientists, and it was not until recently at the 1992 United Nations Conference on Environment and Development that a working definition was discussed, negotiated, and accepted. Consequently, desertification is now defined as land degradation in arid, semi-arid and dry sub-humid areas, resulting from various factors, including climatic variations and human activities. The phenomenon assumes different forms, and is induced by several factors. Some of the factors are human-induced, which predispose land to degradation: deforestation, over-grazing, and inappropriate agricultural activities (slash and burn, reduced fallow period, soil nutrient mining). The combined effects of these factors is the removal of soil vegetative cover, soil physical and chemical impoverization, a decline in productivity and a light to moderate degree of degradation. In addition, they expose the soil to the action of wind which completes and aggravates degradation to strong and extreme levels (UNEP/ISRIC/GLASOD, 1990).

In the Sahel, non-human induced factors such as erratic and limited rainfall with strongly marked annual dry season may also come into play, and act as powerful predisposing factors. Intrinsic soil characteristic such as physical and chemical properties is another predisposing factor. Table 11.1 shows topsoil composition of different sites from four countries. Sand component values are similar to that reported by Sivakumar *et al.*, (1990) who indicated that Sahelian top soil is about 90 per cent sand. These sandy soils are particularly vulnerable to wind erosion because of the limited coherence among the relatively small particles which dry rapidly (Baidu-Forson and Napier 1998). Table 11.2 shows the relative importance of wind erosion compared to other agents of degradation in different ecological zones in the Sudan (Ayoub 1998), and provides additional evidence that land degradation by wind erosion is a serious problem in the Sahel, with devastating socio-economic consequences on the affected populations. Yet, it has not received the attention it deserves at both national and international fora, perhaps because of its insidious nature. This chapter focuses exclusively on wind erosion, the insidious agent of desertification in Sub-Saharan African Sahel, with emphasis on control strategies and surrounding policy issues.

Sub-Saharan African Sahel

Conservative estimates suggest that about 30 per cent of the world's strongly degraded land is located in the Sahel, a region situated between the 100 and 600 mm isohyets (Le Houerou, 1989), and spread over an area of 400-600 km north to south, and 6000 km wide east to west of the African Continent.

Table 11.1: Some Selected Physical Characteristics of Sandy Soils of Four Sahelian Countries

	Selected sites					
	Kamboisé[a] (Byrkina Faso)		Cinzana[a] (Mali) (Mali)		Bambey[b] (Sénégal)	
Particule size distribution (%)	Depth (cm)					
	0-10	10-20	0-10	10-30	0-10	10-35
Sand	68.6	62.2	74.0	68.8	95.6	94.6
Silt	12.2	15.1	20.2	18.5	1.0	0.7
Clay	19.1	22.7	5.9	12.7	3.4	4.7

[a] *Source*: ICRISAT (1987).
[b] *Source*: Charreau (1974).

Table 11.2: Extent of Land Degradation in the Sudan by Different Agents (Million Ha)

Ecological Zone	Wind erosion	Water erosion	Chemical degradation	Physical degradation	Total
Hyper-arid	5.8	2.4	0	0	8.2
Arid	20.0	6.9	3.0	0	29.9
Semi-arid	1.2	7.7	5.3	3.0	17.3
Dry sub humid	0	0.7	3.8	0	4.5
Moist sub humid	0	0.5	3.7	0	4.2
Total	27.0	18.2	15.8	3.0	64.0

Source: Ayoub, Ali Taha (1998).

Although diverse in many ways, the Sahelian countries are all characterized by a very fragile ecosystem caused mainly by poor soils, limited rainfall, frequent drought, and susceptibility to wind erosion that accelerates land degradation. The Sahelian states are all situated at the lower end of the index for human development (UNDP, 1995), ranking between 150 (Mauritania) and 174 (Niger) of the 174 countries classified, with low gross domestic product per capita in the range of $160 (Chad) and $632 (Senegal). The low resilience of the Sahelian natural ecosystem is perhaps the single most important obstacle to sustainable development.

Wind Erosion

The Process

Wind erosion occurs whenever the soil is loose, dry, bare or near bare, and wind velocity exceeds the threshold velocity for initiation of soil particle movement (Koala and Bielders, 1998). The process occurs in several stages from soil movement initiation through transport to final deposition. Lyles *et al.*, 1974 described three distinct modes of soil transport: surface creep, saltation and suspension. Heavy particles are transported by creep, rolling and sliding along the ground, because they are too heavy to be lifted from the surface, while slightly lighter particles are transported by saltation moving in a series of short hops at heights below 1 m. The finest particles move to greater heights and longer distances carried by vertical winds. Creeping particles and those moved by saltation are responsible for on-site consequences, while the suspended particles have off-site impacts which will be described later.

According to Sterk (1996) erosive winds that exceed the threshold wind-speed occur during two periods in the Sahel. During the dry season (October-April) when the area is invaded by dry and rather strong northeasterly winds known as harmattan, moderate wind erosion may result (Michels *et al.*, 1995). The harmattan winds originate over the Sahara desert, and from January to March they usually carry large amounts of dust from remote sources. The second and most important wind erosion period is during the early rainy season (May-July), when rainfall comes with heavy thunderstorms that move westward through the Sahel (Sterk, 1996). Within a fully developed thunderstorm, strong vertical downdrafts occur, causing a forward outflow of cold air that creates the typical dust storms of the Sahel. These events are usually short-lived, approximately 10-30 minutes, but the storms may result in intense soil movement. As indicated earlier, the process has devastating effects in areas where the ecosystem balance has been disturbed by land use systems that reduce soil cover and resilience. These effects could be on or off-site, with accompanying costs of human suffering and misery.

The Effects

The devastating effects of soil erosion can be summarized as: the loss in soil productivity, crop damage and losses, dune formation, air pollution and global climate change. The on-site and off-site effects as well as their consequences are shown on Table 11.3.

Loss in Soil Productivity

An immediate and visible effect of wind erosion is the loss of mineral rich top soil, which if not controlled and reversed leads eventually to long term loss in soil fertility and productivity through nutrients and organic matter losses. On bare soils, soil losses may be in excess of 60-70 t ha^{-1} yr^{-1}, while losses of up to 15 t ha^{-1} have been recorded in farmers' fields during a single wind storm (Bielders *et al.*, 1999). The most accurate estimates of soil loss by wind erosion for Sahelian conditions were reported by Sterk and Stein (1997) who in 1994 recorded a total soil loss of 45.9 t ha^{-1} in just four consecutive storms. According to Bielders *et al.*, (1999) soil losses of similar magnitude have been measured on bare plots in the village of Banizoumbou, some 60 km east of Niamey. They further estimated from mass balance calculations that at least 55 t ha^{-1} of soil were lost during the 1995 rains, and another 24 t ha^{-1} in 1996. Such losses in nutrient equivalents correspond to nutrient uptake by a good average millet crop (600 grain kg ha^{-1}). These data provide a strong evidence of extensive losses of soil from bare fields directly exposed to the erosive action of wind. In practice however, farmers leave some amount of crop residue in the order of 0.3 to 0.5 t ha^{-1} in their fields at the start of the rainy season. Such low levels of crop residue may have very little impact on the intensity of soil erosion by wind, especially at high wind velocities. Only sizeable applications of crop residues in the order of 1.5 to 2 t ha^{-1} are effective at controlling wind erosion. However, the competing use of residues for livestock feed, fuel, and construction make this solution unfeasible. The consequences of such uncontrolled soil and nutrient losses are gradual and progressive decline in soil productivity accompanied by reduced crop yields. In the absence of effective control measures and alternative livelihoods, farmers continue to work such land, until it is completely mined, and has to be abandoned.

Crop Damage and Losses

In a recent field survey in Niger, farmers identified crop damage as one of the more important effects of wind erosion. About 69 per cent of farmers surveyed had observed some damage to their millet crop, damage resulting from sand blasting and/or burial (Lamers, 1998). According to Bielders *et al.*, 1998, direct damage to crops during sand storms can result in the loss of plant tissue and reduced photosynthetic activity as a result of sand blasting, or in the burial of seedlings by deposited sand. Abraded leaves, decreased crop survival rate, poor crop establishment, slower development and lower crop yields were some of the effects observed when plants were subjected to wind and windblown sand (ICRISAT, 1986). The extent of such damage to plants is influenced by such factors as species, wind speed, amount and kind of abrasive material, storm duration and frequencies.

Table 11.3: Effects and Consequences of Land Degradation by Wind Erosion

Effects	Environmental consequences	Socio-economic consequences
ON-SITE 1) Loss of top soil nutrients and soil organic matter	Reduced yield loss of biodiversity	Increased production risk. Loss of medicinal plants or income diversification options requirement
2) Long term loss in soil fertility	Loss of land, reduced crop yields. Increased pressure on remaining land	Increased food insecurity
3) Sandblasting and seedling burial	Reduced crop yields or crop failure	Increased food insecurity Increased labor (resowing)
4) Sand encroachment	Loss of land and shelter. Burial of infrastructures (roads, railways, canals)	Migration (loss of structural investments)
5) Surface crusting	Increased runoff/loss water Loss of yield	Increased labour
OFF-SITE Dust emission	Air pollution Respiratory problems Air traffic at risk Reduced visibility Climate change	Climate related disasters

Dune Formation

Wind blown sand often results in the formation and movement of sand dunes that may encroach on cultivated and inhabited land. Forced migration with all of its associated human misery and tragedy is the major consequence of such land losses, particularly in marginal and poor areas with little or no alternative livelihood.

Air Pollution and Global Climate Change

On- and off-site air pollution from the movement of light weight soil particles or dust is common in the Sahel, which apparently is one of the more important sources of global dust load. Reduced visibility and respiratory health problems are two important consequences. Surveys of farmers in Niger, indicate that dust related health problems were of more concern to farmers than soil loss and crop damage (Lamers, 1998). Evidence also exists that there has been an increase in the number of low visibility days over the last several years in Niger (Mohammed, 1998).

Apparently, more than 100 million tons of dust per annum is blown westward over the Atlantic from Africa (Middleton *et al.*, 1986). It has recently been speculated that the super-saturated, high-energy Sahelian winds may have contributed to the increased frequency and intensity of Atlantic hurricanes observed in recent times. It has been suggested that the dust laden Sahelian air transported during the summer monsoon into the eastern Atlantic by the easterly winds in association with the Inter Tropical Convergence Zone contain sufficient energy to accelerate the formation of severe storms and energize the formation and maintenance of hurricanes (Macleod, 1998). Therefore, without the Sahelian airborne dusts, there would be no super-saturated atmosphere flowing from the east across the Atlantic, and no source of energy to maintain hurricanes force storms, hence fewer and less intense Atlantic hurricanes.

This rapid overview of on- and off-site effects of land degradation by wind erosion, and its associated socio-economic consequences underscores the gravity of the problem in the Sahel. Thus, there is an urgent need to develop and/or promote sustainable control measures that will effectively protect land resources from erosive effects of wind.

Control Strategies

Many techniques and strategies have been shown to have the potential to protect land resources from erosive effects of wind. Bielders *et al.,* (1998) grouped these control measures into three categories: tillage and planting systems such as ridge planting, crop residue management such as mulching with cereal stovers, and planting vegetative green barriers. The latter can be in the form of shelter belts and wind breaks. Shelter belts are long rows of trees aligned in several layers to break the impact of prevailing winds, and constitute the first line of defense. Wind breaks are similar to shelter belts but with fewer rows of trees, and/or perennial grasses such as *Andropogon gayanus, Sporo-bolus festivus*, or *Setaria pallidesfusca*, around smaller cultivated land.

All of these methods function by decreasing wind speed at the soil surface by increasing surface roughness and/or increasing the threshold friction velocity needed to initiate particle movement by wind (Bielders *et al.*, 1998). Although many of these approaches have been shown to be technically feasible and effective by several workers (Renard and Vandenbeldt, 1990; Michels, 1994), their adoption and large scale utilization by farmers has not been encouraging, mainly because of constraints related to the farmers needs, resource and skills demand. Two successful case studies (Boxes 11.1 and 11.2) and one other example of land regeneration through the use of green barriers are summarized below.

Casuarina equietifolia (Filao) was chosen after it was identified by extensive testing as the most suitable for stabilizing the coastal dunes along the Cape Verde peninsula of Senegal. The initial efforts starting in 1905 did not involve the population who consequently did not participate. New funding from USAID and Canadian International Development Agency (CIDA) in the 70s extended the windbreak which stabilized the dunes, thus allowing some measures of agricultural activity, particularly horticulture, in the leeward side of the stabilized dunes. Exploitation and ageing of the Filao which does not regenerate naturally now threaten these gains. The response has been in the form of a new approach with funding from CIDA and the International Development Research Centre (IDRC). An approach that focuses on increasing popular awareness of the problem, stimulating community participation in the management of the trees, developing new income generating activities built around by-products of the tree for women, and renewed research activities to develop tree regeneration and management techniques that will ensure long term survival.

Lessons Learned

An analysis of these and other success stories, identifies a number of factors responsible for the success. Awareness building not only within the affected population but also of the policy makers who may have to take decisions and provide enabling environment that may have immediate or future consequences on the ability of the population to implement a process or technique. The importance of identifying and involving stakeholders in all facets of such projects in order to create a sense of ownership and ensure long term support for implementation also feature prominently. The other side of this coin is that the implemented activities must be demand driven, for there is no point consulting and not accommodating expressed needs. The accommodation of expressed or perceived needs, was perhaps responsible for the holistic, comprehensive and multi-dimensional approaches adopted, and which in a number of cases resulted in the development of income generating activities, that provided much needed resources for immediate needs. Bio-physical impacts and socio-economic gains cannot be obtained in short order, especially in land regeneration programmes. Long term commitment in terms of funding, technical, extension and back stopping support, is therefore an important ingredient for success, and donors, national and international policy makers often neglect this aspect of research and development support.

Similar factors have been identified by others. According to Baidu-Forson and Napier (1998), the majority of farmers surveyed in a comprehensive study, identified perceived needs for conservation, personal benefits, simple and low-cost techniques, and access to farm credit as factors absolutely necessary for the adoption of wind erosion control measures. The same authors also suggested that in order to ensure the adoption of wind erosion control techniques by land operators in the Sahel, the following conditions need to be satisfied: required material should be available locally, required skills and knowledge needed to implement techniques should be available in the community, techniques should be low cost and produce short-

Box 11.1

Survival with moving sands: Ed Debba community forestry project.

Location: Ed Debba, northern Sudan.

Problem: Sand encroachment in villages and farms exposed to prevailing north winds. Once fertile farms along river bed gradually buried by wind blown sand, which also damaged crops and machinery, filled wells and canals and buried houses. Limited alternative livelihood and economic base for the population which viewed migration as an inevitable coping strategy.
Underlying factors include reduced vegetative cover from tree cutting, overgrazing and high population pressure on land.

Action: A comprehensive holistic approach built around protection against wind by planting double or triple rows of Mesquite shelter belts around farms, and Eucalyptus and Leucaena wind break within farms, and complemented by a strong community involvement in project planning and implementation to stimulate a sense of ownership. Other complementary activities include women's programme built around income generating activities, school's programme to create awareness and the spirit of self-help, research activities to diversify tree selection for shelter belts.

Achievements: Efficient cost-effective community based system of wind erosion control put in place, with a high adoption rate, regenerated agricultural land growing at a rate of 25% per annum, 50-100% increase wheat yield at household level, secured settlement for thousands of people, empowered and self-reliant community ready to control and manage their environment in a sustainable manner.

Source: *Mohammed Abdel Mahmoud Ibrahim 1997. Survival with moving sands: the northern province community forestry project Ed Debba – Sudan. Desertification Control Bulletin-UNEP Number 30 pp65-73.*

Box 11.2

Afforestation Project in Nigeria: A success story in desertification/land degradation control.

Location: Kano and Jigawa state in Northern Nigeria.

Problem: Severe land degradation with sand dune formation in areas subjected to extremely high deforestation, livestock overgrazing, increasing agricultural intensification, as well as a hot, dry climate that baked exposed soil, rendering it powdery and easily blown away by the wind.

Action: A multi-faceted approach built around a biophysical approach of shelter belt and windbreak establishment, to achieve an environmental objective, supported by social forestry activities to tackle socio-economic objectives.
Woodlots and orchards were established, roadside planting and nurseries encouraged and established to provide seedlings fuelwoods and poles, thus generating income. A strong community participation was stimulated through the development of a Community Based Organization and Women in Forestry Programmes to ensure project implementation and ownership by such self-help organisations.
Research was undertaken to generate knowledge and inform policy.

Achievements: Effective soil conservation and vegetative cover from well maintained shelter belts. Increased level of awareness of desertification, the benefits of afforestation, and the value and uses of indigenous and introduced tree species. Enhanced existing economic and income generating activities.
Strengthened community organizations, and a high rate of adoption, and spin off benefits to the wider community.

Source: *Clare Westaneys and Ellen Woodley 1998. Afforestation and social forestry in northern Nigeriia: a success story in desertification/land degradation control. Desertification Control Bulletin – UNEP Number 32 pp 62-72.*

term economic benefits, access to capital needed to purchase input should be guaranteed, and land operators must be in a position to capture returns to investments in soil conservation measures.

Policy Implications

The complex nature of the problem and the solutions is evident, hence the need for a coordination of efforts, and the exploitation of past experiences. On the one hand, donor efforts must be coordinated to avoid duplication, waste of resources and counter-productive effects. On the other hand, a lack of effective collaboration among the various government departments responsible for resource management could be counter productive. The importance of public participation in the design and implementation of control measures has been highlighted, and by the same token, the formulation of land regeneration policies and intervention strategies should involve local beneficiaries to ensure policy and programme appropriateness and relevance. The removal of structural barriers with respect to access to land (land tenure), economic resources (subsidies, credit and loan schemes) needs to be tackled decisively, to ensure an enabling environment indispensable to the development and adoption of efficient control measures. Finally, a major policy issue is long-term commitment and support to research and research activities that will ensure the development of simple, low-cost conservation techniques that have been repeatedly shown to be effective and sustainable.

Conclusion

Although this chapter has focused on a particular type of land degradation, that caused by wind erosion in the Sahel, because of its relative importance in this region, we recognize the danger of compartmentalizing problems and solutions related to desertification. Apart from the specific technical aspects of windbreaks and shelter belts construction, many of the broader solutions that tackle the indirect negative impact on people are applicable to other types of land degradation. As we emphasized the relevance, pertinence and effectiveness of a holistic approach to combating wind-induced land degradation, in the same way we recommend a holistic approach to combating the many-faceted challenges of the desertification process.

References

Aubréville, A. 1949. *Climats, Forets, et Desertificationde l'Afrique Tropicale*, Societe d'editions Geographiques, Maritimes et Coloniales, Paris.

Ayoub, T. A. 1998. 'Extent, severity and causative factors of land degradation in the Sudan', *Journal of Arid Environments 38*, 397-409.

Baidu-Forson, J. and Napier Ted L. 1998. 'Wind erosion control within Niger', *Journal of Soil and Water Conservation 55*, 2, 120-125.

Bielders, C.L., Michels, K. and Rajot, J. L. 1999. 'Evaluation of soil losses by wind Erosion under different soil and residue management practices in Niger, West Africa', in *Wind Erosion: An International Symposium/Workshop*, Proceedings of an International Symposium, Manhattan, Kansas, 3-5 June 1997.

Bielders, C.L., Rajot, J.L. and Koala, S. 1998. 'Wind erosion research in Niger: the experience of ICRISAT and Advanced Research Organizations', in M.V.K. Sivakumar, M.A. Zoebisch, S. Koala and T. Maukonen (Eds), *Wind erosion in Africa and West Asia:Problems and Control Strategies*, pp 95-124, Proceedings of the ICARDA/ ICRISAT/UNEP/WMO Expert Group Meeting; 22-25 April 1997, Cairo, Egypt. International Center for Agricultural Research in the Dry Areas, Aleppo, Syria.

Charreau, C. 1974. Soils of tropical dry and dry-wet climatic areas of West Africa and their use and management. A series of lectures at the Department of Agronomy, Cornell University, IRAT, Nogent, France, Agronomy Mimeo 74-26, Department of Agronomy, Cornell University, Ithaca, N.Y. 14850 Spring Semester, 1974.

ICRISAT 1986. *Annual Report 1986*, Patancheru, A.P 502 324, India: ICRISAT.

ICRISAT, 1987. *Alfisols in Semi-Arid Tropics: A Consultants' Workshop*, Proceedings of the consultants workshop on the art and management alternatives for optimizing the productivity of SAT alfisols and related soils, 1-3 December 1983 ICRISAT Center, India.

Koala, S. and Bielders C. L. 1998. 'Extent and severity of wind erosion in West and Central Africa', in M.V.K. Sivukumar, M.A. Zobisch, S. Koala and T. Maukonen. (Ed), *Wind Erosion in Africa and West Asia: Problems and Control Strategies*. Proceedings of the ICARA+DA/ICRISAT/UNEP/WMO Expert group meeting, 22-25 April 1997, Cairo, Egypt. International Centre for Agricultural Research in the Dry Areas (ICARDA), Aleppo, Syria.

Lamers J.P.A.1998. 'Wind Erosion in Niger: Farmer Perceptions, Traditional Techniques to Prevent and Combat Wind Erosion, and Farmer Adoption of Modern Soil Conservation Technologies' in M.V.K. Sivukumar, M.A. Zobisch, S. Koala, and T. Maukonen. (Ed.), *Wind Erosion in Africa and West Asia: Problems and Control Strategies*, pp.173-183, Proceedingsof the CARA + DA / ICRISAT / UNEP / WMO Expert group meeting, 22-25April 1997, Cairo, Egypt. International Centre for Agricultural Research in the Dry Areas,(ICARDA), Aleppo, Syria.

Le Houerou, H.N. 1989. *Grazing Land Ecosystems of the African Sahel*, Springer Verlag, Berlin.

Lyles, L., Schrandt, R.S. and Schmeidler, N.F. 1974. 'How aerodynamics roughnee elements control sand movement', *Trans. American Society of Agricultural Engineers*, 17, 1, 134-139.

Macleod, N.H. 1998. 'The role of airborne Sahelian dust in the frequency and intensity of hurricanes formed in the Eastern Atlantic Ocean', *Third World Foundation Paper*, Washington DC.

Michels, K. 1994. *Wind Erosion in the Southern Sahelian Zone: Extent, Control, and Effects on Millet Production*, PhD dissertation, Verlag U. Grauer, Stuttgart, Germany.

Middleton, N.J. Goudie, A.S. and Wells, G.L. (1986): 'The frequency and source of duststorms', pp.237-260, in W. G. Nickling (Ed), *Aeolian Geomorphology*, Allen and Unwin, Boston.

Mohammed, B. 1998. 'Wind erosion in Niger: Extent, Current Research, and Ongoing Soil Conservation Activities' in M.V.K. Sivakumar, M.A. Zoebisch, S. Koala and T. Maukonen (Eds), Wind erosion in Africa and West Asia: Problems and Control Strategies. *Proceedings of the ICARDA/ICRISAT/UNEP/WMO Expert Group Meeting*; 22-25 April 1997, Cairo, Egypt, pp 125-133.

Renard, C. and Vandenbeldt, R.J. 1990. 'Bordures d'Andropogon gayanus Kunth comme moyen de lutte contre l'erosion eolienne au Sahel', *Agronomie Tropicale* 45, 3, 227-231.

Sivakumar, M.V.K., Renard, C. Klaij, M.C., Ntare, B.R., Fussel, L.K. and Bationo, A. 1990. 'Natural resource management for sustainable agriculture in the Sudano' – *Prospects in the Semi-Arid Tropics 1*, Indian Society of Agronomy, New Delhi, India.

Sterk, G. and Stein, A. 1997. 'Mapping wind blown mass transport by modelling variability in space and time', *Soil Science Society of America Journal.*

Sterk, G., Herrmann, L. and Bationo, A. 1996. 'Wind-blown nutrient transport and soil productivity changes in southwest Niger', *Land Degradation and Development 7*, pp. 325-335.

UNDP 1995. *Human Development Report 1995*, Oxford University Press, Inc. United Nations Development Program, 1 UN Plaza, New York, NY, 10017 USA: UNDP.

UNEP/ISRIC/GLASOD (1990). *World Map of the Status of Human-Induced Soil Degradation*, UNEP, Nairobi.

Chapter 12

Agriculture and Biodiversity in the Drylands of Africa

M. B. K. Darkoh

Introduction

This chapter provides a regional overview of the impact of agriculture on biodiversity in the drylands of Africa. For some time, conservation efforts in Africa have tended to emphasise the international, scientific values of biological diversity and focus on areas of high species richness and endemism (places where there are species and varieties that are found nowhere else). The primary areas of interest have been the lush tropical forests and coral reefs. Drylands have been considered as areas of limited interest in biodiversity. Now, however, this perception is gradually changing as research is proving that drylands are home to unique species with special characteristics, the origin of many important crops and a significant source of genes. Millions of people in Africa depend directly or indirectly on biological resources for basic subsistence needs, and commercial use of natural resources makes an important contribution to national economies. In many traditional African societies, agricultural practices and natural resource use have tended to cause little damage to biodiversity in part because of low population density. In addition, these societies have fostered belief systems as well as social norms that encouraged or even enforced limits to exploitation. Such production methods underlie traditional agricultural practices such as gathering, shifting cultivation, and pastoralism. The values that underpin and reinforce them were well adapted to local conditions in the past, although in some cases at least they are now becoming obsolete and perhaps counterproductive. How best to retain the better attributes of traditional systems under conditions of rapid urbanisation and modernisation must be considered one of the most pressing challenges for African nations.

The bulk of the chapter focuses on regional perspectives on agriculture and biodiversity by examining the situation in Northern Africa, the West African Sahel, East Africa and the Horn, and Southern Africa. The conclusion reached is that agriculture accelerates loss of biodiversity in all these regions because of attempts by farmers to increase crop and animal production to feed the increasing population and to contribute to the growth of the national economies. In all these regions, harmful agricultural practices such as overcultivation, overgrazing, selective grazing, bush fires, cultivation of marginal and easily eroded land, widespread use of chemicals and pesticides, destumping to introduce mechanisation on fields that are often too large, lopping and leaf-thinning of trees for fodder and human use have intensified the degradation of the soil and vegetation and led to alarming decline of species types and their numbers. The drylands of Africa have the potential to be more productive but they are fragile and must be wisely managed. The single most important activity that will enhance biodiversity conservation in the medium to long-term is to raise well informed future generations with a strong commitment to sustain management of natural resources and biodiversity.

Definitions

Biodiversity is a concept that is used to describe the variety of life forms. It encompasses biomes (e.g. tropical moist forests), ecosystems, species, and genetic varieties (McNeely *et al.*, 1990). It can also be defined as the variety and variability among living organisms and the ecological complexes in which they occur (OTA, 1987; USAID, 1993). The interdependent relationships that occur within and among ecosystems, species and genetic varieties are complex, but it is important to remember that these are essentially different ways of looking at the same thing. Biodiversity has captured the world's attention since the United Nations Conference on Environment and Development (UNCED) in 1992.

The term 'drylands' is one that is likely to be easily understood in a general sense but it is not so easy to define in precise terms. Although precise definitions of dryland areas vary, general agreement exists on a number of characteristics. The principal defining characteristic is that mean precipitation is less than mean evapotranspiration for a significant part of the growing season (Hassan and Dregne, 1997). This means that rainfall is deficient for maximum plant production. In this chapter, drylands are considered as regions where annual potential evaporation and plant transpiration exceed annual precipitation. The salient feature of dryland climates is the low average rainfall and the variability in rainfall patterns.

Characteristics

Geographically, drylands are divided into four climatic zones: hyper-arid, arid, semi-arid, and dry sub-humid (UNEP, 1992a, b). Hyper-arid areas are natural deserts that are relatively uninhabited, with the exception of sparse tiny date palm oases that provide habitats where very ancient life styles and traditional ways of managing natural resources are still being practised (Ghabbour, 1997). Arid regions are almost exclusively used for extensive grazing. Semi-arid lands are largely pastoral, but include extensive rain-fed cropping in the wetter parts. Dry sub-humid zones are woodlands and forested lands where intensive cropping is practised, along with livestock production.

Water deficiency is the principal limiting factor affecting soils and plant growth in all the four climatic zones. Dryland soils are generally characterised by low fertility associated with low levels of moisture, organic matter and nitrogen. Plant growth is inhibited by lack of water with vegetative cover varying from forests, woodlands and grasslands in the dry sub-humid and semi-arid zones to virtually nothing in the hyper-arid zones except in oases and stream channels.

Dryland vegetation generally has been notable for its resilience, often recovering rapidly from frequent incidence of drought and fires. Because of this resilience, a re-thinking of ecology and management of natural resources in Africa's pastoral drylands has taken place (Behnke *et al.*, 1993; Scoones, 1996). The new so-called 'disequilibrium theories for range ecology' (DTRE) state that dryland ecosystems do not follow equilibrium dynamics. Instead, such systems are characterised by high levels of temporal and spatial variability in biomass production. A basic assumption of the disequilibrium theories is that livestock grazing causes insignificant changes to arid and semi-arid lands with underlying highly variable climatic conditions. However, recent research seems to point out that this situation may be the case for plant productivity in certain arid and extremely variable environments with limited grazing pressure, but it is not a case for floristic composition and vegetation structure which are important aspects of pastoral strategies (Lykke, 2000). In most of the arid and semi-arid areas, livestock-induced vegetation modification and degradation have been described during past decades in both scientific studies and literature on local knowledge systems (see for example, Hjort af Ornas, 1996).

The dominant biome in Africa's drylands is the savanna. The term 'savanna' has been widely applied to plant communities in the tropics and sub-tropics characterised by various combinations of trees and grasses. They are typified by continuous ground layer, consisting mainly of grasses (*hemicryptophytes*) and a discontinuous woody tree or shrub overstorey of *phanerophytes* (Stock, *et al.*, 1997; Meadows, 2000). In West Africa three types of savanna are biogeographically identifiable. These are the Sahel savanna, Sudan savanna and Guinea savanna which are described in the section on West Africa in this chapter. Huntley (1982) has provided a fairly comprehensive description of the flora and fauna of the Southern African savannas. The African savanna biome has not only provided a home for the majority of humans, livestock and wildlife in the continent, they have been the richest grasslands in the world with a high incidence of indigenous plants and animals and the world's greatest concentration of mammals (UNEP, 1997).

For some time, many large parts of these savannas and adjoining arid and hyper-arid areas in Africa have been protected by their remoteness, their vastness, and their marginal direct usefulness for agriculture or other economic pursuits. This situation is rapidly changing. Now these drylands are under siege. Their biological systems are under threat from recurrent droughts, population and migration pressures, inappropriate agricultural methods and overharvesting of natural resources.

Agricultural Practices and Emerging Threats to Biodiversity

Discussion on biodiversity used to focus mainly on conserving natural habitats, preserving their variety and preventing the extinction of species. For some time, biodiversity has been associated with lush, tropical forest and coral reefs. Drylands have been considered as areas of limited interest for biodiversity. This perception is now slowly changing as research is now generally proving that drylands are home to unique species with special characteristics, the origin of many important crops, particularly cereals and pulses, and a significant source of genes. Dryland ecosystems have provided many of the plant and animal species that have shaped the development of many cultures and civilisations (Hassan and Dregne, 1997). Crop species such as wheat, barley, sorghum, millet and cotton have all originated from dryland ecosystems, as did animal species such as the horse, the sheep, the goat, the cow, and the camel. Drylands harbour many drought resistant species and varieties of plants and animals.

Millions of people in Africa depend directly or indirectly on biological resources for basic subsistence needs, and commercial use of natural resources makes an important contribution to many national economies. However, Africa is undergoing unprecedented losses of biological diversity as a result of habitat alteration, over-harvesting, and other causes. Many forms of human activities including agriculture continue to deplete these resources, some to the point of extinction. This trend in biodiversity loss has adversely affected ecosystem functions in some cases, increasingly posing a threat to the livelihood and well being of many Africans. Habitat change is the greatest threat to biological diversity because plants and animals may fail to adjust to the new ecological setting in the modified environment.

Africa is a diverse region with a dominantly rural economy centred on agriculture that for several generations has been adapted to the dynamics of the biological and physical environment. One of the rudimentary forms of agriculture is gathering among the agro-pastoral societies. This practice has been a response to drought and is common in areas of marginal agricultural potential. A striking feature of the drylands in Africa is that food collecting and agriculture still play a significant role in the subsistence economy of several countries. Dryland ecosystems have a wide range of wild food plants. Grivetti (1979) for example, notes that the principal factor contributing to Tswana nutritional success at the peak of the drought in the Kalahari is a diversified food base with emphasis on wild food plants. It is the removal of the trees and baring of landscape under high human and animal population densities that decreases biodiversity in these areas considered as marginal.

Shifting cultivation is another form of traditional agriculture that is widely practised in the drylands of Africa, which is an efficient way of manipulating environments that would otherwise be unproductive under arable farming. It is undertaken on the basis of intimate knowledge and understanding of the environment. According to Schusky (1989), this practice is carried out by between 25 per cent and 33 per cent of African farmers. To him, there are two kinds of shifting cultivators: those that practise shifting cultivation in order to supplement other, more permanent cropping activities, and those that have been forced out of their 'traditional' homes with little option but to try to make a living by encroaching forests and other environments. It is mainly the latter that have disastrous effects on the natural resources. This is because they lack accumulated knowledge of many generations of shifting cultivators, and at the same time, they do not have access to sufficient land so as to have long fallow periods.

In the past, the sparse population operated the system for subsistence needs. It was not intensive, which did little or no harm to the environment as the long fallow periods restored soil fertility. At present, however, increased population density combined with intensive cultivation of cash crops in addition to subsistence crops has led to a much more intensive shifting with very short fallow period. Under these conditions, shifting cultivation is unstainable as it leads to disastrous decline in soil fertility. As it is essentially suited to the maintenance of subsistence economy of sparse population, it is certainly incapable of yielding adequate returns to provide for the increasing population. The system is inefficient when the population becomes too large to support. Farmers in bid to produce more food crops encroach on forests. Rapid population growth is pushing farmers to extend farming into areas that are agro-ecologically unsuited to these forms of land use. The most pressing problem is the high rate of deforestation, mainly as a result of conversion of forestland and savanna to croplands, which has led to loss of biodiversity in these ecosystems. The continued use of fire under shifting cultivation also inhibits recolonisation and may eventually lead to eradication of woody species. Consequently, the ecosystem's biogeochemical cycles could be permanently altered; the storage component in the biomass may be considerably reduced and even the grasses themselves may be slow to grow (Mannion, 1995).

Pastoralism is another form of agriculture that is carried out in land that cannot be readily cultivated to produce crops or from which it makes economic sense to generate animal products. Nomadic pastoralism occurs in arid and semi-arid environments, which necessitates the migration of animals and herders on a regular basis from one area to another. Like shifting cultivation, nomadic pastoralism, when well managed is ecologically balanced through the varying requirements of the livestock components. This makes the agricultural system stable in a fragile environment. Grazing, however, becomes a threat to biodiversity because it not only displaces wildlife but also disrupts plant-species composition. When the population increases, there tends to be permanent replacement of naturally occurring biota, with lasting changes in the output, input and components of the system. This system is associated with major reduction in species diversity as just a few domesticated animals replace the wide range of wild species (Mannion, 1995). Farmers often remove trees from grazing lands with the intention of improving grass growth by reducing competition for water and soil nutrients, but this has an effect on the biodiversity.

Like traditional pastoralism, modern animal rearing through the privatisation of rangelands and establishment of ranches is a threat to biodiversity. Many farmers in these ecological zones that are rich in biodiversity look at the wild animals as a threat to their domesticated livestock as well as their own lives. As a result, fences have been erected to keep wild animals away from the farms. These and other veterinary fences erected to control the spread of diseases account for a considerable loss of wild life as the animals are denied access to their grazing areas. For example in Botswana, because of the outbreak of diseases, the semiarid rangelands are criss-crossed by veterinary cordon fences emanating from the European Union's (EU's) precautions against importing meat from foot and mouth stricken areas. There are many references in the literature on Botswana to the damaging effects of these fences on the capacity of some highly mobile species (for example, wildebeest, hartebeest, buffalo, zebra, and eland) to cope with drought and the loss of their range (Perkins and Ringrose, 1996; Williamson and Williamson, 1983; Patterson, 1987). Fence damage is in the form of barring of seasonal movement for water and nutrition, trapping and high mortality of animals. Species like wildebeest and hartebeest, which form the bulk of the large herbivore biomass, suffered as much as 90 per cent mortality (Murray, 1988; Patterson, 1987) (see also Chapter 14). Modern animal rearing has also taken lands away and margin-alised indigenous communities in rangelands by depriving them of their grazing and cultivation rights (Sporton *et al.*, 1999).

The majority of farming enterprises that characterise Africa are subsistent mixed farms which also support the majority of the population. Mixed farming may be rain-fed or reliant on irrigation. Innovative methods of crop production involve the use of more productive hybrids, the use of crop plants resistant to various diseases which require active crop breeding policies and institutions, improved irrigation and efficient use of artificial fertilisers. In West Africa, approximately 80 per cent of cattle are reared in crop-livestock systems to fully integrated systems (Jabbar, 1993). This is causing environmental change as forests are cut down to expand the area of crop production into which livestock are then introduced.

Government policies that have been adopted to promote agriculture also play an important role in habitat change and the loss of biodiversity. This can be experienced in countries that are trying to promote export crop production or to be self-sufficient in food production with preferential support given to large-scale monocropping systems often using imported seeds. This has resulted in the abandonment of the indigenous plant varieties and species. In Sudan, for example, extension services and donor agencies have promoted sorghum monocultures and discouraged traditional methods even though they have proved far more resistant to pests and drought and are more sustainable in the long-term (Bedigian, 1991).

In the savanna lands of northern Ghana, promotion of alternative cash crops such as yams has resulted in unanticipated deforestation. Farmers clear land for planting yams and also cut down trees to obtain poles required to support the growing yams. This promotion of cash crops as a strategy in many African countries has had adverse effects on biodiversity. In areas such as central Tanzania where fuel is needed to cure tobacco, a lot of wood is consumed every year. As a result, such areas like Isikizya village has long exhausted its principal source of wood supply, the natural Miombo woodland.

Unnecessarily high levels of useage of fertilisers and pesticides have sometimes been stimulated through inappropriate price subsidies (Repetto, 1985, 1989). These have had adverse effects on biodiversity as they and kill other living organisms together with pests.

A World Bank report states that the dependence on pesticides in the Gezira Province of central Sudan led to a massive and permanent reduction of beneficial invertebrates and that an aid programme linked to multinational chemical companies inhibited rational pest control in Gezira (World Bank, 1991).

The status of the environment is a crucial aspect in the development process of Africa; however, the rich biodiversity, together with the time tested, diversified indigenous systems of managing them, have recently come under increasing pressures caused by population growth, social change, migration in search for new agricultural lands and the degradation of the once frontier regions, shifts in cash cropping and the impact of policies adopted by the post colonial governments.

In Africa, about two-thirds of the land that supports habitats for wild plants and animals is now used for other purposes (MacKinnon and MacKinnon, 1986). Agriculture and other human activities are reducing biological diversity at a rate that is unprecedented in the history of the continent. It is not easy to assess, with our limited knowledge, the consequences of the disappearance of species for the stability of the continent's environment or the economic value lost because of extinction. McNeely (1995), citing Cary and Mooney (1990), notes that since the beginning of this century alone, on a world-wide scale, about 75 per cent of genetic diversity of the most important crops has disappeared from farmers' fields. This has increased agricultural vulnerability and reduced the essential variety of diets of rural people.

Traditional and local species and races of domesticated plants and animals vital for the nutrition of the poorest people are either neglected or being displaced by exotic species in Africa. Once exotics establish a beachhead in a given ecosystem, they often proliferate, suppressing native species. 'Bioinvasion' poses one of the greatest threats to Africa.

Modern farming technology is now removing innovation from the farm and placing it instead in the laboratory (McNeely, 1995). The uniform varieties produced at the research centres with their dependence on fertilisers and pesticides are fast displacing farm-bred varieties. Once these traditional varieties are gone, the knowledge of their cultivation and use is also lost. Traditional agriculture is currently threatened by the new global consumer culture that is spreading through television, trade and other means.

Management systems that were effective for thousands of years in natural resource utilisation and conservation have become obsolete in a few decades, replaced by systems of exploitation that bring short-term profits for a few and long-term costs for many (McNeely, *ibid.*). With her doors now open to the influx of global corporations, whose far-flung operations transcend national borders and allegiances, and with private capital surges giving international investors a growing stake in her natural heritage, Africa now faces an environmental assault from globalisation, with ominous consequences for biodiversity.

Regional Perspectives

Northern Africa

The drylands of North Africa occupy about 6 million km^2. They experience the Mediterranean climate that is characterised by warm-wet winters and cool dry summers. The semi-arid to humid zones have a native vegetation dominated by sclerophyllous evergreen trees and shrubs. Commercial farming dominates. The drier areas have a steppe type of vegetation with perennial grasses and/or dwarf shrubs. Subsistence farming and grazing dominate. In the hyper-arid zones in the Sahara, native vegetation is restricted to depressions in waterways. The density of species or 'areal species richness', when calculated for the whole desert is of course very small (Ghabbour, 1997). Date-palm agriculture is practised in the oases. Irrigation is on the increase causing major changes in the natural drylands habitats.

Biological diversity varies in complex ways depending on local moisture regimes, topography and soil type. The number of flowering plants of the sub-region is approximately 6000, with an endemism of about 25 per cent, with the highest endemism in the west and the east of the sub region (IPED, 1994; Hassan and Dregne, 1997). Presently, about 200 species are endangered (3.3 per cent) (*ibid.*). A few crops are thought to be native: the date palms, olives and primitive wheat, barley from the oases. Some of these are known to have genetic traits of potential use for breeding programmes world-wide.

Many large mammals have become extinct over the past 150 years. The only common survivors are the jackal, wild boar, and the fox. Others are on the verge of extinction: the cheetah, leopard, striped hyena, mountain gazelle, the arbary sheep and the baboon. About 100 species of small mammals remain, 200 resident species of birds and some 80 species of reptiles.

The original breed of cattle, Atlas brown (*Bos taurus*) has been excessively crossed with European and Indian cattle (*Bos indicus*) and it is doubtful whether they now remain in their original state.

Several major breeds of sheep have retained their genetic identity, and little breeding work has been done with camels.

Increasing human population has caused removal of large areas previously covered by forest and scrubland, partly for agricultural cultivation, and for stocking. Traditionally, livestock raising has been the major agricultural activity of the people of this region with sheep being the most important as a source of wool and meat. Following closely is the cultivation of grains particularly wheat and barley in the more humid areas. This has been greatly encouraged by the introduction of mechanised agriculture and exploitation of underground water sources. Aboriculture based on fruit trees such as olives, almonds, figs, citrus trees, date palm (naturally linked with oases) and grape vines are important along the Mediterranean coast and other suitable hinterland areas. In Libya and Algeria, increased incomes from oil and natural gas have enhanced the power to invest in capital development and agriculture. This has had a telling impact on grazing lands and woody vegetation formations alike in the arid zones.

In Egypt, a major constraint to biodiversity is expansion of agriculture into more marginal areas and the loss of fertile arable land to urban expansion and development (about 25,000 feddans-10,500 hectares-lost in this way every year). In the arid and semi arid zones of Algeria, Morocco and Tunisia, rain-fed croplands have been degraded mainly through extension of cropping into dry, sandy soils and use of inappropriate heavy machinery. A further constraint in all these countries is the conversion of agricultural land to other uses which has had great impact on whole ecosystems, destroying the ecological equilibrium and accelerating wind and water erosion, thus contributing to desertification and decline in biodiversity. In Libya, degradation of land resources and therefore, desert encroachment and erosion of biodiversity are directly linked to overgrazing and conversion of grazing and other rangelands to cultivation. Wind erosion is particularly serious in the cultivated areas of Jeffara plain (110,000 hectares) and the Gharian Plateau (92,000 hectares) where dry farming is practised. Shifting sand dunes and sand drifts are of particular concern. Irrigated agriculture is expanding and as a result decreasing the area available for grazing.

There has also been widespread destruction of woodlands as a result of clearing for cultivation, or grazing and fires. In Morocco, between 1973 and 1983, bush fires destroyed an estimated annual average of 3000 hectares of forest and alpha grassland (especially cereals). In 1980 reforested and alpha grasslands of more than 6000 hectares were burned causing an estimated loss of US $ 1.4 million (UNEP, 1983). If present deforestation rates (0.5 to 1.0 per cent per year) continue, it is unlikely that forests will exist by the year 2050. This will have a strong impact on total biodiversity.

The removal of perennial vegetative cover through human mismanagement of land causes desertification. As this author has stressed in his chapter on desertification in this book, the consequences of desertification have been particularly severe on the fauna. The flora is likely to become severely affected over the next 50 years, unless preventive measures are taken. On the whole, biodiversity is under severe threat in North Africa, largely because of changes in land use and human mismanagement of land.

The West African Sahel

The 'Sahel' refers geographically to the narrow band of territory bordering the Sahara desert and lying between the Sahara and the Sudan and Guinea savanna lands to the south. It is characterised by the strong seasonality of the climate, with a short rainy season and long dry season. Ecologically, however, the countries commonly referred to as the Sahel countries of West Africa straddle the different savanna types mentioned above. For example, as Table 12.1 on Mali's situation shows, there are six agroclimatic zones which are, from north to south: Saharan, North Sahelian, South Sahelian, North Sudanian, South Sudanian and Guinean.

Table 12.1: Republic of Mali: Agroclimatic Zones

Zone	Climate	Rainfall (mm/Yr)	Length of Agricultural Season (d/Yr)
AS- Saharan	Arid	<150	<25
NH- North Sahelian	Semi-arid	150- 350	25- 45
SH- South Sahelian	Semi-arid	350- 550	45- 90
NS- North Sudanian	Semi-humid	550- 750	90- 120
SS- South Sudanian	Semi-humid	750- 1150	120- 150
G- Guinean	Humid	1150- 1450	150- 180

Source: Republic of Mali (1987: 15).

The Saharan Zone is characterised by highly irregular haphazard precipitation amounting to less than 150 mm. Drought is a limiting factor to plant growth and its impact is strengthened by the wind which increases evaporation. The soil is covered with a green herbaceous carpet upon which the nomadic herds and wild fauna feed. There are a few stunted bushes, especially in depressions. Surface water resources are limited to the Niger River at its bend, oases, temporary wadis and a few ponds. Particularly adapted wild fauna can be found in this zone (Dorcas and Dama gazelles in the south, Barbary sheep in the mountains, and addax); they live off the meagre, ephemeral vegetation.

The fundamental characteristic of the Sahelian Zones (North and South) is the long dry season, with a short rainy season of about three to four months (June-July to August-September). Plant formations are very discontinuous in space and fleeting in time. The plant cover consists of small, stunted, often thorny trees, with spreading crowns like parasols. The herbaceous carpet is often sparse and grows rapidly with the first rains. The long dry season forces the plants to have an extended period of dormancy every year, often characterised by falling leaves. *Acacia albida* whose cycle is reversed is an exception. This zone is the normal habitat for many species of fauna: Dorcas, Damas, and Red-fronted gazelles, and some oryx. The Sahelian zone is a land of mostly transhumant shepherds. Herding is in fact the only way of life that ensures permanent exploitation of these regions through the mobility from one pasture to another.

The Sudanian Zones (North and South) are best described as savanna woodlands. The zone deteriorates progressively from south to north into a savanna that is still rather rich in woody perennials. The herbaceous carpet is made up of perennials and annual grasses. Water resources are abundant: there are large rivers (Niger and Senegal), small rivers and lakes and ponds. This Sudanian Zone is an area whose activities revolve around food and cash crops (cotton, groundnuts), and livestock, especially sedentary livestock. There are also transhumants coming from the south heading north.

The Guinean Zone is a mosaic of savanna woodland and open woodland forest but these two formations are regularly victims of bush fire. The herbaceous vegetation contains many perennial species. The river valleys are fringed by dense vegetation made up of gallery forests. This is the home of the tse-tse fly (which causes sleeping sickness) and filarial worms (which cause onchocerciasis or river blindness). This zone is characterised by cereal growing, industrial plants (e.g. cotton), fruits, and tubers. There is traditional sedentary livestock, mainly with trypanotolerant races that are smaller than Sahelian ones.

While the flora of the West African Sahel region as a whole has relatively low species diversity (less than 1800), the genetic diversity of these species is known to be high with considerable resilience in relation to the periodic droughts and fire disturbances that affect the region (Hassan and Dregne, 1997). Nevertheless, increasing population pressures and cyclical drought stresses have led to serious reductions in regional animal populations.

Approximately 100 million people of the West African Sahel rely largely on the primary production potential of the area. The population increases vary between 2.2 to 2.8 per cent p.a. with urbanisation rates at around 7 per cent p.a. Based on considerations of weather patterns alone, the vegetation of the Sahel adapts regularly to shifts in the climatic gradients, and species loss is, therefore, limited. However, the increase of the human and animal population and clearing of land for agriculture has had major impacts on biodiversity. Man is responsible here more than elsewhere for the alarming decline of species types and their numbers. Harmful agricultural practices intensify the degradation of the soil and vegetation. These include overcultivation, overgrazing, selective grazing, bush fires, cultivation of marginal and easily eroded land, reduction of fallow lands which are not compensated by organic fertiliser, total elimination of trees from the open woodland forests (destumping) to introduce mechanisation on fields that are often too large and bordering fragile lands; lopping and leaf-thinning trees (African Rosewood, baobab) for fodder and leaves for sauces and fruit.

Longer wet periods have allowed cropping to penetrate into grazing areas. Reverses in precipitation patterns have made these lands highly vulnerable. Although it has often been assessed that the heavy grazing pressure may lead to lowering of the biodiversity of Sahelian rangelands, recent evidence indicates that, for example, bush encroachment naturally follows. The Sahelian vegetation appears very unstable but resilient to pastoral stresses due to the strong dynamism of annual seed production, dispersion and germination cycles, and to the spatial equilibrium between herbaceous and woody plants.

Animal populations seem seriously reduced over the last decades: many major mammals have largely disappeared for example, the leopard, the cheetah, the elephant, the Nile crocodile and the hippopotamus. Birds of prey are less common. Animal reserves have not been very effective, and civil strife has reduced the protection offered by legislation and designated areas.

In recent years, considerable efforts have been mounted to rescue the genetic diversity of native crops and of local varieties. The international research centre ICRISAT, with facilities in Niger and Mali now has significant *ex situ* collections of seed of sorghums millets, and some pulses.

Limited work has been performed on identifying and monitoring the types of biodiversity of the Sahelian sub-region. There is a very serious need for careful surveillance of the biodiversity of the region, which seems particularly threatened due to significant increases in population pressure and long spells of increasing unfavourable weather conditions.

East Africa and the Horn

The dryland countries of Djibouti, Eritrea, Ethiopia, Kenya, Somalia, Sudan and Uganda occupy just less than 5 million km^2 around the Horn of Africa. Of these about 80 per cent (4 million km^2) receive less 600 mm precipitation per year on average. The population is around 120 million, with an overall annual growth rate of 3.2 per cent. The dryland areas have slightly lower population growth rates. The countries are all among the economically poorest of the world.

The drylands of this region are drought prone and there have been severe droughts in recent years. As a result, there have been significant variations in cropping areas, in livestock productivity and in natural biomass available for harvesting.

Although the number of studies on biodiversity of the region are few and monitoring of biodiversity difficult due to the strong variations in inter-annual rainfall, several authors have identified significant reductions in beta and gamma biodiversity over the last thirty years. The expansion of croplands, particularly in the Sudan and Kenya, has altered habitats of both wild plants and animals significantly. In both Kenya and Tanzania, there are widespread movements of landless poor from the more humid and densely populated high potential areas to the adjoining peripheral dryland areas in search of farming lands (Darkoh, 1982, 1996a). This permanent immigration, which is a recent phenomenon, has been problematic in terms not only of desertification as described by the author in one of the chapters in this book but also of biodiversity loss because of land mismanagement and land use conflicts. Due to land-hunger and agricultural and settlement expansion, even protected areas and their adjacent lands are heavily encroached upon, with serious consequences for the ecosystem survival of the majority of large mammals and carnivores because of the loss of critical habitats and the blockage of animal migration corridors. Half of Tanzania's protected areas have been encroached upon by land-hungry people. The size of the Maswa Game Reserve was reduced three times in 25 years, and Nkomazi Game Reserve was reduced twice, because of people moving into the area (SARDC/ IUCN/ SADC, 1994). Over the sub-region as a whole, large human and animal populations, with few alternatives for economic activities outside agriculture, cause increased pressure on natural resources at times of drought.

Pastoralism has been going on for centuries over a large portion of especially Djibouti, Kenya, Sudan, Tanzania and Somalia. Nomadic pastoralism occurs in arid and semi arid environments, which necessitates the migration of animals and herders on a regular basis from one area to another. Nomadic pastoralism, when well managed is ecologically balanced through the varying requirements of the livestock components. This makes this agricultural system really stable in a fragile environment. Grazing, however, becomes a threat to biodiversity because it not only displaces wildlife but also disrupts plant-species composition. In eastern Sudan, since the 1950s, the natural pastures have undergone visible changes, inevitably destabilising animal production systems. Overgrazing and climate-induced changes have triggered off the homogenisation of the floristic composition of the natural pastures. The result has been an increase in drought resistant species and low quality grazing plants in the heavily exploited grazing lands (Akhtar-Schuster *et al.,* 2000).

Many countries of the sub-region have suffered from frequent civil strife, international conflicts and general political instability during the last decades. The ensuing forced migration and concentration of refugees in local areas at levels well beyond local capacities to provide land for farming and fuelwood for cooking has exacerbated loss of biodiversity from overcultivation and overharvesting of biomass fuel. The widespread availability of firearms has also contributed to severe reductions in wild animal populations. There is significant poaching of key species and limited enforcement of international conventions, e.g. CITES. In some Protected Areas in Tanzania, poaching has either eliminated entire wildlife populations or reduced them to very low levels (Mwalyosi and Sosovele, 1999: 122).

Although East Africa has been intensively studied ecologically, little is still known about the biodiversity of the area except for large mammals and monitoring procedures have been difficult to undertake except in selected places. East Africa's low lands are predominantly arid, within which pockets of humid environments have formed from highlands, wetlands and the rift valley. This has created a strong diversity of habitats in the region. These drylands have much to offer in terms of plant resources with possibilities of wider cultivation and commercial exploitation. It is widely believed that long periods of dry years, combined with mounting human and livestock pressure on land, cause severe strain on the biodiversity of the countries around the Horn of Africa. There are therefore, indications of links between dryland degradation and biodiversity.

Southern Africa

The eleven countries of the Southern Africa sub-region have been severely hit by drought at least five times in the 20th century including the latest droughts in 1981-87 and 1991-92 (Tyson 1986, Vogel 1995). Historical and geological records indicate that incidents of below-normal precipitation in all parts of the sub-region recur at intervals. The effects of these droughts have varied. In general the vegetation cover has been severely reduced and fauna put under severe stress. The biodiversity of these dry lands is seemingly in a state of flux. Southern Africa has one of the fastest growing populations in the world and faces the challenge of trying to increase food production by 3 per cent a year. This is normally done by bringing more land under cultivation.

The vegetation of Southern Africa has the highest species richness of any floral region in the world (Gibbs Russell, 1987). Cowling and Hilton-Taylor (1994) have identified Southern Africa's eight centres of plant diversity (hot spots) (Figure 12.1) which include about 3.5 per cent of the world's flora on just 0.2 per cent of the earth's surface area. The 'hottest' of these hot spots (Myers, 1990) is the Cape Floristic Region, where biodiversity results in almost 9000 species being squeezed into just 90,000km^2 (Meadows, 2000). Hot spots according to Myers (1990) are areas that feature exceptional concentrations of species with high levels of endemism, and face exceptional threats

Figure 12.1: Southern African Biodiversity Hot Spots

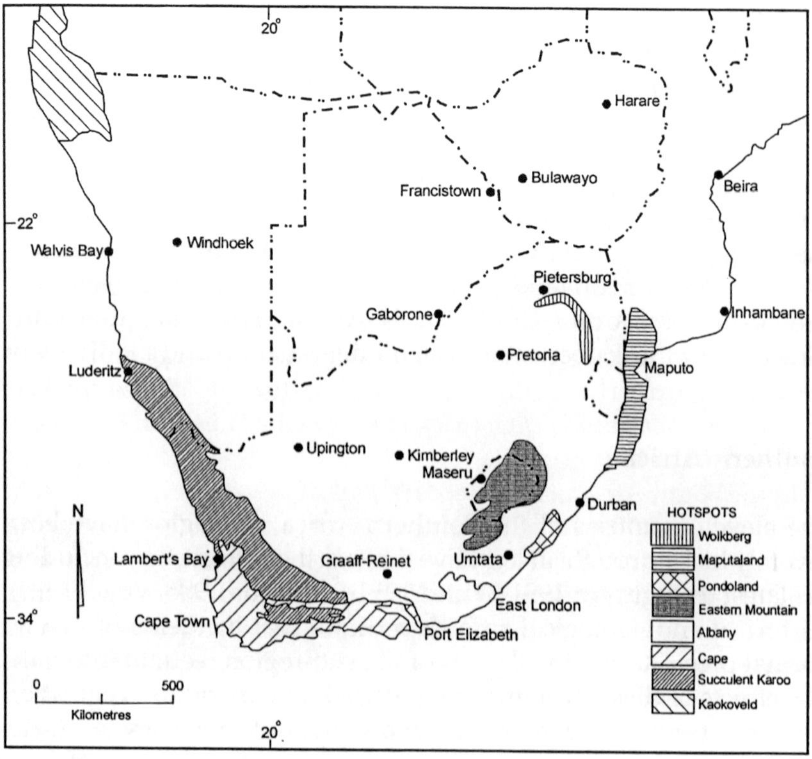

Source: Cowling and Hilton-Taylor, 1994.

of destruction. The rich floral biodiversity could be attributed to the region's morphological evolution and contemporary ecological conditions. The region's transitional location relative to the subtropical summer rainfall and temperate winter rainfall climates combined with a varied topography results in ecological gradients along which many species can be found.

Grazing is the most extensive form of land use in Southern Africa, and about two-thirds of the region is estimated to be suitable for grazing. The drylands of the region have supported pastoral communities for hundreds, perhaps thousands of years, probably without excessive grazing until the 20th century. The mixed indigenous herbivore populations, comprising a high proportion of browsers have been largely replaced by high densities of grazers, principally cattle. The effect of overgrazing and browsing has led directly or indirectly to the transformation of several millions of hectares of formerly productive natural pastures to closed thickets of *Acacia mellifera* and *Dichrastachys cinerea* in the Northern Transvaal, northern Cape and Namibia (Huntley, 1982).

In the South African karoo ecosystems of low and erratic rainfall, Tainton *et al.*, (1989) have noted that overgrazing, selective grazing and lack of fire have caused major changes in the composition and structure of the vegetation. While overgrazing and selective grazing lead to a sward dominated by unpalatable pioneer grasses such as *Aristida congests* subsp *barbicollis, Tragus racemosus, Sporobolus nitens,* and shrubs, overgrazing and the exclusion of fire have resulted in the dramatic encroachment of karoo and savanna species into arid grasslands (Tainton *et al., ibid.*).

The most important impact of introduced plants on the biodiversity in the region is their effects on the endemic flora. This invasion poses the most serious threat to the survival of the numerous endangered plant species occurring in the biomass. Huntley (1984) states that an area of some 66 000 km^2 of former grassland has been encroached on by species such as *Chrysocoma tenuifolia, Pentzia incana, Felicia muricata* and *F filifolia.* The replacement of grassveld by karoo is undesirable because it is accompanied by reduced cover, increased runoff, and erosion (Snyman and Van Rensburg, 1986). In some cases areas invaded by karoo are so severely eroded that regeneration to productive grassland may not be possible although overgrazing reduces the competitiveness of the grass layer and produces a more favourable environment for the germination and establishment of shrubs and trees. *Acacia karro* is the most widespread of the encroaching species, having encroached into the arid and semi-arid grasslands of the Transvaal, central Orange Free State, eastern Cape and Natal (Acocks, 1975).

In drier and more fragile zones such as Kalahari with persistent famine and drought, many people are living through a period of rapid and dramatic changes in land use patterns, economic conditions and natural environment. This pace of change often exceeds the capacity of local institutions to design new land use practices that support both the natural resources and people's welfare. With more technological developments (e.g. dams and boreholes), more marginal areas have been opened up. Livestock numbers now amount to about 45 million cattle and 71 million sheep and goats (SARDC/ IUCN/ SADC, 1994). The recent expansion of arable cultivation in the Kalahari has been seen as one of the major causes of land degradation in the region (Ringrose *et al.*, 1997b), resulting in loss of biodiversity.

One of the most significant hot spots in the Kalahari is the Okavango Delta, a large wetland located in Ngamiland district of Northern Botswana. In this wetland, water is the primary factor controlling the environment and associated life of plant and animals. The Okavango is the habitat for diverse species of plants, arachnids (spiders, scorpion, ticks and mites), large herbivores and birds. Land use changes in the Okavango Delta are having tremendous impacts on the ecosystem of this wetland. While it is not known how many rare or threatened species of flora and fauna exist in the Delta, the wetland ecosystem as a whole is a critically endangered environment of international significance. The potential threats to the Okavango are the possible expansion of cattle in the area following tsetse eradication; pollution from pesticide spraying; repeated uncontrolled burning of vegetation; siltation resulting from vegetation cutting, especially papyrus and reeds; and large-scale water removal for development programmes. The proposed construction of a national water project in Namibia would represent the first major diversion of water, posing the biggest threat to the Okavango. The growth of tourism in the area offers prospects for conservation but if not properly managed, this may exacerbate the threat to sustainable use of resources in this wetland area in the long-run.

Cultivation of crops in the drylands too has an effect on the biological diversity in the region. Unfortunately, cultivation is not only a major factor in reducing biodiversity in what would be considered legitimate arable land. There are large tracts of land in semi-arid areas (that cover the largest area of the subcontinent) that are being cultivated. There are unconfirmed reports that in the Namaqualand agricultural region (68 719 km^2), for example, 2 449 km^2 had been cultivated by 1983 and only 180 km^2 was under irrigation (Macdonald, 1989). For Karoo agricultural region (290 600 km^2), 1 920 km^2 is cultivated whereas a mere 200 km^2 is considered suitable for rain-fed crop cultivation (Schoeman and Scotney, 1987). Such examples show that there are some drylands that are being unwisely cultivated under dry conditions.

This has an effect on the biological diversity in these areas because such drylands are unlikely to revert to the original plant communities' characteristic of uncultivated sites. Probably the most important effect of cultivation on ecological processes in the sub-region is the effect of soil erosion as illustrated *par excellence* by the situation in Lesotho (Darkoh, 1987).

The loss of woodlands in particular is not only a loss of biodiversity in the drylands but also has implications for climatic changes. This is because trees act as sinks of atmospheric carbon dioxide. The loss of trees induced during initial preparations for land cultivation diminishes the flux of carbon dioxide from the soil via photosynthesis and reduces the storage capacity of the biosphere for carbon. Burning as used by farmers in Zambia and other countries to clear land releases stored carbon into the atmosphere along with carbon from the litter and soil, which contributes to the enhanced greenhouse effect.

Increase in agricultural activities can be attributed to government policies. In Botswana, the post independence government agricultural policy encouraged people to go into arable farming by providing agricultural extension services and other incentives. This has resulted in the abandonment of the indigenous plant varieties and species. In the case of the Basarwa in the Kalahari, the government's settlement policies have curtailed their movement. The provision of permanent water such as wells and boreholes has resulted in concentrated settlements around the water points. This has had adverse effects on the surrounding ecosystems.

The Future of Africa's Biodiversity

The economic and social forces that are driving the transformation of the natural habitats of Africa's biodiversity are such that they cannot be easily stopped as long as the stated goals of most people and their political representatives are to increase production of food for the growing population at all costs.

It has been established that agriculture accelerates loss of biodiversity in the drylands and this comes as a result of attempts by farmers to increase crop and animal production to feed the increasing population and to contribute to the growth of the national economies. In order to avoid a situation where agricultural activity leads to conflicts over resource use and loss of biodiversity, a broad landscape planning approach to conservation is required to ensure environmental sustainability. Multiple-use areas bringing benefits to local people that combine conservation with sustainable land use need to be considered as practical and possible alternatives to the establishment of protected areas and buffer zones that are often in conflict with local people over rights of access to land and other natural resources. It is crucial that biodiversity conservation be extended beyond buffer zones and protected areas to include all elements of the landscape and all ecosystems (USAID, 1993). The drylands of Africa have the potential to be more productive if wisely managed.

Successful biodiversity conservation strategies must respect and incorporate African values, knowledge systems and priorities. With regard to ecological knowledge, the traditional people occupying the marginal environments such as semi-arid areas tend to possess a lot of knowledge on how to conserve and use the resources sustainably (McNeely, 1995; Darkoh, 1996b). Any biodiversity conservation projects must involve local people in the management and use of biological resources. This will make the local people take the biological resources as their own and will fight to protect them against depletion. Successful conservation strategies take many forms, both traditional and modern. In Africa, the emphasis on mordern approaches and formal education in general have been at the expense of traditional knowledge systems. Most schools use curricula and teaching materials based on those of Europe. The single most important activity that will enhance biodiversity conservation in the medium to long-term is to raise well-informed future generations with a strong commitment to sustain management of natural resources and biodiversity. Biodiversity conservation should be an integral component of environmental education throughout the formal education systems. Both traditional and modern knowledge systems and values need to be integrated into school curricula. We are not talking here about simply going back to 'traditional' practices, whatever that means in the context of ever-evolving agricultural systems. A blend of modern science and indigenous knowledge will be required to face the challenges of increasing agricultural production and lessening damage to the environment in the decades ahead.

There is need to use agricultural production systems that are likely to intensify production so as to reduce the rate at which more land is cleared for cultivation. Such production systems need to be self-sustaining such that land could be used for a long time without the need to shift to other places. Agro-forestry, the planting of trees on and around the farm should be recommended because these trees will serve as windbreaks and shelterbelts. They also provide other needs like fuel wood, timber, poles, fodder, fruits and medicines. This will cut down the rate at which trees are harvested to meet these needs.

The sustainable use of drylands depends upon the development projects that take into consideration the fact that all the components of drylands systems directly or indirectly affect all other components. However, development activities in drylands have historically been less successful. Their failure could be attributed to the fact that the past projects often did not consider the entire system that includes the local people in planning, implementing and making a follow up of the projects.

There is need to assess the agricultural potential of the semi-arid environments and develop specific agricultural policies or programmes to enhance the sustainable utilisation of the ecosystem. For such policies or programmes to succeed, they have to be in line with the needs, aspirations and abilities of the users. In other words, they have to be formulated with the participation of the beneficiaries.

So far, only limited research on biological diversity, species composition, characteristics and endemism has been conducted in Africa's drylands. Only a few countries such as Botswana, Kenya, Namibia, South Africa and the countries of North Africa have been adequately researched to contribute significantly to understanding of dryland ecology. For the majority of Sub-Saharan African countries, drylands remain a largely unknown domain. Knowledge of dryland ecosystems and their genetic characteristics is an important tool in efforts to mitigate the effects of drought and combat desertification. Research is, therefore, needed to fill the gaps in our knowledge of biodiversity across regions, countries and local areas in the drylands of Africa. Efforts should be made to identify and strengthen activities that transfer or disseminate lessons, learned for example, through regional sharing of information. The transfer of ideas and technologies is an important aspect of the challenge to strengthen the capacity of African societies to take responsibility for the management of biodiversity in the ways that are sustainable in the long term.

There is also a serious lack of natural resource inventories and other baseline data that are of fundamental importance for monitoring biodiversity trends. Inventories provide information on existing levels and patterns of biodiversity. Baseline inventories can, if carefully carried out provide the basis for ecological monitoring. Above all, there is an urgent need for regular ecological monitoring to identify both positive and negative trends in biodiversity in the drylands of Africa.

Finally, African countries, along with countries in the North with a stake in Africa's natural resources, should find ways of managing and curbing globalisation to alleviate problems of biodiversity loss as well as other forms of environmental and human insecurity. Mechanisms and institutions are needed to change globalisation to become conducive to sustainable and more humane development (Opschoor, 1986; see also Chapter 21 in this book).

References

Acocks, J. P. H. 1975. 'Veld types of South Africa', (2nd Ed), *Memoirs of the Botanical Survey of South Africa* 40, 1-128.

Akhtar-Schuster , M., Kirk, M., Gerstengarbe, F.W. and Werner, P.C. 2000. 'Causes and impacts of the declining resources in the eastern Sahel', *Desertification Control Bulletin*, 36, 35-42.

Bedigian, D. 1991. 'Genetic diversity of traditional sesame cultivators and cultural', *Conservation, and Ecodevelopment*, Westview, Boulder, Colorado.

Behnke, R., Scoones, R. and Kerven, C. 1993. *Range Ecology at Disequilibrium: New Models of Natural Variability and Pastoral Adaptation in African Savanna*, Overseas Development Institute, London.

Cowling, R. M. and Hilton-Taylor, C. 1994. 'Patterns of plant diversity and endemism in Southern Africa: an overview', in Huntley, B. J. (Ed), *Botanical Diversity in Southern Africa*, National

Botanical Institute, Pretoria.

Darkoh, M. B. K. 1982. 'Population expansion and desertification in Tanzania', *Desertification Control Bulletin*, 6, 28-33.

Darkoh, M. B. K. 1987. 'Socio-economic and institutional factors behind desertification in Southern Africa'. *AREA*, 19, 1, : 25-33.

Darkoh, M. B. K. 1996a. ' Environmental problems in Kenya's arid and semi-arid lands', in Eden, M. J. and Parry, J. T. (Eds), *Land Degradation in the Tropics*, Pinter, A. Cassell Imprint, London, pp 73-99.

Darkoh, M. B. K. 1996b. ' Towards an adaptive and community-based approach to the management of natural resources in the drylands of Sub-Saharan Africa', in Hjort-af-Ornas, A. (Ed), *Approaching Nature from Local Communities*, EPOS, Linkoping University, Sweden, pp 73-99.

Darkoh, M. B. K. 1998. 'The nature, causes and consequences of desertification in the drylands of Africa', *Land Degradation and Development*, 9, 1-20.

Ghabbour, S. I. 1997. ' Threats to biodiversity in Arab countries', in Barakat, H. N. and Hegazy, A. K. (Eds.), *Reviews in ecology Desert Conservation and Development*, Metropole, Cairo, Egypt, pp. 129-157.

Gibbs Russell, G. E. 1987. 'Preliminary floristic analysis of the major biomes in southern Africa', *Bothalia* 17, 213- 327.

Grivetti, L. E. 1979. 'Kalahari agro-pastoral hunter-gatherers: the Tswana example', *Ecology of Food and Nutrition* 7, 235-256.

Hassan, H. and Dregne, H. E., 1997. *Natural Habitats and Ecosystems Management in Drylands: An Overview*, World Bank Environment Department Papers, Paper No. 51, World Bank, Washington DC.

Howell, D. 1976. 'Observations of the role of grazing in revegetating problem patches of veld', *Proceedings of the Grassland Society of Southern Africa*, 11, 59-64.

Hulme, M. and Kelly, M. 1993, ' Exploring the links between desertification and climate change', *Environment*, 35, 6, 1-11, 39-45.

Huntley, B. J. 1982. 'Southern African savanna', in Walker, B. H. (Ed) *Ecology of Tropical African Savannas*. Ecological Studies No. 42, Springer-Verlag, Berlin pp 101-109.

Huntley, B. J. 1984. 'Characteristics of South African biomes', in Booysen, P de V. and Tainton, N. M. (Eds), *Ecological Effect of Fire in South African Ecosystems* Springer, Berlin, pp 2-17.

IPED, 1994. *Effects of Desertification and Drought on the Biodiversity of the Drylands*, International Panel of Experts on Desertification, Preliminary Executive Summary Report, United Nations, Geneva.

Jabbar, M.A. 1993. 'Evolving crop livestock farming systems in the humid zone of west Africa: potential and research needs', *Outlook on Agriculture* 22, 13-21.

Kelly, M. and Hulme, M. 1993. 'Desertification and climate change', *Tiempo*, 8, 1-7.

Lykke, A. M. 2000. 'Refining the ecological aspects of disequilibrium theories for Africa's pastoral drylands', *Desertification Control Bulletin* 36, 23-33.

Mannion, A. M. 1995. *Agriculture and Environmental Change: Temporal and Spatial Dimensions*, John Wiley and Sons Ltd, Chichester.

Mcdonald, I. A. W. 1989. 'Man's role in changing the face of Southern Africa', in Huntley, B. J. (Ed), *Biotic Diversity in Southern Africa: Concepts and Conservation*, Oxford University Press, Cape Town, pp 51-77.

MacKinnon, J. and MacKinnon, K. 1986. *Review of the Protected Area System in the Afrotropical Realm*, IUCN, Gland, Switzerland.

McNeely, J. A. 1995. 'Biodiversity conservation and tradition agrosystems', in Saunier, R.E. and Meganck, R.A.(Eds), *Conservation of Biodiversity and the New Regional Planning*, IUCN, Gland, Switzerland, pp. 21-31.

McNeely, J. .A., Miller. K. R., Reid, W. V., Mittermeier, R. A. and Werner, T. B. 1990. *Conserving the World's Biological Diversity*, IUCN, Gland, Switzerland.

Meadows, M. E. 2000. 'The ecological resource base: biodiversity and conservation', in Fox, R. and Rowntree, K. (Eds), *The Geography of South Africa in a Changing World*, Oxford University Press Southern Africa, Cape Town, pp. 361- 389.

Murray, M. 1988. *Management Plan for Central Kalahari and Khutse Game Reserve*, Kalahari Conservation Society, Gaborone, Botswana.

Mwalyosi, R. B. B. and Sosovele, H. 1999. 'National environmental policies in Tanzania', in Salih, M. and Shibru, T. (Eds), Environmental Planning, Polices and Politics in Eastern and *Southern Africa*, Macmillan, London, pp. 121-140.

Myers, N. 1990. 'The biodiversity challenge: expanded hot spot analysis', *The Environmentalist*, 10, 243-256.

Nicholson, S. E. 1978. ' Climatic variation in the Sahel and other African regions during the past five centuries', *Journal of Arid Environments,* 1, 3-24.

Office of Technology Assessment (OTA), 1987. *Technologies to Maintain Biological Diversity,* US Government Printing Office, Washington, DC.

Opschoor, J. B. 1996. 'Institutional change and development towards sustainability', in Costanza, R., Segura, O., and Martinez-Alier, J. *Getting Down to Earth: Practical Applications of Ecological Economics,* Island Press, Washington DC pp. 327-351.

Patterson, L. 1987. 'Cordon fences', *Kalahari Conservation Newsletter 18,* 10-11.

Perkins, J. S. and Ringrose, S. 1996. *Development Cooperation Objectives and the Beef Protocol: The Case of Botswana: A Study of Livestock/ Wildlife/ Tourism/ Degradation Linkages,* University of Botswana, Gaborone.

Repetto, R. 1985. 'Environmental ethics and global imperatives', in Repetto, R. (Ed), *The Global Possible: Resources, Development and the New Century,* Yale university Press, New Haven.

Repetto, R. 1989. *Wasting Assets: Natural Resources in the National Income Accounts,* World Resource Institute, Washington DC.

Republic of Mali, 1987. *National Plan of Action to Combat Desertification,* Ministry of Natural Resources and Livestock, Bamako, Mali.

Ringrose, S., Vander Post, C. and Matheson, W. 1997. ' The use of image processing and GIS techniques to determine the extent and possible causes of land management/fenceline induced degradation problems in the Okavango area, Northern Botswana' *International Journal of Remote Sensing,* 18, 2337-2364.

SARDC/ IUCN/ SADC, 1994. *State of the Environment in Southern Africa,* The Penrose Press, Johannesburg, South Africa.

Schoeman, J. L. and Scotney, D. M. 1987. 'Agriculture potential as determined by soil, terrain and climate', *South African Journal of Science,* 83, 260- 268.

Schusky, E. L. 1989. *Culture and Agriculture,* Bergin and Garvey Publishers, New York.

Scoones, I. 1996. *Living with Uncertainty: New Directions in Pastoral Development in Africa,* Intermediate Technology Publications, London.

Snyman, H. A. and Van Rensburg, W. L. J. 1986. 'Effect of slope and cover on run-off, soil loss and water use efficiency of natural veld', *Journal of the Grassland Society of Southern Africa,* 3, 4, 117-121.

Sporton, D., Thomas, D. S. G. and Morrison, J. 1999. 'Outcomes of social and environmental changes in the Kalahari of Botswana: the role of migration', *Journal of Southern African Studies,* 25, 3, 441-459.

Stock, W. D., Allsopp, N., van der Heyden, F. and Witkowski, E. T. F. 1997. 'Plantform and function', in Cowling, R. M., Richardson, D. M. and Pierce, S. M. (Eds), *Vegetation of Southern Africa,* Cambridge University Press, Cambridge, pp. 376-398.

Stuart, S. N., Adams, R. J. and Jenkins, M. D. 1990. *Biodiversity in Sub-Saharan Africa and Its Islands: Conservation, Management, and Sustainable Use,* Occasional Paper of the IUCN Species Survival Commission No. 6, IUCN, Gland, Switzerland.

Tainton, N. M., Zacharias, P. J. K. and Hardy, M. B., 1989. ' The contribution of veld diversity to the agricultural economy' in Huntley, J. B. (Ed), *Biotic Diversity in Southern Africa: Concepts and Conservation: Concepts and Conservation.* Oxford University Press, Cape Town, pp. 107-120.

Tyson, P.D. 1986. *Climate Change and Variability in Southern Africa,* Oxford University Press, Cape Town.

UNEP, 1983. *Promotion of Exchange of Information and Expertise on Desertification Control and technology in Africa, Volume 1: Actions,* Desertification Branch of UNEP, Nairobi.

UNEP, 1992a. *World Atlas of Desertification,* Arnold, London.

UNEP, 1992b. *Status of Desertification and Implementation of the United Nations Plan of Action to Combat Desertification,* United Nations Environment Programme, Nairobi.

UNEP, 1997. *Global Environment Outlook,* Oxford University Press, New York.

USAID, 1993. 'African Biodiversity: Foundation for the Future', Professional Printing Inc, Beltsville, Maryland.

Vogel, C.H. 1995. 'People and drought in South Africa: reaction and mitigation', in Binns, T. (Ed), *People and Environment in Africa,* John Wiley and Sons Ltd, Chicheter, pp. 249–256.

Williamson, D. T. and Williamson, J. E. 1983. 'An assessment of the impact of fences on large herbivore biomass in the Kalahari', *Botswana Notes and Records,* 13, 107-110.

World Bank, 1991. *World Development Report 1991*, Oxford University Press, Washington, DC.

This chapter is a slightly expanded version of an article appearing in *Sowing the Seeds for Sustainability*, pp 38–50, Proceedings of the Eighth Interactive Session held at the Second IUCN World Conservation Congress, Amman, Jordan, 7th October 2000, edited by R. Wiseman and L. Hopkins (2001), IUCN – The World Conservation Union.

Chapter 13

Indigenous Systems of Land Evaluation: Some Examples from Eastern and Southern Africa

H. M. Mushala

Introduction

Land evaluation is a process involving the gathering of environmental, technological and economic information (data) including its integration in a spatial context, and synthesis for specified purposes such as land use planning (Young, 1981). Land evaluation seeks to assess the suitability of land for sustainable production taking into account inherent characteristics of land and the socio-economic milieu of production. The integration of economic and environmental interests requires that environmental concerns be given equal importance with economic performance in evaluating the impacts of development projects, and that reliable indicators of environmental performance be developed. Usually land is considered in its broadest context to encompass all elements of the physical and biological environment that influence land use.

These procedures are critical in both modern and indigenous approaches to land evaluation which are applicable mainly in subsistence agriculture. In indigenous farming systems the farmers accumulate knowledge of their immediate environment and make adaptations for their survival under diverse conditions, although modern scientists have tended to ignore this fact (Gata, 1994). As a result, forms of land evaluation developed in non-subsistence farming economies have been used in small-holder developing economies without capturing the immense knowledge of the indigenous stakeholders, and have invariably proved a poor or misleading guide to the improvement of subsistence agriculture. This has happened because the evaluation methods in use have not put conditions of production in their proper context, and ignored the fact that the farmers' understanding of diverse and complex ecological relationships is the essence of their lives. In other words, indigenous farming systems have always matched land uses to constraints of local environments. Studies have shown that land management strategies recommended to small holder farmers without incorporating their basic knowledge of the local environment have either been rejected or abandoned or only implemented under fear of coercion (Msanya *et al.*, 1998). In a review of the role of indigenous knowledge in development projects Warren (1991) noted that indigenous knowledge provided an improved approach to managing natural resources compared with proposed project technologies; and projects in which indigenous knowledge was ignored were characterised by failures.

The objective of this chapter is threefold. First, it seeks to analyse some of the characteristics and shortcomings of the orthodox methods of land evaluation. Secondly it identifies the salient features of the indigenous systems of land evaluation that can be incorporated amicably into modern approaches to land evaluation. Third, it advocates the incorporation of relevant indigenous land evaluation schemes in modern systems of land evaluation.

Orthodox Systems of Land Evaluation

Conventional systems of land evaluation comprise three inter-related phases: description of the physical environment; appraisal (evaluation) phase; and development phase. During the descriptive phase natural resource surveys are undertaken in order to describe the climate, geology, geomorphology, hydrology and soils of the area in question. The data obtained during this phase is used in the evaluation phase by combining it with data from technology (crop requirements, farming systems, socio-economics) to determine the resource potential of the area. The end product of this phase is usually a land capability classification map or a land suitability map. The final phase is essentially an implementation stage, during which the resource potential is translated into production. The process of land evaluation, therefore, describes promising land use types; determines the requirements for each land use type e.g. for water, nutrients, avoidance of erosion; describes physical properties of land mapping units from resource surveys, e.g. climate, slope, soils; and compares the requirements of the land use types with the properties of the land units to arrive at a land suitability classification.

Examples of common systems of land evaluation include the US Department of Agriculture (USDA) system of land capability classification and its variants and the FAO Framework for Land Evaluation (FAO 1976). The USDA system is a qualitative, general purpose classification designed mainly for soil conservation, and is intended for determining the maximum intensity of land use consistent with low erosion risks and sustained productivity (Landon, 1991). It is an interpretative grouping of *land* mapping units based on soil characteristics. The system is based on restrictions upon the type of land use or the land potential i.e. limitations. Land is classified mainly on the basis of permanent limitations and the level of management assumed is that 'within the ability of a majority of the farmers'.

The USDA system comprises a three-category structure based on soil mapping units. Soil is in fact taken in its wider meaning of land as it incorporates slope angle, climate and frequency of flooding (Young, 1978). The simplest unit in the system is the capability unit, which is a grouping of soil mapping units that have the same potential, limitations and management responses. It is assumed that all soils within a given capability unit can be used for similar crops, require similar management responses and have a comparable productive potential.

A capability sub-class is the second highest unit, which is a grouping of capability units that have the same kinds of limitation or hazard. The original system identified four limitations as being crucial i.e. erosion hazard (e), excess water (w), soil root-zone limitations (s) and climatic limitations (c). Subsequent modifications of the scheme use additional limitations such as stoniness, low fertility and salinity.

A capability class is the highest unit in the hierarchy and is a grouping of capability subclasses that have the same relative degree of limitation or hazard. Classes are indicated by roman numerals, the limitation to the type of land use and the risks of damage to the environment increasing from Class I to Class VIII.

The USDA system:

- is based on negative limitations rather than positive potential;
- the class to which land is allocated is strongly influenced by considerations of soil conservation;
- assumes (implicitly) a decreasing order of value from arable use, through grazing and forestry to recreation, wildlife conservation and water catchment uses; and
- is only valid where these priorities hold good (Young, 1981).

At the subclass level, the original system does not give precise limits for subclass and class allocations. This lack of precise criteria for the environmental parameters permitted in each class could lead to subjectivity and looseness of definition. It is not clear how the climatic limitation is to be interpreted, nor does it deal satisfactorily with wetlands. Wetness is classed on the basis of 'continuing limitations after drainage', using drainage measures considered practical at the present day.

Other systems developed on the basis of the USDA model include the Land Use Capability Classification of the UK Soil Survey (Bibby and Mackney, 1969), and the system of the US Bureau of Reclamation (USBR). The UK system reduces the number of classes to seven by the omission of the USDA Class V (special purpose lands including wetlands). Class definitions are similar to those of the USDA system. The division between arable and non-arable land again comes between classes 4 and 5. At the capability sub-class level, five kinds of limitations are recognized: wetness (w), soil limitation (s), gradient (g), liability to erosion (e) and climatic limitation (c). Guidelines for the permitted range of each limitation within each capability class are given. The land class becomes lower as rainfall increases. The advantages associated with this scheme are that:

- the purpose for which it is used is quite clear;
- it is adaptable to local conditions (of the West);
- it provides specific guidelines for application;
- application is flexible; and
- it is presented simply and concisely.

The USBR system is specifically for the classification of potentially irrigable land. It is also based on the limitations of the land rather than its positive qualities, but contains precise local specifications for the permitted range of values within each land class. Costs of management and amelioration measures are taken into account. Land

Classes 1–3 have progressively less repayment capacity for development under irrigation; Class 4 is special-purpose land; Class 5 is non-arable at present but could become arable if some major works were undertaken; and Class 6 is non-arable.

The FAO Framework

The FAO Framework for Land Evaluation (FAO, 1976) is a standard set of principles and concepts on which national or regional land evaluation systems can be constructed. It is explicit in terms of the intended land use and the level of management envisaged. It recommends the use of reconnaissance surveys, which are usually qualitative, and economics is incorporated only as broad background assumptions. Semi-detailed and detailed surveys include comparisons of costs and returns quantitatively for a more specific 'land utilisation type'. The degree of suitability for a given use may be expressed in terms of net income per unit area.

The framework's structure (Table 13.1) is compatible with other systems but allows greater flexibility. There are two orders, suitable (S) and not suitable (N). Conditionally suitable (Sc) is a 'phase' of the order suitable, but its extent must be small with respect to the total study area. Land classes, subclasses and units are similar categories to those in the USDA system. The framework employs several terms to describe land features, especially land quality and land characteristics (Table 13.2). Land qualities rightly emphasise the combination of land characteristics which affect crop growth, but there is a danger of constructing an over-complex system when their effects could be simply indicated in terms of just one or two land characteristics such as soil texture or depth. Although these latter may not be the land qualities which directly affect crop growth, their effects are sufficiently known by users of major land suitability classifications not to require further explanation and incorporation into a formal (and unfamiliar) system. In some cases the use of land qualities rather than land characteristics may obscure the user's immediate appreciation of which lands he can and cannot improve.

Table 13.1: Structure of the FAO Land Suitability Classification

S SUITABLE	The land can support the land use indefinitely and benefits justify inputs
S1 Highly suitable	Land without significant limitations Include the best 20–30% of suitable land as S1. This land is not perfect but is the best that can be hoped for.
S2 Moderately suitable	Land that is clearly suitable but which has limitations that either reduce productivity or increase the inputs needed to sustain productivity compared with those needed on S1 land
S3 Marginally suitable	Land with limitations so severe that benefits are reduced and/or the inputs needed to sustain production are increased so that this cost is only marginally justified
N Not suitable	Land that cannot support the land use on a sustained basis, or land on which benefits do not justify necessary inputs
N1 Currently not suitable	Land with limitations to sustained use that cannot be overcome at currently acceptable cost
N2 Permanently not suitable	Land with limitations to sustained use that cannot be overcome
Examples of classes in the third category S2e Land assessed as S2 on account of limitation of erosion hazard S2w Land assessed as S2 on account of inadequate availability of water N2e Land assessed as N2 on account of limitation of erosion hazard	

Source: FAO, 1993.

The use of standard national or international land evaluation schemes has some problems. For example, according to Landon (1991), the quantified limits for distinguishing features of subclasses may not coincide with the values measurable or mappable in practice. These limits are usually modified for local use; or manipulated to coincide with the naturally occurring, mappable limits of soil and land features. It becomes problematic when small holder development is being considered because individual holdings are very small (down to 3ha, or less) and it is essential to ensure that they are not sited on unsuitable land.

Table 13.2: Terms Used in FAO Land Suitability Classification

FAO term	Meaning	Examples
Land quality	Complex attributes which directly affect specific kinds of land use; derived from land characteristics	Erosion resistance, water availability, flood hazard. Crop yield levels are aggregate land qualities
Land characteristic	Measurable or estimated parameter, use for land resource mapping	Slope angle, rainfall, soil texture/type/ salinity
Diagnostic criterion	Land characteristic or quality which determines suitability of land for a particular use. Expresses the limitations of land for the given use, and the requirements of that use.	Usually a combination of land qualities and/or characteristics

Additional problems may occur when different land evaluations are applied to large areas. The use of different diagnostic criteria in class definitions within the two (or more) systems may prevent simple interrelations of classes, and will very often preclude simple recommendations on development priorities for the region based on land suitability criteria alone. At this level a whole range of outside factors must be taken into account, including government policy, sociology, demography, market forces, financial and economic analyses and the exercise necessarily becomes one involving a multi-disciplinary approach. As we have also seen for the USDA system, subjectivity is common especially when precise criteria for environmental parameters are lacking. For this reason some land evaluation systems are not suitable for some ecosystems (e.g. wetlands).

Most of the approaches for data collection often adopted comprehensive data collection techniques using formal survey methods governed by inflexible terms of reference. Due to the differences in environmental and socio-economic conditions in different project locations, much of the information collected is often never used, and some potentially useful data is often not collected (FAO, 1998).

Indigenous Systems of Land Evaluation

Many local communities have used indigenous systems of land evaluation for a long time although most of the data/information is not recorded. These data are in the form of 'passed-down knowledge' about the local environment of the farmers by their ancestors, or knowledge pragmatically gained through working the land from early childhood (Osunade, 1993). Since they have not been systematically documented, the indigenous methods are not cited widely in the literature to compete favourably with modern methods.

However, the indigenous systems of land evaluation are effective because they depend on the dual experiences of the individual farmer. First, the farmer accumulates knowledge that has been informally passed on by ancestors. Normally the family land is inherited and passed on to subsequent generations. Secondly, the farmer acquires experience as a result of directly working the land from childhood. Over time one differentiates the soil types within the farming area and crops suited thereto; develops and adopts appropriate conservation techniques based on available technology and constraints of the natural environment; and will automatically adapt to changing situations imposed by elements of climate. The above experiences assist indigenous farmers to determine the suitability of a given parcel of land for sustainable cultivation based on observations made on their farmland of specific land attributes.

Swaziland is one of the countries where the application of indigenous systems of land evaluation are documented. Agriculture in Swaziland comprises both small-holder cultivation mainly on Swazi Nation Land (SNL) and estate farming on privately owned Title Deed Land (TDL). The main crop on small-holder land is maize usually inter-cropped with beans and pumpkin. Osunade (1992, 1993, 1994) documents studies undertaken in Swaziland on the application of indigenous methods in land evaluation. In these studies it is observed that in soil determination, the attributes of particle size (texture) and colour are used respectively to classify all soils into soil classes and to distinguish soil classes into types. The textural soil names form the classes. The addition of carefully selected colour names helps to differentiate several soil types of the same textural class. From these studies it is clear that farmers have amassed knowledge in which soil classes can be determined by the recognition of respective soil and slope unit systems and by using other land attributes as indicators of capability. It was established that in Swaziland, the recognition of the soil types in the field formed the basis for the soil unit system. A soil unit can be defined as an area of land with a particular soil type, which can be used as a uniform piece of land. The modern soil series, therefore, is the equivalent of the indigenous soil unit. The Soil Series concept is accepted by modern science as the basic soil-mapping unit and assumes uniformity of soil texture, profile, colour and drainage characteristics, as does the indigenous soil unit. There are suitable terms in siSwati available to describe the different soil units. In this way a ready-made alternative nomenclature can be derived from the indigenous soil knowledge system (Osunade, 1992). Table 13.3 shows the various soil classes and soil types applied in Swaziland.

In terms of the slope unit system, farming communities use land slope in the determination of the capability of a given parcel of land. The slope unit system delimits the entire landscape into segments which are recognised by the indigenous farmers as uniform in terms of other physical characteristics. In this case the slope unit system can be conceived as analogous to the land facet, applied in the orthodox systems of land evaluation. Table 13.4 below shows the capability ratings of slope/soil units as applied in Swaziland.

Table 13.3: Soil Unit System and Capability Rating in Swaziland

Soil Class	Soil Unit	Rating
Matshe (stony soils)	*Matsheni* (area of stony soil)	Good to fair
Sihlatsi (Sandy soils)	*Esihlabatsini* (area of sandy soil)	Fair
Lubumba (Clay soils)	*Elubumbeni* (area of clay soil)	Fair
Lubhubhushi (Loamy soils)	*Elubhubhushini* (area of loamy soil)	Good

Source: Osunade (1992).

Apart from Swaziland, comparable systems of land evaluation have been applied in Ethiopia (Ridgway, 1997), Tanzania (Msanya *et al.*, 1998), Zambia (Dolva and Renna, 1990) and Zimbabwe (Gata,1994) of which examples are given in Tables 13.5-13.9. In terms of traditional soil classification and land use systems in Zimbabwe, Gata (1994) notes the recognition by the Shona of the

Table 13.4: Capability Ratings of Slope / Soil Units

Slope/soil units	Soil type	Rating
Ligquma (hill crest)	sand/clay	unfit
Elukhalweni (hill slope)	coarse sand	unfit
Phansentsaba (footslope)	sand/clay	good
Egcekeni/Litsafa (lowland)	Clay/loam/sand	good
Ogwini (valley slope)	loam/sand	good
Umfula (valley bottom)	silt/loam	fair if well drained

Source: Osunade (1992).

salinity of mopani soils and the relation that exists between soils and indicator vegetation. It is indicated also that two systems of traditional soil classifications are in use. The first system uses specific names for specific soil types based on colour and texture (parameters used by scientists in field descriptions of soils) and includes descriptive expressions such as 'red clays', 'brown sandy loams', common terminology in soil science. The second one is an ecological one that describes soils in terms of their edaphic environment, such as 'vlei soils' or 'mopani soils'. The most reliable indicators of the agricultural possibilities of a region in this system are found in the native vegetation.

The factors of soil and slope units have been combined with other land attributes to evaluate the capability of a given piece of land. These other land attributes relate to the general site characteristics of an area, presence and behaviour of certain animals, nature and characteristics of the vegetation, and the presence of particular plant species (Osunade 1993). Table 13.10 indicates how site characteristics, animal life and vegetation elements are used as indicators of Land Capability in Swaziland. Gata (1994) also notes the use of other attributes in the evaluation of land in Zimbabwe. She indicates that soil-vegetation associations are widely used by traditional farmers as indicators of the agricultural potential of soils.

It is further indicated that chemical characteristics of salinity and organic matter content are usually expressed qualitatively on the basis of the taste, smell and appearance of the soil. Some of the relevant examples are given in Table 13.11 which correlates traditional Shona and Ndebele terms to modern scientific terminology.

Table 13.5: Soil Classification by Small-Scale Farmers

SOIL CLASSES	SOIL TYPES
Matshe (stony soils)	*Umgubane* (lateritic soil) *Matshe* (gravelly or stony soils)
Sihlabasi (sandy soils)	*Sihlabatsi lekhatekile* or *lukhetse* (coarse sand soil) *Sihlabatsi lesicosakele* (fine sandy soil) *Sihlabatsi lesibovana* (reddish sandy soil) *Sihlabatsi lesimnyama* (blackish sandy soil) *Sihlabatsi lesinsundvu* (brown sandy soil) *Sihlabatsi lesimphunga* (grey sandy soil) *Sihlabatsi lesimhlophe* (white sandy soil)
Libumba (clay soils)	*Sidzaka* (dark clay soil) *Libumba lelibovu* or *libovu* (red clay soil or red Aochre@) *Lelinsundvu* (brown clay soil) *Lelimphunga* (grey clay soil) *Lesimhlophe* (white clay soil) *Lubhubhushi* (loamy soils) *Lubhubhushi lolubovu* or *sibovu* (red loamy soil) *Lubhubhushi lolumnyama* (black/dark loamy soil) *Lubhubhushi lolusundvu* (brown loamy soil) *Lubhubhushi lolumphunga* (grey loamy soil) *Lubhubhushi loluphuti* (yellow loamy soil) *Sihlabatsi lesidzaka lesilubhubhushi* (sandy clay) *Sihlabatsi lesimphunga lesilubhubhushi* (sandy soil) *Sidzaka lesiphunga lesilubhubhushi* (grey clay loamy soil)

Source: Osunade (1993).

In summary, as it is illustrated above a majority of indigenous land evaluation systems:

- use soil physical characteristics as the basis of classification;
- can identify soils locally and these could easily be translated into common scientific nomenclature;
- soils can be related to other environmental characteristics such as slope, climate, landforms based on surface or subsurface properties as is common with orthodox land evaluation mapping units procedures;

Table 13.6: Indigenous Soil Types and Crop Suitability: An Example from Southern Ethiopia

Gobo –	Red soil with low moisture holding capacity. Suitable for ensete, coffee, tef and taro.
Tella –	Black soil which retains moisture. Suitable for maize, tef and beans.
Anchurwa –	Sandy soil with low moisture holding capacity. Suitable for groundnuts.
Cherae –	Swampy soil, suitable for taro and tef.
Boqinta –	Heavy clay soil. Suitable for sweet potato, taro and other tubers.

Source: Ridgway (1997).

Table 13.7: Some of the Main Features of the Clay Soil (*We Bumba*) in Zambia

Characteristics	Farmer's Descriptions	Technical Descriptions
Physiography	Crest/upland	middle and lower slope
Drainage	Water sinks slowly, runoff	well drained
Texture topsoil Subsoil	clayey (light) clayey	clay, clay loam, clay
Colour topsoil Subsoil	grey red	dark-brown, brown, red

Source: Dolva and Renna (1990).

- limitations can be determined on the basis of the soil physical characteristics which can be translated into modern scientific terminology; and
- land users perceive soil types in terms of their uses.

Table 13.8: Common Local Names for Soil Types and their Meaning in English Used in Karatu and Mbulu Districts of Tanzania

Local Name	English equivalent
Hhaper Daaten or Yamadat	Red soil
Hasang=w or Kichanga	Sandy soil
Hhaper Bo=o or Yamaboi	Black soil
Hhaper Bo=o Hhsang=w	Black soil with sandy texture
Sasagwan or Changarawe	Gravels (normally in black soils)
Harki or Chumvi (Magadi)	Soil rich in salts
Hhaper Burki (Tifutifu)	Forest soil (soil rich in organic matter)
Tleii or Mfinyanzi	Clayey soil
Hhaper Baraduxa	River deposited soil (floodplain soil)
Hhaper Bulgar Bo=o	Heavy cracking black soil

Source: Msanya *et al.* (1998).

Table 13.9: Local Soil Names in Zambia

Local Soil Name	**Language**	**Translation**
Aliumba	Namwanga	Clay soil for pot making
Busenga	Bemba	Sandy soil
Busenga bunono (unono)	Bemba	small, sandy soil
Changalawe/Chilambe	Namwanga	soil found on top of rocks
Chimembe	Namwanga	white soil which is bitter
Chipama	Namwanga	hard soil
Chipau	Namwanga	white soil
Chitipa	Namwanga	muddy soil
Chundu	Namwanga	loose and soft, whitish or blackish and/or fertile soil
Ilambo/Ilambwe	Bemba	black soil with no sand
Ilambo chilobaloba	Bemba	black, sandy soil black soil
Ilambo mo muchanga mo	Bemba	with some sand
Imbulo	Mambwe	clay soil for pot making
Inkama	Mambwe	hard soil
Inkama	Mambwe	small, hard soil
munono	Bemba	sandy soil with dust
Kafuche wa muchanga	Bemba,	hard soil
	Namwanga	hard soil

Table 13.9 concluded

Local Soil Name	Language	Translation
Kapama	Bemba	red soil
Katondo	Bemba	sandy soil
Lusenga	Bemba	sandy soil
Muchanga	Bemba	slightly sandy soil
Muchanga panono	Bemba	soil mixed with gravel
Musakalabwe (nsakalabwe)	Bemba	soil that can be moulded
Nkenda	Bemba	soil that can bemoulded
Nkananka / Chkanka	Bemba	soil that can be moulded
Nkanka wa kashika	Bemba, Namwanga	red soil
Senga / Musesenga	Bemba	really red soil
Senga usekemuke	Mambwe, Namwanga	sandy soil
Utifi (utife)	Mambwe	loose, sandy soil
Wa buta	Nyanja	black soil
Wa fita	Bemba	white soil
Wa matipa	Bemba	black soil
We bumba	Bemba	muddy soil
We bumba, wa fita	Bemba	clay soil
We bumba, wa nkande	Bemba	black, clay soil
Ukalamba	Bemba	big, clay soil which can be moulded

Source: Dolva and Renna (1990).

Table 13.10: Other Land Factors as Indicators of Land Capability

Land Factors	Capability perception
SITE CHARACTERISTICS	
Swampy Area	Infertility. Could be improved by adding Akraal manure@ (cow dung) to absorb the water. Water loving plants especially grasses can do well on it.
Water-logged area	Infertile. Usually underlain by impervious clay layer. Difficult to work.
Alluvial deposits	Highly fertile along big rivers. Subject to occasional flooding. Valley bottom cultivation prohibited in some places.
Stones and boulders	Areas with mixture of black red aggregates could carry fertile soil while white and grey soil are infertile.

Table 13.10 concluded

Land Factors	Capability perception
ANIMAL LIFE	
Earthworm casts	Found on nutrient-rich workable fertile soil. Never found on acidic soils. Earthworm cries signify imminent rainfall. Presence are harmful to potatoes and groundnuts.
Burrowing animals	Fertility. Help to improve soil structure especially where moles and crickets are abundant.
Insects	*Litfuku, umswenya* are fertility insects found near kraals. *Tinhluva* insect found in warm soil indicates fertility. Appearance on ground surface indicates imminent rainfall.
Termite hills	Fertility. Removed and spread on fields especially for growing pumpkins and mushrooms.
VEGETATION ELEMENTS	
Luxuriant vegetation cover	Fertility. Any land area that supports rich vegetal cover should support cultivated crops.
Decomposed organic matter	High fertility.
Fungi under grasses	Fertility. Presence of *sihlangu* indicates high fertility.

Source: Osunade (1993).

Table 13.11: Correlation of Traditional Terms and Scientific Knowledge

Shona	Ndebele	Descriptions and Remarks
shapa/sapa	*ihlabathi*	Light sandy soil. Associated vegetation: (Uapaca kirkiana), *mutsatsati* (Faunea spp.) *bungure mutunguru/munhunguru* (Flacourtia indica).
Rusekenya	*unknown*	Extremely sandy, usually very deep, *ruzekete* infertile soil.
ndoronya	*unknown*	Descriptive of its rheological properties: no strength, and will fail under minimal stress. Light textured, gleyed, hydromorphic soil. Associated vegetation *mukute* (Syzygium huillense).

Table 13.11 concluded

Shona	Ndebele	Descriptions and Remarks
tsangarahwe	*unknown*	Shallow, gravelly, well drained soils which most respondents described as good for *rukangahwe rapoko* (finger millet). The preference of this soil for *rapoko* is related mainly to the drainage and possibly is also because there is less competition from weeds on these stony soils, making them ideal for broadcasting *rapoko* and practising minimum tillage.
Chishava	*unknown*	Gravelly red soil, non-arable good for building on.
Mpunzo *chiombwe* *rondo* *rondo*	*ibumba*	Clays that are Ainfertile@ but good for ceramics. Light grey in colour. Considered to be good indicators of sites suitable for location of wells. Associated vegetation: *mukute* (Syzygium spp.), *muroro* (Annona stenophylla), *mtasamviringa* (Ficus ingens).
Bukutu jiho	*isibomvu*	Red clays, usually very fertile, very slippery when wet. Associated vegetation: *musasa* (Brachystegia spiciformis), *mutukutu* (Pliostigma thonningii).
Gova	*isidhaka*	Vertisol (black cracking clay). Associated vegetation: *muunga* (Acacia nilotica), *mubaya mhondoro* (Acacia polycantha), *kananga* (Acacia nigrescens).
Chivavane	*isikwaka*	Very hard sodic to saline soils, usually a medium grey colour. Poor permeability, as indicated by standing water in the rainy season.
Gokoro	*isimunyu*	Much paler in colour and more saline than *chivane/isikakwa.*

Source: Gata (1994).

An Integrated Approach

One of the main criticisms of orthodox methods of land evaluation systems is that they have ignored existing indigenous systems although both have common attributes which can be integrated. It should be emphasized that the development of land capability classification methods for use by small holder farmers should draw strength and initial impetus from the local skills, experience and experimental ability already present within the farming communities (Richard 1985). This will strengthen the link between organized science and indigenous knowledge to maximize efficiency.

If a realization is reached that problems of agriculture as they affect most African countries can only be defined and solved at the local level, then there is first a need for universality of knowledge and second to bridge the gap between the indigenous and modern systems of land evaluation (Osunade, 1987). The scientific knowledge should be made available to the farmers in a manner that makes it relevant to them. After all, as has been recommended by Msanya *et al.* (1998), local people have substantial knowledge of land evaluation that suits their land use practices. Also soil types identified by local people can be correlated to those identified by professionals based on the main features of the identified soil. As such both systems can be used in an integrated manner. The FAO (1998) has proposed a feasible integrated approach from which some of the ideas presented here are derived. Land evaluation at the grassroot level would involve a village or local community with a defined area of land at its disposal, and a land resources management group whose main function is to ensure that the village land resources are managed in a sustainable manner. It is assumed that the objective of land evaluation at this level takes into consideration the needs and desires of the community at large and that there are no conflicting objectives. In many cases the main objective of the community is to supply basic needs of food, fibre, fuelwood and building materials as opposed to maximisation of financial or economic returns, which may be a primary objective of the better off farmers in the community.

The systematic evaluation of land requires information (data). Data collection should be geared to gaining an understanding of how the land/land use ecosystem functions. Data collection should be efficient, focusing on minimum data sets, and flexible, to allow collection of any additional data which may be relevant (FAO, 1998).

Evaluation may include land mapping units and land types as used in the orthodox land evaluation systems, data on existing and proposed land utilisation types, production systems and general socio-economic data at the household or community level. The important thing is that data collection should be efficient and that the data acquired is sufficient to do the job. These data may not necessarily have been recorded before, but need to be recorded at this stage. Particular attention should be given to maximising use of the information resources of the local communities. Local people usually understand the operation of the land use systems on which they depend better than outside experts. Base maps are necessary and should definitely be prepared. Aerial photographs may be used to obtain some of the base data. Ridgway (1997) reports on how airphotos were used in Ethiopia by peasant associations to identify natural resources and as a basis for land use planning. Other data can be obtained from secondary sources but consultations should be made with the community to obtain data based on indigenous knowledge. Rapid Rural Appraisal (RRA) and Participatory Rural Appraisal (PRA) techniques are particularly important in obtaining the essential data. These techniques provide an efficient way of filling some of the information gaps, and can also be used in conjunction with conventional mapping techniques. RRA and PRA techniques are particularly well suited to collection of socio-economic information and data on agricultural and other land use practices at the household level. Although emphasis is given to data collection by RRA methods which are generally more efficient and have the advantage of direct involvement of villagers, some formal surveys are necessary to fill information gaps. In Ethiopia, techniques based on airphoto interpretation with field checks, carried out by trained extension agents, have been successfully applied to mapping land resources in villages or groups of villages (Ethiopian Red Cross Society, 1988).

Evaluation of land resources involves a comparison of the properties of land with the requirements of possible types of land use. Evaluation is based on an inherent understanding of the interactions between land and land use. The procedure of comparing the land characteristics and qualities with the requirements of land use may be carried out either by simple matching, based either on previous experience or scientific knowledge, or by modelling of crop or animal growth and development in the particular environment. At the simplest level, matching is carried out based on transfer of experience from observations of similar production systems on similar soils. A farmer re-cultivating an area of fallow will draw analogies from his experience of similar soils elsewhere and from his previous experience on the same plot in deciding which crops to grow and which management practices and inputs to apply. This approach does not require specification of individual land use requirements and land characteristics but is simply based on experience and intuitive reasoning.

Matching procedure for FAO is fairly easy to understand and operate but can be extremely time consuming to carry out manually when large numbers of LMUs and production systems are involved. At this stage the use of modelling can be considered. According to FAO (1998):

- Empirical models are derived from statistically valid relationships of observed data. They do not make pretence to represent the processes which may link these observations, and should only be used over the particular range of conditions where they have been validated.
- Deterministic models simulate the processes involved in biological or physical systems, such as crop growth and development in specific environmental conditions. These models are more difficult to develop and require more understanding of systems. They usually contain more detail on physiological processes than is normally required for land evaluation, and do not usually include the complete range of soil and land factors which may affect yield in an actual field situation.
- Heuristic models are based on experience and expert knowledge, which does not necessarily imply an understanding of underlying processes. The sequence of decisions taken by the farmer on what crops to plant, where to plant them and at what time, may be regarded as a heuristic model. Heuristic models can be computerized given an adequate knowledge base. ALES is a commercially available software which provides the tools for building an expert system.

Models used for land evaluation of crop based production systems include:

- Agro-ecological zoning (FAO, 1996) and
- Crop Yield Simulation and Land Assessment Model for Botswana (CYSLAMB).

Agro-ecological zones are defined on the basis of combinations of soil, landform and climatic characteristics. Each zone has similar combinations of constraints and potentials for land use. When combined with an inventory of land utilisation types and their requirements, zoning can then be used as the basis of a methodology for land evaluation.

As a result of land evaluation, a series of physically suitable land use options are derived for each LMU. These options must be appraised according to financial and economic viability, potential impact on the various stakeholders, and potential impacts on the environment. Financial analysis is based on market prices and costs. Economic analysis takes into account shadow prices and opportunity costs. Whereas a land use plan of a large farm or group of commercial enterprises may rely on financial analysis to determine feasibility, a land use plan covering a broader range of activities within a nation, a district or a community, would normally use more complex procedures of economic analysis to evaluate alternatives.

Based on the land inventory, an evaluation is made of the suitability of each LMU for each production system. Preferably the evaluation may be carried out by qualitative matching procedures which are understandable and easy to apply. Matching procedures can be based on a limited number of key land characteristics which are mapped with the corresponding requirements of the LUT or production system. It may not be necessary to distinguish different classes of suitability, but simply to distinguish areas of suitable or non-suitable areas. As we have seen indigenous methods of evaluating land resources according to their suitability for certain uses exist. These can be applied to LUTs or production systems with which the community has direct experience. Such methods can be elucidated and further developed through topical RRA or PRA techniques, with extension agents or NGOs acting as facilitators.

It should be noted that in the described procedure there is no assessment of the risks associated with a particular land use. It is possible to build historical profiles of the occurrence of natural disasters from farmer recall. Focused interviews with key community members may relate the incidence of such occurrences to famine, times of plenty, or events in political or social development.

Conclusion

The chapter has attempted to characterise modern and indigenous systems of land evaluation as a basis for proposing an integrated approach. It has been shown that modern systems of land evaluation have their inherent weaknesses that make them ineffective in solving problems of the small-scale farmers. Some problems also exist within indigenous systems of land evaluation, but these systems are more realistic to the farmer because they are based on gained experiences, which, unfortunately is not found in most literature. However, there are attributes from both evaluation systems which complement each other. An integrated approach is proposed taking into account the simplicity of the indigenous systems and effective methods of collecting data. It is emphasized that the experiences of the local farmers should be utilized to make scientific knowledge more relevant to affected farming communities.

References

Bibby, J. S. and Mackney, D. 1969. 'Land use capability classification', *Soil Survey, Technical Monograph 1,* Rothamsted.

Dolva, H. and Renna, R. 1990. *Indigenous Soil Classification: A Study among small-scale farmers in the Northern Province of Zambia,* NORAGRIC, Agricultural University of Norway.

Ethiopian Red Cross Society 1988. *Rapid rural appraisal: A closer look at rural life in Wollo,* Ethiopian Red Cross Society/IIED, Addis Ababa and London.

FAO. 1976. 'A framework for land evaluation', FAO *Soils Bulletin,* 32, FAO, Rome

FAO, 1996. 'Agro-ecological zoning guidelines', *FAO Soils Bulletin,* 73, Rome.

FAO,1998. Integrated Land Use Planning: the key to sustainable land resource management in the 21st century, (Draft), FAO, Rome.

Gata, N. R. 1994. 'Indigenous science and technologies for sustainable agriculture, food systems and natural resource management with special reference to Zimbabwe', in J.Z.Z. Matowanyika, V. Garibaldi, and E. Musimwa, (Eds), *Indigenous Knowledge Systems and Natural Resource Management in Southern Africa,* pp 85-100, Africa 2000Network/IUCN Zimbabwe.

Landon, J.R. 1991. (Ed) *Booker Tropical Soil Manual: A Handbook for Soil Survey and Agricultural Land Evaluation in the Tropics and Subtropics,* Longman, New York.

Msanya, B.M., Mwaseba, D., Ole-Meiludie, D.J., Kimbi, G.G., Massawe, A.R. and Kimaro, D.N. 1998. 'Indigenous knowledge in soil classification: the case study of Mbulu and Karatu districts in Tanzania,' *Uongozi – Journal of Management Development* 10, 2, 237-260.

Osunade, M.A.A.1992. 'Soils and the small farmers of Swaziland', *UNISWA Research Journal,* 6, 71-82.

Osunade, M.A.A. 1993. 'Modern and indigenous land evaluation approaches in Swaziland,' *International Journal of Environment Studies,* 33, 32 - 45.

Osunade, M.A.A. 1994. 'Community environmental knowledge and land resource surveys in Swaziland,' *Singapore Journal of Tropical Geography,* 15, (2), 157-170.

Richard, P. 1985. *Indigenous Agricultural Revolution*, Hutchinson London.

Ridgway, R. 1997. 'Applications of large scale aerial photographs in participatory land use planning in rural Ethiopia', *The Land* 1, 1, 67-74.

Young, A.T. 1978. 'Recent advances in the survey and evaluation of land resources,' *Progress in Physical Geography,* 2, 3, 426-479.

Young, A.T. 1981. *Tropical Soils and Soils Survey*, Cambridge University Press, Cambridge.

Warren, D.M. 1991. *Using Indigenous Knowledge in Agricultural Development*, Discussion Paper No. 27, World Bank, Washington DC.

Chapter 14

The Livestock Industry and the Subsistence Economy of Rangelands in Africa with Special Reference to Botswana

B. T. Kgabung

Introduction and Background

Increasing demands on the land for goods and services are placing considerable pressure on traditional management systems in developing countries (Child *et al.*, 1987), as well as on the resources themselves. Most of the lands in these countries are classified as arid and semi-arid (*ibid.*). Arid or semi-arid rangelands fall within the broader biome of tropical savannas. The latter are widely recognised as forming a transition zone between closed tropical forests and open desertic steppe (Bourliere and Hadley, 1983; Cole, 1986) and according to UNEP (1997: 5), they occupy 1286 million ha (43.4 per cent) of Africa, 663.3 million ha (37.5 per cent) of Australasia and 516 million ha (29.2 per cent) of South America. A climatic gradient is inherent within the savannas from humid to arid (Sinclair, 1983), depicting a continuum of vegetation types ranging from dense moist woodlands and thickets through open shrub lands to extensive treeless grasslands (Huntley and Walker, 1982).

An important distinction can be made between semi-arid savannas (which are the focus of this chapter) with an average annual rainfall of 350-600mm and their humid equivalents in areas ranging up to 1300mm (Huntley and Walker, 1982). A marked dry season is common in these two tropical savanna systems. In the humid savannas it is manifested by a short dry period (Cole, 1986), and in the arid and semi-arid savannas may extend up to seven or more months (Huntley, 1982).

Although annual arable dryland agriculture is not an ideal form of land use (Walker, 1979), semi-arid savannas constitute a major part of the world's grazing rangelands (Coppock *et al.*, 1986) which represent an optimal habitat for grazing mammals, wild and domestic, large and small. The semi-arid flora also provide a variety of other wild foods, medicinal plants, energy resources, and construction materials.

These semi-arid lands were until recently dominated by low-density usage by pastoral nomads and hunter- gatherers. Over time resource use has intensified and new patterns of use have emerged. In Africa, for example, these changes have been striking for the past forty to fifty years resulting in encroachment of traditional modes of utilisation by a variety of sectors, including wildlife conservation but more especially, more intense forms of livestock and arable production (Warren, *et al.*, 1996). These new methods of use have far reaching implications on several spheres of the environment and particularly on rural livelihoods and land capability.

Not only do semi arid environments contain important flora and fauna, they also occupy a large part (about a third) of the earth's land surface which is inhabited by some 500-600 million people (Sandford, 1993; Western & Finch, 1996). This, as noted by Heathcote (1983), serves to emphasise that an understanding of the ecological and socio-economic impact of development in these regions has implications for vast tracts of land and large numbers of people. Of this total number of arid and semi-arid human inhabitants, 30-40 million are believed to have 'animal based' economies and the majority of these are pastoralists (Sandford, 1983). Within the 30-40 million, 50-60 per cent are found in Africa (*ibid.*).

This chapter first reviews trends in resource management in semi-arid rangelands of Africa. The aim is to unravel both at the regional and national levels, past and present management strategies or policies in order to determine how the relationship between livestock rearing and hunting and gathering have evolved amidst resource use changes. Second, the chapter outlines the ecology of semi-arid environments in order to provide an understanding of their functions and dynamics which, according to Walker (1997) is an inevitable prerequisite for ecologically sound management and optimal use of resources in semi-arid rangelands. Finally, the impact of the livestock industry on the subsistence use of wildlife and veld products in the semi-arid rangelands of Botswana is examined as an example to illustrate the effect that more intensive forms of agriculture are having on ecologically marginal environments in Africa.

Resource Use in Africa's Semi-arid Rangelands

Savannas are among the oldest ecosystems used by people (Solbring, 1993) and African rangelands are no exception. These people existed by hunting wildlife and gathering of wild plant materials, with the situation in Africa changing about 6000 years ago when domestic animal production and crop agriculture were introduced (Solbring, 1983). Although the pattern and intensity have changed since the introduction of domestic animals, prior to colonization by Europeans utilisation remained light because both human and domestic animal populations were small and most of the inhabitants were nomadic (Walker *et al.*, 1981).

African savannas, notably in the southern and eastern parts of the continent, support a great wealth of wild herbivores and their carnivorous predators, (McNaughton, 1985). Significantly, the largest concentrations of wildlife occur in areas where there is a juxtaposition of different types of savanna - each of which provides grazing and browse at different periods of the year. This sustains the migrations of game that are such a characteristic feature of areas like the Serengeti plains of Tanzania and the northern Kgalagadi, Etosha pan and Okavango area of southern Africa. These savannas also provide conducive conditions (grass and browse) for domestic herbivores and a wide variety of veld products.

The use of these range resources for wildlife utilization, gathering and livestock production has over time either changed in importance or intensified due to pressure exerted by high human population growth (which at present stands at 2.8 per cent per annum in Africa). Rapid increase in livestock numbers, colonization and modernization, land use changes, governments policies, international influence and other factors, have all contributed to this growth. As Scoones *et al.*, (1993) elaborate, because of its location between arid lands and higher potential forests and agricultural lands, the African savanna zone has had to absorb increasing numbers of people. Pastoralists from the drylands and agriculturalists from the sub humid zones are both expanding into these areas, as the traditional land available to them diminishes.

Trends in Landuse and Utilisation of Natural Resources

Livestock Production

The drier savannas of Africa have, for a long time since the introduction of domestic herbivore in the continent, been the main livestock rearing areas (Cole, 1986). Moist Savanna woodlands (in areas of more than 600 mm mean annual rainfall) are avoided for this use because this is where tse-tse flies act as vectors for trypanosomiasis.

African livestock farming is based on the rearing of cattle, camels, sheep and goats, with draught animals playing a minor role. Products obtained from livestock include milk, meat, hides, manure, blood, draught power and social security. African pastoralism is characterized by the collective ownership of land and individual pastoralist family ownership of livestock (Solbring, 1993), therefore allowing open access to grazing land. Mobility is yet another vital aspect of pastoralism in arid and semi-arid zones of Africa, as well as in Southwestern Asia, where pastoralists are either nomadic or migrate in order to effectively exploit the available grazing and water resources (Sandford, 1983). In fact, three distinct traditional livestock management strategies have evolved in Africa since the introduction of domestic herbivores. These are nomadic pastoralism, transhumance and sedentary animal husbandries (see Table 14.1).

It is frequently pointed out in the literature that the practices of traditional pastoralism represents highly adaptive strategies which enable pastoralist societies to exploit grazing lands and to survive periods of extreme stress (Lamprey, 1983). As Sinclair and Fryxell (1985) put it, migratory pastoralists have traditionally lived with their cattle in balance with the vegetation. Despite this, Solbring (1993) points out that nomadic pastoralism in its traditional application probably no longer suffices to meet the needs of an expanding African human population. Neither can it meet the aspirations of the pastoralists population for a more sedentary and material lifestyle. The question that then arises is whether modern livestock management systems are capable of sustaining such changes in land use practices. The past forty to fifty years has witnessed pressure and changes in the use of rangelands by livestock farmers. This has been accompanied by erosion of the traditional systems of pastoral resources management and emergence of new ones based on modern technology which in most cases if not all are adopted from the developed world.

Table 14.1: Traditional Livestock Management Strategies

Aspect/ Management System	Nomadic Pastoralism	Transhumance Animal Husbandry	Sedentary Animal Husbandry
Form	Constant movement from place to place, e.g. the Tuareg nomads of Niger/ Algeria	Regularly repeated migration between wet & dry season pastures, e.g. Sahelian pastoralists move to the north in the rains & southwards in the dry season (Sinclair & Fryxell 1985)	Practised throughout the year in the vicinity of perma-nent settlements. Small-scale transhumance (a few days) practised
Objectives	To cope with marked spatial & temporal variability of forage & water resources (Solbring, 1993)	Same as nomadic pastoralism	Mixed farming (pastoralism & subsistence agriculture)
Advantages	Mobility allows environment to recuperate & regenerate	Same as nomadic pastoralism	Animals feed on plant remnants & wees in the fields while in turn they provide manure & draught power for arable farming
Limitations	Cannot cope with expanding human population	Cannot cope with expanding human population	Cannot cope with expanding human population

Source: Kgabung (1999).

Table 14.2 summarises developments that have been and continue to be undertaken within the domain of the livestock industry, or that affect it.

With regard to sedentarization, governments especially in East Africa and the Sahelian region believe that if pastoralists can be settled in well defined areas, centralised control can be increased, taxes be levied, herd sizes be limited in line with some predetermined idea of 'carrying capacity' and herders encouraged to produce for market (Graham, 1988). In the arid zone of Northern Kenya the rate of sedentarization around water points greatly increased since the drought from 1968 to 1976 when many destitute pastoralists became dependents upon famine relief food (Lamprey, 1983).

Table 14.2: Developments in the Livestock Industry

Developments	Reasons / Objectives	Implications
sedentarization	• government induced to settle pastoralists who were otherwise thought to be uncontrollable because of their mobility • sometimes occurs spontaneously to lay claim to grazing land	• concentration of people & livetock in one area (in contrast to nomadic & transhumance animal husbandries) = over exploitation = range degradation • disruption of mobility & seasonal grazing patterns • reduce area formerly occupied by pastoralists
privatization & commercialization	• to intensify production • to effectively personalise range degradation costs (Sandford, 1983)	• accentuated problems of inequality in distribution of resources (livestock & land) and stratification of pastoralists in terms of wealth • break down of the entire community-based property systems (Scoones et al., 1993) • loss of traditionally used rangelands = marginalization of subsistence users • concentration & degradation in communal lands
borehole	• provide permanent surface water for livestock & people	• encourage sedentarization = curtailment of movement, concentration = over exploitation • allows spatial expansion of livestock holdings • increase in livestock numbers
veterinary services (vaccinations, artificial inseminations, cordon fences & quarantine camps)	• to eradicate diseases • fences erected to control diseases • to increase production to increase production	• increase in livestock numbers • loss of mobility over large areas due to fences
subsidies	• international aid • national government policy & support for privatization & commercialization	• more investment in the livestock industry (Perkins & Ringrose, (1996) • increase in livestock numbers

Table 14.2 concluded

Developments	Reasons/Objectives	Implications
loss of traditional pastures to other uses	• need for more land for other uses e.g. for conservation & agriculture	• land shortage & landuse conflicts e.g. between the Amboseli Masai pastoralists & conservationists/Kenyan Government
road development	• development in general	• rangelands became accessible to more people = over exploitation (Squires, 1981)

Source: Kgabung (1999).

Because of sedentarization, and the imposition of political boundaries, the Turkana (Northern Kenya) pastoralists, for example, can no longer move their herds to the mountains of Ethiopia as they have done for centuries, if not millennia (Solbring, 1993). In the Sahel region the wet season grazing area in the north is no longer available to pastoralists while the southern Sahel became grazed year-round with no rest period for reproduction (seeding) and with consequent destruction of grasslands (Barnes, 1972; Kelly and Walker, 1976). In short, denied access to traditional dry season grazing reserves, herdsmen have either been pushed onto marginal lands or become concentrated on the more productive land. Overuse of such restricted areas has eroded the basis for their livelihood and initiated a downward spiral of marginalisation, impoverishment and environmental degradation (Darkoh, 1996).

MacDonald (1989) notes that with the advent of the borehole and windmill pump it very soon became possible to graze domesticated herds virtually throughout South Africa on a year-round basis. As acknowledged by Western (1976) and Campbell (1979), some livestock populations are too large to be supported by their traditional pastoralist economies due to developments in the livestock industry such as the introduction of veterinary facilities and boreholes.

The move towards privatization is particularly noticeable in Eastern Africa, Kenya, Somalia and the Sudan. In Southern Africa, Botswana, through a series of livestock development projects and land reform policies such as Tribal Grazing Land Policy (TGLP), and now through its new agricultural policy (fencing of communal lands) is following the same course. In Kenya during the 1960s, pastoralist

Masai were encouraged to register land individually and become commercial beef ranchers (Graham 1988). As Darkoh (1996) asserts, in recent years, large sections of the nomadic livestock economy in sub-Saharan Africa have experienced rapid commercialization. The main causes are the explosive growth in the size of both the rural and urban population (Warren *et al.*, 1996) and associated steady increase in the levels of meat consumption, coupled with new opportunities for export offered by more developed and affluent neighbouring as well as distant countries.

Privatization is based on the concept of 'Tragedy of the Commons' (Hardin, 1968). This is:

> held to be the inevitable result of grazing of individually-owned livestock on communally-held rangelands, wherein the benefits of an extra animal accrues entirely to the owner, while the costs, reduced grazing and range degradation are shared by the community (Perkins, 1991: 75).

That is, the open access grazing regime is believed to result in overstocking, overgrazing hence range degradation.

Since both privatization and borehole drilling are capital-oriented ventures which poorer farmers cannot afford, typically most of the land and livestock is held by a small class of wealthy and influential cattle owners (Sandford, 1983). In the process, this may force the poorer farmers out of the system completely or they remain as herders of rich livestock owners (Graham, 1988). It is also common in some African countries that those with private ranches or enclosures still maintain their livestock on the open range, and use their enclosures for fodder production and reserve grazing (hence over exploitation in communal areas), for example, in Botswana and Somalia through such 'dual grazing' practices. Today in some areas there is almost no common land left, and those who missed the bandwagon and did not enclose may have been permanently excluded (*ibid.*).

As depicted in Table 14.2, traditional pastures have been lost to other uses. For example, large areas of pastoral rangelands have been expropriated for exclusive wildlife conservation (Little, 1984) under the pretext that pastoralists overstock, overgraze and damage their range while wildlife are seen as existing in harmony with their surroundings (Hannerwood and Rogers, 1987). The alienation of grazing land for the exclusive use of wildlife and tourism has a very direct impact on pastoralists communities. It has given rise to questions about 'people versus animals' in the formulation of African wildlife policies (Collett, 1987).

Hunting and Gathering

For at least 40 000 years, hunter-gatherer societies in African savannas led nomadic ways of life, moving seasonally within their areas in pursuit of wildlife resources, veld products and water. They existed like that for at least 40 000 years in Southern Africa (Campbell, 1971 and Hitchcock, 1978). Hunting and gathering remained an important activity even after the introduction of cultivation and livestock in Africa. The latter activities were until recently extensive and to a large extent existed harmoniously with the environment because both human and livestock populations were at low densities. Collett (1984b) argues that there is archaeological and historical evidence to show that prior to European colonisation pastoralists tended not to exploit wildlife except during periods of stock loss.

Today, hunting and gathering as a form of land use is in decline all over Africa basically because the resources that support the system have been drastically reduced. However, there are some rural inhabitants who still subsist on hunting and gathering. Cunningham (1989) states that the use of plant resources to meet human needs is an important aspect of the multiple-use of land in many parts of Southern Africa in which indigenous plants, for example, are a source of fuel, medicine, building material, craftwork materials, dye, income and food supplements. Various factors attribute to the decline in hunting and gathering system:

- increase in the human population;
- changes in the land uses, notably encroachment into hunting and gathering areas by livestock or crop production;
- harvesting and hunting pressure; and
- poaching.

Historical records show that the resultant effects of these include reduction or displacement of wildlife and as Cunningham, (1989) notes reduction in the range of plant resources which provide a buffer against rural poverty during climatic extremes. Huntley (1982), Barnes (1982) and Scholes and Walker (1993) indicate that the mixed indigenous herbivore populations have been largely replaced by high densities of domestic animals, mainly grazers except in game reserves and in areas with tse-tse fly.

During the present century recognition of the value of wildlife resources and the threats to its continued existence posed by increased herds of domestic stock and growing population pressure, has led to the setting aside of large areas as national parks, where the native fauna is protected against hunting and competition from introduced species (Cole, 1986).

In summary, the resource utilisation changes that occurred in the semi-arid tropical Savannas of Africa over the past 40 or 50 years have been striking with traditional modes of utilisation being encroached upon from a variety of sectors, including wildlife conservation but more especially more intensive forms of livestock and arable production (Warren *et al.*, 1996). These as discussed above have far reaching implications of both socio-economic and environmental dimensions.

In general, over exploitation and poverty ensue despite the modern management strategies which have dismissed traditional resource management as primitive, self-destructing, and thus too unproductive to provide goods for the increasing human demands. As Lamprey (1983: 660) summarises:

> in the last fifty years numerous modern influences have undoubtedly exacerbated destructive land use by pastoralists in the savanna rangelands of developing countries and to this extent they are victims of circumstances beyond their control.

It should be emphasised, however, that the applicability of modern technology in arid and semi-arid ecosystems remains a moot question, more especially because these technologies evolved in peculiar socio-economic, political and ecological contexts. In addition, if appropriate solutions are to be found they must be framed within an understanding of the ecology of arid and semi-arid savannas.

Semi-Arid Rangelands Ecology and Ecological Change Theories

As Walker (1979) highlights, a thorough understanding of the function and dynamics of semi-arid ecosystems should be an essential prerequisite for ecologically sound management and optimal use of resources found in them. If this is not appreciated, mankind may continue to promote resource management policies and technical interventions that are unsuitable and detrimental, hence resulting in the degradation of ecosystems.

Factors Determining Savanna Composition, Structure, Physiognomy and Dynamics

Ecological studies in the savannas, for example, Scholes and Walker (1993), Cole (1986), Lamprey (1983), Ellis *et al.*, (1988) reveal that four principal factors: water (plant available moisture, PAM); soil nutrients (plant available nutrients, PAN); fire and herbivory determine savanna distribution, structure and function. The occurrence and interplay of these factors result in long- and short-term changes constantly modifying savanna composition, physiognomy, dynamics and ecological processes. Because of this, savannas, particularly arid and semi arid ones are said to be dynamic and non-equilibrial (Scholes and Walker, 1993; Coppock *et al.*, 1996; Walker and Noy-Meir, 1982).

The consensus among researchers as exemplified by Tinley (1982) and Huntley (1982) is also that the overwhelmingly important factor determining the distribution of savanna is soil water availability which is determined by the availability of rainfall. As Scholes and Walker, (1993:80) put it, 'water availability acts as a switch on savanna function'. To this effect it should be realised that rainfall in savannas is highly seasonal or variable both in space and time, coupled with the incidence of drought. The seasonality becomes increasingly marked as one proceeds from the edge of the rain forests to that of the desert (Bourliere and Hadley, 1983). The direct linear correlation between rainfall and primary biomass is well-established especially in dry savannas (Coe *et al.*, 1976; Scholes and Walker, 1993). Since rainfall is erratic, it is such that as it increases so does the amount of vegetation cover which also decreases when it decreases.

Soil nutrient content in dry savannas is mostly related to availability of nitrogen and phosphorus to plants (Scholes and Walker, 1993). It interacts with rainfall (higher rainfall leading to leached soils) and results in a continuum of savanna vegetation types from arid / eutrophic (non leached) on the fringes of the semi deserts through to moist/dystrophic savanna woodlands (Huntley and Walker, 1982). In short the plant available moisture and plant available nutrients determine above-ground biomass in savannas. This structure is then modified by herbivory and fire which are often seen as competing. The latter refers to the fact that if primary biomass is not eaten by the late dry season, it may well constitute a fuel load.

Fire, which has been practised for at least 50,000 years in Africa (Gillion, 1983) and herbivory among other things are viewed as disturbances that contribute to the functional stability of semi arid ecosystems. The changes they exert should be seen as part and parcel of the whole component of the system. Westoby *et al.*, (1989) note that, burning, if properly done transforms bush encroached landscapes into grassland savanna. Herbivory on the other hand, particularly grazing and depending on the intensity can transform savanna grassland into bush thicket. Fire can then be used to revert the situation to grassland savanna. As a result of the influence of these factors, it has been pointed out that savanna regions can be regarded as areas of natural or semi-natural vegetation (Gillion, 1983).

Drought, Resilience and the 'Boom and Bust' Nature of Semi-Arid Ecosystems' Flora and Fauna

As noted earlier, drought (hence erratic rainfall) is a recurrent feature of arid and semi arid savannas. Fluctuations in rainfall determine the availability of both primary and secondary production of vegetation matter, with the distribution and numbers of animals limited by the confined availability of water and forage (Walker, 1979; Ellis *et al.*, 1988; Perkins, 1991 and MacNaughton, 1985). Thus, irregularity in rainfall in arid and semi arid zones result in fluctuations in the production of vegetation which in turn causes animals to fluctuate. For these reasons, semi-arid savannas are said to be characterised by 'boom and bust' cycles whose animal populations follow a pattern of expansion in wetter periods and decline and essentially retreat to 'refuge areas' in drought periods (Perkins and Ringrose, 1996).

These 'boom and bust' cycles are evident and well documented throughout Africa. For example, during the drought from 1968 to 1976, cattle populations in the Sahelian Zone and north-eastern Africa were reduced by 50 per cent in several countries. The same happened in the Turkana District of Kenya during the drought from 1979 - 1980 (Ellis *et al.*, 1988). It is argued that domestic and wild ungulates populations approach but never reach extinction because droughts do not last long enough to completely eliminate them. Recovery is thus possible with the onset of rains, in many cases being relatively rapid with some herds returning to pre-drought levels (*ibid*). Wild populations follow the same trends but recovery to the pre-drought levels is prevented by man's activities.

According to many researchers (e.g. Walker and Noy-Meir, 1982; Ellis *et al.*, 1988), these 'boom and bust' cycles reveal the resilience of semi arid systems, which as highlighted earlier, are continually subjected to considerable and frequent disturbances (e.g fire, drought, grazing and disease) but still continue to exist and re-establish themselves, even in cases where such shocks are severe. As further asserted by Walker and Noy-Meir (1982: 585):

> the savannas are not only adapted to survive the temporal variations (disturbances) but in all probability, require them to maintain their resilience.

A resilient system is described as one which can change quite markedly as a result of disturbance, but which then returns towards its original equilibrium condition (*ibid*). Resilience is thus defined as 'the capacity of a system to absorb disturbance without quantitatively changing its behaviour' (Walker & Noy-Meir, 1982: 556).

Besides drought, another factor which clearly shows the resilience of these ecosystems is the rinderpest epidemic which drastically reduced ungulate populations of eastern and southern Africa in 1890's, but whose populations were able to recover thereafter. In short, semi arid ecosystems because of their dynamic nature described above, are said to be non-equilibrial, but persistent or resilient, with system dynamics perhaps affected more by abiotic controls (Ellis *et al.*, 1988).

The Importance of Animal Mobility in Semi-Arid Rangelands

In light of the spatial and temporal heterogeneity of arid and semi arid savannas, one would wonder how the animals and the people they house have survived the recurrent droughts and the accompanying fluctuating production levels to this day. Studies point to migration as a major coping strategy for both wild and domestic animals and subsequently human populations (Walker, 1979; Sinclair and Fryxell, 1985; Barnes, 1982 and McNaughton, 1985). Thus, migration patterns simply tend to follow changes in the amount and quality of the food supply, and in the case of most species, supplies of drinking water.

Walker (1979) observes that in East Africa, migration of large game herds are regular and predictable reflecting the more regular seasonal rainfall and production pattern there as opposed to the irregular migrations and movements of large herbivores in Southern Africa. According to Dumont (1978), from historical and prehistorical evidence, it appears that the migration has been operating for many centuries possibly since domestic cattle first appeared in the Sahel 5000 years BP. This has actually been apparent in all African drylands where pastoralism was or is practised until the fairly recent past.

One conclusion from the studies undertaken is that migration is an ecologically stable strategy because it allows vegetation to recover from grazing each year and it also allows a much larger population of cattle and people to be on the arid and semi arid grasslands compared with a situation of year-round residence on the same land (Barnes, 1979; Sinclair, 1979 and Sinclair and Fryxell, 1985).

It is to be realised, as Walker (1979) notes, that many of the worst problems associated with management of the semi arid ecosystems stem from the fact that increasing populations have stopped the movements of both wild animals, and of people and their domestic herds. As emphasised by McNaughton, (1985:283),

> deterioration of agricultural grazing lands may be less a consequence of stocking density than of the reduced responsiveness of a confined fauna to the temporal and spatial dynamics of grassland ecosystems.

Ecological Impact of Large Herbivore Grazing/Browsing and Ecological Change Theories

There is evidence that the dynamics, stability and resilience of savanna ecosystems can be significantly affected by herbivores. The grazing system as a whole is said to be self-regulating with ungulate populations tending to increase to equilibrium levels determined by primary production and limited competition between herbivore species (Lamprey, 1983). Competition is limited because different wild game species, particularly in Africa and as experienced in the Serengeti, favour different plant species and plants at different stages of growth (McNaughton, 1976, 1985). This, as a result, is said to maintain the stability of the ecosystem. On the other hand, domestic ungulates are not regulated in this manner and frequently increase to levels that are greater than the carrying capacity of their habitats. As already mentioned, this has been promoted by man's modern management activities such as provision of water points, sedentarization, and construction of roads and tracks. His control (or alteration) of fire results in lack of reversion of bush thicket to grassland hence loss of grazing resource (Squires, 1981). As a result of all these, grazing by livestock imposes a more severe strain on the vegetation.

It is widely known that, in semi arid savannas, livestock through their grazing, cause changes in vegetation structure, composition and dynamics (Barnes, 1979; Cole, 1986 and Squires, 1981). Heavy grazing is said to result in reduction of palatable species, hence livestock may be forced to graze upon progressively less palatable species or browse upon shrubs normally avoided. Literature shows that the ultimate outcome of heavy grazing is bush encroachment, which Lamprey (1983) attributes to lack of burning as the fuel load is consumed by livestock. The other explanation of bush encroachment is what Walker and Noy-Meir (1982) refer to as 'a two layer soil-moisture competition model'. This demonstrates that owing to the nature of their root systems, grasses out compete trees for water in the upper layers (about 30cm from the surface). Reduction of grass due to overgrazing therefore leads to less evapotranspiration, hence water penetrates to deeper levels for exclusive use by shrubs – thus bush encroachment.

Livestock induced vegetation changes or impacts are evident wherever semi-arid savannas occur (Walter *et al.*, 1981). They are well documented in Africa's semi-arid savannas, for example, in northern Tanzania and northern and eastern Kenya (Lamprey, 1983) in the Sahel (Sinclair and Fryxell, 1985) in Southern Africa (Barnes, 1979; Huntley, 1982) and west Africa (Cole, 1986). These changes are perceived differently by various people but generally they have consistently been viewed or framed within a range degradation or desertification framework, in the on going debate about the impact of the livestock industry on the environment.

Thus there is in the minds of many individuals, governments, international organisation alike, a belief that semi arid rangelands are undergoing severe degradation and desertification as a result of poor management and overstocking hence overgrazing. This belief has been described as the 'Mainstream View' by Sandford (1983) and he singles out pastoralists as not being in agreement with it. To be noted also is that this view has to a large degree governed range management policy and it continues to do so. Abel and Blaikie, (1989) point out that to be the case in Botswana and Zimbabwe.

The mainstream view is based on the 'Range Succession Model' summarised by Westoby *et al.*, 1989. This model advocates that livestock stocking rates should be restricted to recommended carrying capacity which will be applied on a continuing basis so as to avoid range deterioration. The assumption in this model is that plants and animals reach an equilibrium in the 'predator-prey' relationship (Warren *et al.*, 1996), such that increased stocking rates leads to range degradation while destocking improves the range. The model therefore sees primary production as a function of stocking rates. This is in spite of the account given above that in semi arid ecosystems vegetation is a function of rainfall which varies both in space and time hence variation in primary production.

It is evident that this model has worked in temperate zones of North American and European rangelands where a relatively stable climate has influenced the setting up of a carrying capacity to match particular stocking rates on a continuous basis (White, 1993 and Sandford, 1983). However, the applicability of the model in semi-arid ecosystems, where climatic variation is the norm, remains challenged (Moleele, 1994; Abel and Blaikie, 1989 and White, 1993). This leads to the recent alternative range management model referred to as the 'Alternative Multiple Stable State or State and Transition Model' by Westoby *et al.*, (1989).

The model encourages a more flexible approach in the management of semi arid savannas. It fundamentally heeds the fact that savannas are continuously changed by man and natural factors. Therefore, under this model as Westoby *et al.* (1989: 238) argues:

> managers would see themselves as facing an oncoming stream of events, a mixture of opportunities and hazards and their objective should be to seize the opportunities and evade the hazards as far as possible.

In contrast to the conservative stocking strategies adopted under the Range Succession Model, Abel and Blaikie (1989) assert that the great instability but intrinsic resilience of semi arid rangelands should encourage governments to more favourably regard the opportunistic stocking strategies.

These ecological change theories are discussed not to enter into a range degradation debate but to provide insights as to which range management strategy may be appropriate under the ecological, socio-economic and political conditions or constraints experienced in semi arid environments.

In summary, understanding the savanna dynamics discussed in this chapter could form a baseline to enable managers to come up with appropriate means of exploiting the system. However, as shown in the account of recent African experience with regard to livestock production and wildlife utilisation, such has not been the case. Mankind continues to adopt ideas or models which disregard the most important aspects of semi-arid savannas such as mobility, spatial and temporal climatic variations, amongst others (Lewin, 1986). It should also be realised that the issue of livestock impact does not lie solely with their direct influence but owes a lot of explanation to the management strategy or theoretical model followed. Hence Abel and Blaikie (1989) assert that in addition to spatial and temporal variation in rangeland resources, social equity and institutional problems should as well be considered in the establishment of range management strategies. Indeed the current management approach which to a large extent is based on modern technology and commercialization, puts the resilience and sustainability of semi-arid savannas at stake.

Rangelands in Botswana

Botswana is a land locked semi-arid country located in Southern Africa. Like all such environments (semi-arid), the country is naturally characterised by vast rangelands which are predominantly grass cover with scattered shrubs and trees. These rangelands are mainly used for livestock production, wildlife utilisation and gathering of veld products.

Livestock production has been of paramount importance in the country. As White (1993) notes, it was the heart of the national economy in both the pre-colonial and colonial times, and remains the heart of the rural economy today. The industry, mainly beef production, accounts for the largest portion of the agricultural sector's contribution to GDP (Ministry of Finance Development Planning, 1991).

In spite of its economic significance, the livestock industry is alleged to be expanding at the expense of other uses of the rangelands as well as the very environment that supports it. This as White (1993) indicates has resulted in concern being expressed for many years over the apparent degradation of the communal rangelands due to overgrazing.

Botswana's Physical and Biological Environment

Botswana is geographically divided into the 'Kalahari Sandveld' which extends over some 75-80 per cent of the country, and the 'hardveld' which is confined to the eastern most part (Cooke, 1983). These areas differ in a number of respects such as rainfall, soil properties and ground water recharge (Veenendaal and Opschoor, 1986). The hardveld is characterised by varied relief and geology with more fertile soils and slightly higher and more reliable rainfall, and has surface water provided by rivers dissecting it (Cooke, 1983). The Kalahari sandveld on the other hand, is devoid of surface water and is characterised by low and highly variable annual rainfall and infertile Kalahari sands which support extensive cover of grasses and shrubs (Cooke, 1983 and Perkins, 1991). Although lacking surface water, the sandveld is dotted and veined with pans and ancient river beds which, in addition to their high nutrient content and better grasses, hold surface water during the rainy season and for varying periods into the dry season (Cooke, 1983).

The Kalahari environment consists of extensive grazing, ephemeral surface water and in addition to these, a large variety of wild animals which are dependent on this grazing and water resources (*ibid.*). Like elsewhere in arid and semi arid environments, the animals notably wildebeest, hartebeest, eland, springbok and gemsbok have developed physiological and behavioral strategies such as movement to survive the harsh conditions. In the very large area of Botswana, as summarised by Cooke (1983: 7):

> it is the presence or absence of water in rivers, swamps and pans which has been and still is of critical importance to both man and his domestic stock and to all forms of wildlife.

Background to Human Activity

Archaeological research and information has established that the entire area now known as Botswana was occupied by hunter-gatherers for more than half a million years (Campbell, 1971) and by agro-pastoralists for the past 2000 years (Campbell, 1983).

Bantu pastoralist/cultivator groups are believed to have always lived more in the eastern hardveld (Veenendaal and Opschoor, 1986). According to Campbell (1971), the Tswana were, by 1820, well established in this area, living in large villages. Their cattle were kept in grazing lands normally at quite a distance from the villages and crop fields (Veenendaal and Opschoor, 1986). There, semi nomadism was practised with cattle being moved in an east-west pattern between wells, dams, springs and rivers in the east and temporary rainfed water supplies in the western sandveld (*ibid.*).

From the mid 19th century, as noted by Cooke (1983), increasing political stability and powerful external cultural and technological influences led to a rapid rise in the human and livestock populations. This exerted great pressure on the eastern hardveld resources, especially on grazing, woodland and water. Environmental degradation resulted from this, which according to Cooke (1983), was noticed as long ago as 1943.

The Kalahari sandveld, though devoid of surface water and characterised by harsh conditions, has for a long time been extensively used by hunter-gatherer populations variously known as the San, Bushmen or Basarwa and pastoralists all of whom were fairly mobile (Campbell, 1971; Cooke, 1983; Hitchcock, 1978). During the 19th century, the latter began to settle permanently particularly on pans where best pastures and water, either from spring or shallow wells, were available (Campbell, 1983). These people were tied to water by their very numbers, their livestock and to the land by their need for subsistence agriculture. Though settled around permanent water for most part of the year, the agropastoralists moved out during the rainy season to exploit fresher resources in the outlying pans hence permitting a form of rotational grazing (Campbell, 1971 and Devitt, 1971).

The whole economy of the San depended on gathering veld foods and hunting (Campbell, 1971), activities which for most pastoralist people tended to be minor occupations in comparison to livestock keeping (Devitt, 1971). Both the Kgalagadi and Tswana pastoralists traditionally turned to wildlife when cattle and crops failed

(Campbell, 1983). They also traded in wildlife products especially after the arrival of the colonialists (*ibid.*). Altogether, Murray (1978) found that in western Botswana, wildlife utilisation was an important economic activity for at least 39 per cent of the total resident population (48,556 people) most of whom own few or no cattle.

The above summarises a traditional way of utilisation of range resources particularly in the sandveld where aspects such as low population density, mobility and open access to resources then reigned. The question that now arise is how can the Kalahari environment cope in the face of the greater demands of the society in terms of land and water resources for agriculture, ranching, mineral exploitation, wildlife conservation and tourism and a host of other uses.

Livestock Industry in Botswana

Livestock farming in Botswana involves the rearing of cattle and small stock such as goats and sheep with the former being the major livestock production activity. Livestock ownership ranges from small subsistence farmers to fully commercial farmers who aim at profit maximisation. There are two distinct livestock production systems, viz, the traditional cattle post system in the communal areas and the ranch system in the freehold or leasehold commercial ranches. The bulk of the livestock (about 85 per cent) is in communal areas while the remaining 15 per cent is in the ranches (Mannathoko, 1992).

At Independence in 1966, both livestock and people were mainly confined to the Eastern Hardveld, in the freehold farming areas and in those parts of the country where surface water was available. According to Machacha (1992), the cattle population was then about 1.3 million and that for small stock was about 1.0 million. The late sixties and the early seventies saw a rapid increase or expansion of livestock activities to the west into the sandveld. Several factors are widely regarded as having driven this expansion:

- The introduction of the borehole technology opened up increasingly larger parts of the sandveld to livestock utilisation which overcame the 'age-old' problem of a lack of surface water

(Cooke, 1985). The major borehole drilling started in 1956 (Campbell, 1983) and was at government expense as part of a drought relief effort (Hitchcock, 1980). Subsequently, individuals sank their own boreholes including many people from the eastern hardveld who had suffered for years from denuded pastures (Campbell, 1983).

- Government policies have provided extensive support for beef production hence the industry became very attractive to farmers. As summarised by Arntzen *et al.*, (1996), this attractiveness has taken the form of artificially high beef producer prices, input subsidies given by the National Development Bank, a lax and lenient tax system, provision of veterinary services (such as quarantine camps, vaccination campaigns, boreholes, trek routes and artificial insemination) through Livestock Development Projects (LDP) I & II and a change in land tenure (through the establishment of the TGLP.
- The livestock industry has also been made profitable, hence prompted to expand by the European Economic Community (EEC) subsidy (90 per cent) on beef exports and increasingly privileged access to their markets (Perkins and Ringrose, 1996 and Veenendaal and Opschoor, 1986). Access to this export market was/is at a cost for the government and indeed the environment because it required stringent disease and hygiene requirements to be met. This led to the erection of the veterinary cordon fences to control Foot and Mouth Disease, hence safe guarding the market.
- Extensive road building programme throughout the country is yet another factor that encouraged the expansion. Immediately after independence, such developments were confined to the eastern hardveld. Today, however, the Kalahari sandveld is transversed by major roads many of which are updated to bitumen standard (e.g Trans Kalahari Road, Kang-Hukuntsi Road and Sekoma-Tsabong Road) while many are of secondary standard. This road development made the recently very minimally exploited Kalahari rangelands accessible.

The above factors have indeed made livestock production a viable enterprise in Botswana in general and in the sandveld in particular. To be noted is that factors influencing the development of the livestock sub-sector in Botswana resemble those evident throughout Africa.

General consequences of the above factors are summarised below:

- The post independence period witnessed not only remarkable geographic expansion of livestock holding into the sandveld pastures, but also cattle numbers swelling to levels detrimental to the environment (Campbell, 1985; Hitchcock, 1980; White, 1993 and World Bank Supervision Mission, 1986). Cattle statistic trends for the country are summarised in Table 14.3. Sandford (1977) asserts that both livestock numbers and the area accessible to grazing increased by about two and half times between 1965 and 1976. Mannathoko (1992) further claims that 64 per cent of the cattle are in the sandveld whilst 36 per cent are in the hardveld, and both areas have a 50 per cent share of the goat population. Altogether, Arntzen *et al.*, 1996 indicate that permanent access to the rangeland has increased from 2.4 per cent in 1936 to at least 45 per cent of the country in 1986.
- The advent of the borehole changed the traditional system of semi-nomadist with cattle now being kept at cattle posts near fixed water points and grazing freely on the surrounding communally held land (Veenendaal and Opshoor, 1986). Thus, whereas in the past cattle were brought to the sandveld from the east only seasonally, with boreholes, year round grazing was permitted in the Kalahari (Hitchcock, 1980 and Cooke, 1985). Campbell (1983) notes that the actual number of boreholes in the Kalahari is very high, and that outside national parks and game reserves, few areas remain where they do not exist (see Table 14.4).

Table 14.3: Cattle Trends with Time in Botswana

Year	National Cattle Herd
1934	1 189 000
1939	671 000
1947	966 871
1954	1 140 000
1957	1 310 000
1965	1 481 000
1970	1 350 000
1975	2 017 000
1980	2 390 000
1985	2 459 000
1990	2 696 000
1993	1 820 700

Source: adapted from White, 1993 & MOA, 1993.

Table 14.4: Borehole Statistics over Time in Botswana

Year	Number of Boreholes
1954	1 350
1960	3 050
1965	4 400
1971	4 750
1976	4 500
1979	4 000–6 000
1986	6 000

Source: Moleele (1994).

- Overutilisation hence range deterioration results in communal grazing lands due to concentration triggered by among other things some government policy loopholes such as 'dual grazing rights'. This give TGLP ranchers exclusive rights to graze in their ranches while at the same time they retain their rights to graze in the communal lands.
- Boreholes and the ranching system require large capital investments (high cost of boreholes and ranch equipment) which smaller farmers cannot afford. Therefore benefits of the livestock development subsidies accrue mainly to the wealthy because they are, for example, able to monopolise the development of the new water resources and thus obtain an ever-growing share of the national herd (White, 1993; Veenendaal and Opshoor, 1986). Many of the boreholes in the sandveld though drilled at government expense ended up in private hands (Hitchcock, 1980). Thus, these developments have led to skewed distribution of resources. For example, 45 per cent of the rural families in Botswana own no cattle, while about 7 per cent of the cattle owners own half the national herd (McGowan International and Coopers Lybrand, 1987). Because of the capital inputs that go with ranching, access to extensive grazing resources (ranches) is therefore concentrated in the hands of wealthy farmers.
- There is a problem of assuming exclusive grazing rights on land around privately owned boreholes in communal lands hence restricted grazing by others, often small farmers (Hitchcock, 1980).

Impact of Livestock Development

Development in the livestock sub-sector is widely held to have far reaching adverse implications. First, there are the direct impacts of livestock itself and its related infrastructure on the environment (notably on vegetation and wildlife). Secondly, there are the trickle down adverse effects on those using the said environmental resources to earn a living.

To be borne in mind about the livestock industry in Botswana, as is generally the case throughout Africa, is that two major foreign undertakings have been imposed on the semi arid environments: *boreholes* (permanent water) and *fences*. The direct impact of the livestock on the environment should therefore be assessed in relation to those innovations (see Figure 14.1).

Ecological Impact of Boreholes

Water in arid and semi-arid rangelands, as highlighted earlier, is the principal focus of attention and grazing distribution is governed by it (Squires, 1981). Thus, the influence of the grazing animal on the vegetation is determined by the existence of water points and by the capacity of the animal to forage away from it (*ibid.*). Squires (1981) further reports that a concept referred to as 'piosphere' recognises the importance of the location of water points as a determinant of dispersion and patterns of movement. Andrew, (1988) describes the 'piosphere effect' as the combined grazing, trampling and excretion effects of large herbivore that are concentrated at the water point itself, and decrease with increasing distance into surrounding rangeland.

Large concentrations of livestock around boreholes in Botswana are reported in a lot of literature, particularly during dry periods. This, as highlighted by Campbell, (1983) and Skarpe (1983) is due to the fact that cattle are more or less tied to boreholes, hence bound to one area restricted by their requirements for water at least every second or third day. Unable to move far from water and since the pasture is now grazed permanently rather than seasonally, cattle concentration results in heavy grazing and the accompanying vegetation changes in the vicinity of boreholes. During drought, the area around boreholes deteriorates more because of the continued

Figure 14.1: Summary of the Impact of Livestock Development

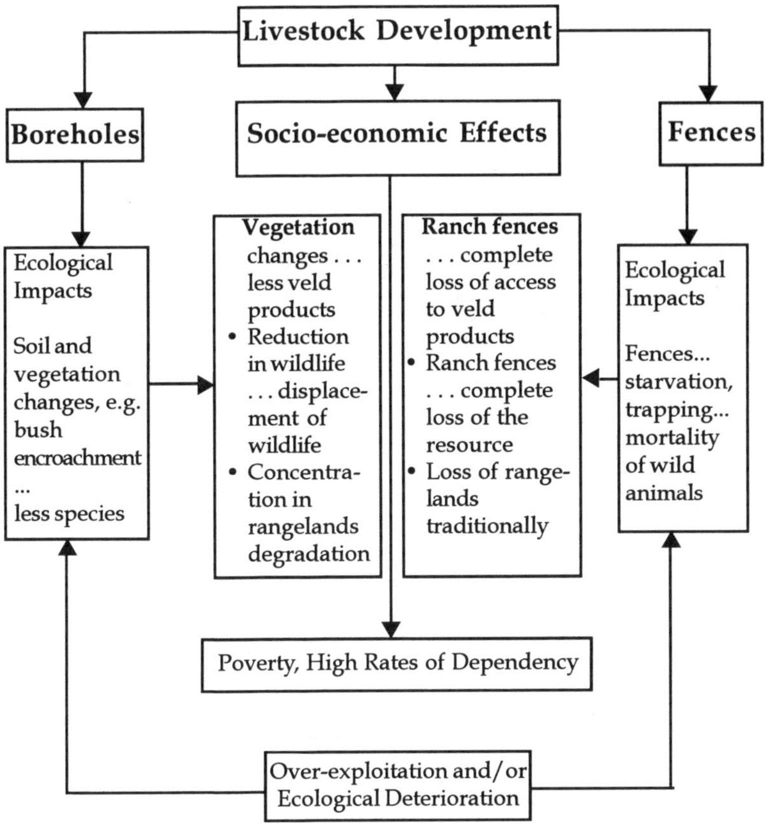

Source: Kgabung (1999).

presence of cattle near it, preferring to drink than to eat (Campbell, 1983). It is widened more by people driving their stock further away than normal, in areas not used in good years. Sometimes livestock boreholes are drilled in these new areas, hoping to alleviate drought effects (*ibid.*). Piosphere effects have been studied in Botswana (e.g. Bogman & Lammers, 1995; Moleele, 1994; Perkins, 1991; Perkins and Thomas, 1993 and Tolsma *et al.*, 1987). In all, some vegetation gradient or changes are realised with distance from the borehole. Perkins, (1991) indicates that the impact of cattle around water sources or borehole may be summarised into five major zones (Figure 14. 2).

Together with Moleele (1994), Perkins (1991) notes that the zones

Figure 14.2: Piosphere Zonation around Borehole-led Cattleposts in Eastern Kalahari Sandveld

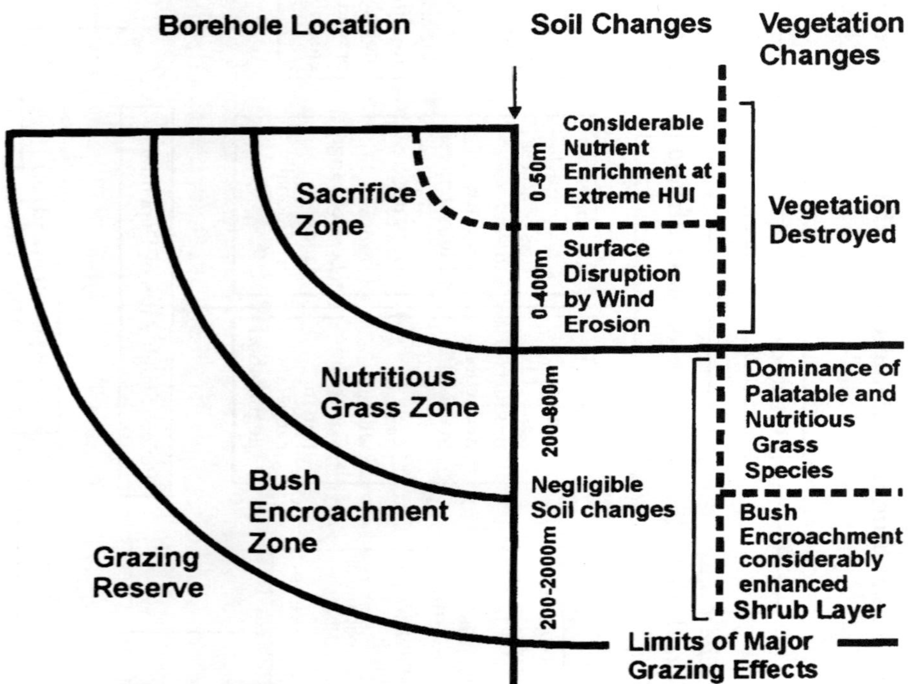

Source: Perkins (1991).

may change due to differences in duration and intensity of use, types of vegetation, soils and of animals using the site. For example, piosphere based research in eastern Kalahari (Perkins, 1991 and Perkins and Thomas, 1993) reveal the following pattern:

- a bare ground or 'sacrifice zone' (0–400m);
- bush encroached zone, where the cover and density of the woody layer has increased (400 -2000m); and
- a grazing reserve, where the impact of large herbivores is not readily apparent (>2000m).

Perkins (1996), also, notes that studies on similar soils, in areas with longer history of cattle - keeping have shown that the bush encroached layer can extend to distance of 7 km from villages. Bogman and Lammers, (1995) piosphere study in Mabutsane indicate a bush encroached area of 5km.

In conformity with these studies, Campbell (1983) and Skarpe (1983) note that more boreholes, cattle posts and villages in the Kalahari exhibit a pattern of widening rings around them and that though the vegetation is poor around the borehole, it improves with distance.

According to Campbell (1985), cattle expansion and subsequently overgrazing endanger traditional plant foods on which, because of regular droughts, all are to some extent forced to rely. He notes that this is especially evident in the vicinity of boreholes where wild plants are disappearing. Hitchcock (1987) also notes that cattle eat some of the same plants as people and also that cattle often browse shoots and fruits hence destroying evidence of some traditional under ground species. In spite of those assertions, it should be realised that although the general ecological changes that follow the introduction of livestock are known, its direct impact upon the distribution, abundance and diversity of veld products is assumed rather than known.

Although cattle are selective grazers, depletion of preferred species caused by heavy grazing prompt them to switch to other species and browsing, hence triggering competition for food resources with wild animals. In the process wildlife, particularly in pans are displaced due to large numbers of livestock, and also the presence of people.

Figure 14.3: Veterinary Cordon Fences and Dry Season Wildlife Movement

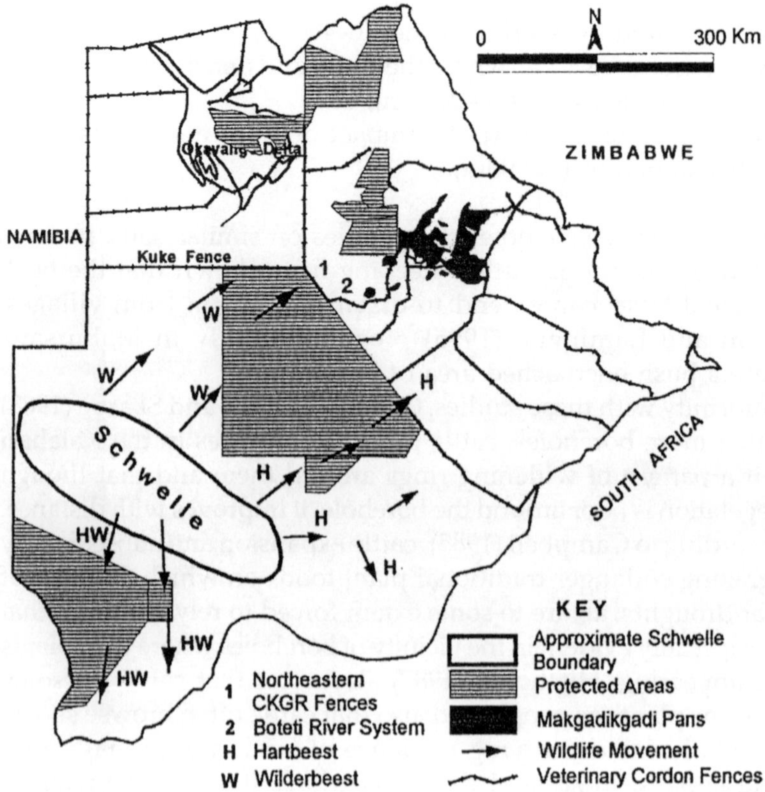

Wildlife Movement Data Derived from Carter, 1983 in Sweet, 1986.

Ecological Impact of Fences

The semi arid savanna of Botswana is today criss-crossed by cordon fences emanating from the EEC,s precautions against importing meat from Foot and Mouth Disease stricken areas (Figure 14.3). The erection of these fences is continuing with the most recent extension resulting from the outbreak of the cattle lung disease in Ngamiland.

There are many indications that these fences have undoubtedly had a crippling effect on the capacity of some highly mobile species (e.g. wildebeest, hartebeest, buffalo, zebra, eland) to cope with drought and the loss of their range (Perkins and Ringrose, 1996; Williamson and Williamson, 1983). Fence damage is in the form of barring of seasonal movement and trapping of animals as has been experienced, for example, with the 'buffalo' fence across the southern and western Okavango tributaries.

During the wet season large numbers of Kalahari species such as wildebeest, hartebeest, springbok, gemsbok and other species frequent the Schwelle area of western Kalahari (see Figure 14.3). The Schwelle is an area of mineralised pans and woodlands and therefore it provides animals with minerals in addition to vegetation of relatively high moisture content, and often surface water in the pans. Many species come to this area to breed at this time of the year, whereby the pregnant and lactating females make use of the available nutritional resources (World Bank Supervision Mission, 1986). As the dry season approaches these species, notably wildebeest and hartebeest leave the Schwelle for water and pasture in the Boteti region and other areas.

Figure 14.3 specifically shows migration patterns of wildebeest which are more regular and well known. Other seasonal movements such as those of hartebeest, gemsbok, springbok, and eland are less predictable and normally more localised (Sweet, 1986). Although less predictable, Carter (1982) in Sweet (1986) demonstrates a more easterly migration towards the Limpopo for hartebeest. Recent surveys show that populations of some species notably wildebeest and hartebeest have been drastically reduced by over 90 per cent during the 1980's drought (Perkins and Ringrose, 1996) (see Table 14.5). The cause of the decline is that these species were and continue to be blocked from their dry season pastures by the veterinary cordon fences such as the Kuke fence and the eastern boundary fence of Central Kalahari Game Reserve (CKGR). Human settlement and cattle accumulation around crucial watering places like Lake Xau prevent access of these migratory animals to water and monopolise the grazing within a large radius (*ibid.*). In addition to displacement by increased livestock and people, the above demonstrates that wildlife distribution patterns and their availability in an area may be due to events or changes very far away.

Table 14.5: Comparison of Kalahari Wildlife Populations in 1978 and 1994

SPECIES	1978	1994	COMMENTS
Wildebeest	315 058	17 934	Widespread populations. Decimated – Longest migration in Africa irrevocably lost. Affected by fences, drought, loss of refugia-cattle post expansion.
Hartebeest	293 462	44 737	Decimated but widespread (recovery unlikely). Affected by fences, drought, loss of refugia-cattle post expansion.
Gemsbok	71 423	85 368	Confined to protected areas – Hunting, habitat replacement e.g. pans lost to livestock.
Eland	18 832	11 757	Fragmented populations – Loss of habitat, hunting.
Springbok	101 408	67 777	Widespread. Concentrated along pan systems. Affected by hunting.
Ostrich	92 286	27 744	Very widespread populations. Affected by fences, hunting, live capture.
Kudu	6 429	7 849	Cryptic browser.
Steenbok	2 122	45 252	Cryptic and solitary. Remains in livestock areas.
Duiker	6 594	13 780	Cryptic and solitary. Remains in livestock areas.

Source: Adapted from Perkins and Ringrose (1996).

The Kalahari species, like elsewhere in Botswana, are as well denied access to pastures by ranch fences resulting from the implementation of TGLP. Hence, Hitchcock, (1978) argues that TGLP overlooked the fact that extensive areas of 'tribal land' zoned for commercial ranches were already occupied by cattle posts and sizeable populations of hunter-gatherers, most of whom have few or no stock. To this effect Perkins and Ringrose (1996: 91) argue that:

> government policy have permitted the widespread substitution of game by livestock in all but protected areas and those regions where the obstacle of finding suitable groundwater cannot be overcome.

Socio-economic Effects

Livestock development through the direct impact of boreholes and fences have resulted in ecological changes (vegetation changes and reductions in wildlife populations). This has reduced the very resource base that has for a long time supported a hunting and gathering and semi nomadic pastoralist subsistence economy. As Hitchcock, (1978: 268) summarises:

> there are more boreholes, more cattle, more people. The number of intact hunter-gatherer groups has visibly dropped. Clearly, there are changes ... bringing about a reduction of hunting and gathering as a viable way of life.

This is widely asserted, but often it is just general statements or assumptions which had not first sought the direct impact of livestock on the said resources and subsequently linking that to the socio-economic effects.

It is noteworthy that, because of the changes or impacts resulting from the development of livestock activities, subsistence users of the range resources have been compelled to seek alternative sources of subsistence (Hitchcock, 1987) or survival strategies. As further noted by Arntzen, *et al.*, (1996: 91):

> people who can not meet their basic needs are particularly adaptive because they have no choice in their struggle for survival.

Literature reveals the following survival means:

- Increased dependence on livestock products and a shift to arable agriculture to make use of the availability of packages such as Arable Land Development Programme (ALDEP),

- Accelerate Rainfed Area Programme (ARAP) and other drought relief packages.
- Most hunter-gatherers and non-stock owners have been absorbed into the cattle post system as lowly paid herders existing at subsistence level (Perkins, 1991). They often get items such as milk and some food rations in exchange.
- Perkins, (1991) also reports dependence on Drought Relief /Recovery food in order to survive. This scheme has been stopped and replaced by the Destitute Relief scheme. With the latter, food rations are confined to those regarded as destitute, while the able bodied can participate in the Labour Intensive Rural Public Works programmes (LIRPW). Opportunities for LIRPW are however, extremely limited in the remote cattle post areas of the Kgalagadi sandveld(*ibid*). The ensuing results are hardship and poverty (Perkins, 1996).
- Increased selling of 'free' natural resources such as fuel wood (Arntzen et al, 1996). This is however, not really possible in the sandveld but in the hardveld.
- Increase in rural-urban migration.

Generally speaking, with modern resource management strategies and the current population trends, the productivity of the Kalahari will steadily decline until it neither supports domestic stock nor wildlife (Campbell, 1983), hence the people, especially those who have no alternative but to rely on its (Kalahari) natural resources for survival.

Range Resource Use Policy/Government Intervention

Government intervention in resource management normally comes in the form of policies and regulations. Most modern legislation that governs resource use today was established after independence. Prior to this, and under the traditional system, as highlighted earlier, all resources were communal property under the jurisdiction of chiefs or with access rights governed by either social ties or permission from the owner.

Livestock Related Legislation

The major livestock development and land reform is the TGLP

established in 1975. It had the objectives of: (1) curbing the problem of range degradation; (2) increasing productivity of livestock; and (3) safe guarding the interest of those citizens owning few or no cattle (Tsimako, 1991). The policy provided for the delineation and development of three distinct categories of grazing land: (1) commercial, where individuals will be granted leases and exclusive rights to areas of land (ranches); (2) communal, where the customary tenure system would remain; and (3) reserve, which would not be allocated but instead set aside for the future (Hitchcock, 1980).

Basically, TGLP was driven by the contention that *:*

> . . . Botswana's rangelands are subject to widespread deterioration, and secondly, that this occurrence stems from the traditional system of land tenure, which must therefore be reformed. (Perkins, 1991: 74).

As is the case throughout Africa's rangelands, the guiding principles behind this modern livestock management strategy (TGLP) are: concepts of the predetermined carrying capacity and 'the Tragedy of the Commons' (Hardin, 1968). The premise with this policy is therefore that reduced stocking rates, fencing and privatisation of land leads to better utilisation of rangelands, hence the establishment of the ranch system. Together with TGLP is the adoption of new Agricultural Policy which seeks to extend the fencing of the land to much larger parts of the communally owned lands.

Hunting Regulations

The present utilization of wildlife centres around policies such as the Wildlife Conservation and National Parks Policy of 1992 (as amended), the Fauna Conservation Act and the Tourism Policy of 1990.

Hunting is administered under the Unified Hunting Regulations issued in 1979. Any citizen can apply for a licence to hunt anywhere in the country, and licenses are allocated by raffles. This means that people who hunt for recreation (mostly urban dwellers) pay the same fees and compete for licenses with rural hunters, most hunting to eke out a living.The issuing of free special licenses have been established under these regulations as well but, this is targeted at Remote Area dwellers.

Consequently many rural people who hunt for subsistence,

including many who do not qualify for free special licenses are unable to get the hunting licenses and are thus deprived of the wildlife resource. White, (1995: 16) therefore rightly argues that:

> while the licensing system may be equal, it is certainly not equitable but contributes to the economic disadvantage of the rural areas. . .

This results in increased levels of poaching and impoverishment of rural subsistence users. White (1995), further notes that many licenses are used to kill more animals than are permitted and that many people apply for licenses with no intention of hunting but to sell them at a higher price.

In short, as Tyler (1995) states, the current wildlife management policies do not give any incentive for people of Botswana to be concerned about the welfare of the wildlife resource. Thus, they do not thoroughly address issues of wildlife conservation and sustainable utilization.

Wildlife Management Areas (WMAs) and Controlled Hunting Areas (CHAs) have been established under the 1986 Wildlife Conservation Policy, within which the aim is to stimulate and sustain utilization of wildlife resources, for the benefit of rural communities and to encourage private sector investment in the resource. In spite of this, livestock holdings are increasingly encroaching into these areas because of lack of enforcement and unclear policy guidelines or objectives.

Policies Related to Gathering of Veld Products

Conservation and management of veld products is vested with the Agricultural Resources Conservation Act of 1972. The Act provides for conservation of all agricultural resources that have been defined to include soils, water, plant and animal life. The Agricultural Resources Board, and district conservation committees have been established as executing bodies. They have specific functions of exercising supervision over utilization of agricultural resources (amongst which are plant resources) and promoting their conservation through issuing of conservation orders and regulations. It should, however, be realised that this Act is very general and has no specific regulations on the use and management of veld products. To this effect, veld products are still regarded as freely available resources to be used at any rate, hence over exploitation. The Agricultural Resources Board has, however, introduced a license system for commercial harvest of veld products such as grapple plant, mainly in the Kgalagadi and Southern districts (Government of Botswana, 1990).

On the realization of over exploitation of veld products and other natural resources, the government also emphasises their sustainable utilization through the 1992 National Policy on Natural Resources Conservation and Development. The policy is still new and has not been fully implemented, hence the continued depletion of veld products. Conservation and sustainable utilization of veld products is also actively promoted by Non-Governmental Organisations (NGOs) such as Thusano Lefatsheng and Veld Product Research. However, their efforts and goals are not fully realised because they are not backed by relevant legislation.

Government policies discussed above (particularly those relating to wildlife and livestock) have generally resulted in changes in resource ownership from common property to state controlled and private ownership by individuals and/or groups. This, to say the least, has restricted access and use of resources by the subsistence or poor communities. There is also the problem of lack of enforcement of legislation hence utilisation of resources continues to be unsustainable.

Conclusion

As indicated in this paper, there has recently been a spread throughout Africa of more intensive agricultural activities into ecologically marginal environments occupied by pastoral people and/or hunter-gatherers. Population growth, modern technology and government policy have been shown to be the driving forces behind this.

In the process, and in respect of the livestock sub-sector, a number of trends have been realised. For example, ownership of animals and the control of rangelands is passing out of the hands of pastoralists and hunter-gatherers. The control over rangelands and water resources is also changing from communal ownership by whole communities or segments of them into the hands of new groups: state, company or private ranches, rich rural or urban investors or new parastatal agencies (Swift, 1982).

Consequently, hunter-gatherers and pastoralists are progressively excluded from the land and water they traditionally used. Not only this but also over exploitation hence an erosion of the very resource base that has sustained them for a very long time. These people are in the final analysis reduced to a destitute level. Thus, while traditional pastoralist societies have coexisted harmoniously with the environment (wildlife and vegetation inclusive), under modern management strategies, livestock production has proved to be destructive, not only to the environment but to the majority of rangelands' human inhabitants. The applicability of modern technology in arid and semi-arid ecosystems therefore remains to be questioned because as it has been shown in this paper, developments under modern management strategies are not sustainable in terms of balancing economic development with sustainability of ecosystems and social equity.

It should however be realised that while the general impact of livestock is known, the extent of this impact on veld products and wildlife, hence on the well being of subsistence users, is scarcely studied.

Recommendations

- Research should be undertaken to investigate how vegetation changes resulting from livestock grazing can be linked with changes in the abundance, distribution and diversity of veld products.
- There is need for documentation of past and present veld products inventories in order to form baseline data that will enable long term monitoring of the livestock impact.
- More research work is needed to investigate the relative impact of the livestock developments on wildlife distribution patterns and their availability to subsistence users, bearing in mind that there are other causes of impacts (e.g. hunting pressure, drought etc). Since most wildlife species are not confined to one area, there is need to place local changes within a regional perspective.

- The present livestock management policy (TGLP), should be replaced by one that recognises or suits the dynamics (ecology) of the semi arid environments and one that also has interests of small farmers at heart. In its formulation, not only should the farmers be consulted but they should as well be involved in the decision-making or conception phase.
- There is need for legislation specifically aimed at promoting sustainable utilisation of veld products and wildlife resources and which takes local communities into account.
- Research on the economic potential of veld products is also necessary.
- Finally, tourism should be recognised as an economically and ecologically viable development option or enterprise in the sandveld.

References

Abel, N.O.J. and Blaikie, P.M. 1989. 'Land degradation, stocking rates and conservation policies in the communal rangelands of Botswana and Zimbabwe', *Land Degradation and Rehabilitation.* 1, 101-123.

Andrew, M.H. 1988. 'Grazing impact in relation to livestock watering points', *Trends in Ecology and Evolution,* 3, 336-339.

Arntzen, J. Fidzani, H. and Tacheba, G. 1996. 'Communal rangelands in Botswana: Less subsistence, more commerce and few beneficiaries', in Odada, E. Totolo, O. Smith, M. and Ingram, J. (Eds), *Global Change and Subsistence Rangelands in Southern Africa: The Impacts of Climatic Variability and Resource Access on Rural Livelihoods,* Report of a workshop, 'Southern African Rangelands Today and Tomorrow: Social Institutions for an Ecologically Sustainable Future', Gaborone.

Bogman, F. Lammers, A.V. 1995. *A Biophysical Research in the Mabutsane Sub-District (Botswana) with Emphasis on Cattle System,* Utrecht University, Netherlands.

Barnes, L.D. 1979. 'Cattle ranching in the semi-arid savannas of east and southern Africa', in B. H. Walker (Ed), *Management of Semi-Arid Ecosystems,* Elsevier Scientific Publishing Company, Amsterdam.

Barnes, L. D. 1982. 'Management strategies for the utilisation of Southern African Savanna', in B.J. Huntley, and B.H. Walker (Eds.), *Ecology of Tropical Savannas,* Springer-Verlag, Berlin.

Bourliere, F. and Hadley 1983. 'Present day savanna - an overview', pp. 312-331, in F. Bourliere (Ed), *Ecosystems of the World, 13 : Tropical Savannas,* Elsevier Scientific Publishing Company, Amsterdam.

Campbell, A.C. 1971. 'Traditional utilization of wildlife in the Kalahari', *Botswana Notes and Records,* Special Edition, 1, 103-113.

Campbell, A. 1983. 'A comment on Kalahari wildlife and the kuke fence', *Botswana Notes & Records,* 13, 107-110.

Campbell, A. 1985. 'The use of wild plants, and drought in Botswana', *Journal Of Arid Environments,* 11, 81-91.

Campbell, A.C. and Child, G. 1971. 'The impact of man on the environment of Botswana', *Botswana Notes & Records,* 3, 91-111.

Child, D. R., Heady, H. F. , Peterson, R. A., Pieper, R. D., Poulton, C. E. and Jones, R. K. (Eds), 1987. *Arid and Semi-arid Rangelands : Guidelines for Development,* Winrock International Morrilton.

Coe, M., Cumming, D.H.M. and Phillipson, J. 1976 'Biomass production of large African herbivores in relation to rainfall and primary production', *Oecologia,* 22, 341-354.

Cooke, H.J. 1983. Environmental change in Botswana: *Which Way Botswana's Wildlife?* Proceedings of the Symposium of the Kalahari Conservation Society, Gaborone, pp. 11-18.

Cooke, H.J. 1985. 'The Kalahari today: a case of conflict over resource use', *Geographical Journal,* 151, 75-85.

Cole, M.M. 1986. *The Savannas: Biogeography and Geobotany,* Academic Press, London.

Collett, D. 1987. 'Pastoralists and wildlife: image and reality in Kenya Masailand', pp. 129-148, in Anderson and Groove (Eds), *Conservation in Africa: People, Policies and Practices,* Cambridge University Press, Cambridge.

Coppock, D.L. Ellis, J.E. and Swift, D.M. 1986. 'Livestock feeding ecology and resource utilization in a nomadic pastoral ecosystem', in *Journal Of Applied Ecology.* 23, 573-583.

Cumming, D.H.M. 1982. 'The influence of large herbivores on savanna structure', pp. 211-228 in B. J. Huntley and Walker (Eds), *Ecology of Tropical Savannas,* Springer-Verlag, Berlin.

Cunningham, A.B. 1989. 'Indigenous plant use: balancing human resources', in B.J. Huntley, (Ed), *Biotic Diversity In Southern Africa: Concepts and Conservation,* Oxford University Press, Cape Town.

Darkoh, M.B.K. 1996. 'The human dimension of desertification in the drylands of Africa', *Journal of Social Development in Africa,* 11, 2, 89-106.

De Leewe, N. P. and Tothill, C. J. 1993. 'The concept of rangelands carrying capacity in sub-Saharan Africa - myth or reality' pp. 77-88, in R. H. Behnke Jr. and I. Scoores, and C. Kerven , (Eds), *Range Ecology at Disequilibrium,* Overseas Development Institute, London.

Devitt, P. 1971. Man and his environment in the western Kalahari: *Botswana Notes and Records,* Special Edition, 1, 50-56.

Dumont, H.J. 1978. 'Neolithic hyperarid period preceded the present climate of the central Sahel', in *Nature,* 274, 356-358.

Ellis, E. and Swift, D.M. 1988. 'Stability of African pastoral ecosystems: Alternative paradigms and implications for development', *Journal Of Range Management,* 41, 6, 450-459.

Fryxell, M. J. and Sinclair, A.R.E. 1985. 'The Sahel of Africa: ecology of a disaster', in Canadian *Journal of Zoology,* 63, 987-994.

Gillion, D. 1983. 'The fire problem in tropical savannas', in Bourliere, F. (Ed), *Tropical Savannas,* pp 617-641, Elsevier, Amsterdam.

Government Paper No. 1 1990. *Botswana National Conservation Strategy, National Policy on Natural Resources Conservation and Development,* Government Printer, Gaborone.

Government of Botswana 1992. *Wildlife Conservation and National Parks Act,* Government Printer, Gaborone.

Graham, O. 1988. 'Disappearing pastures : The privatisation of Africa's rangelands', *Appropriate Technology,* 14, 4, 20-22.

Habitat, 1987. *Global Report on Human Settlement,* Oxford University Press, Oxford.

Hannerwood, K. and Rogers W. A. 1987. 'Pastoralism, conservation and the overgrazing controversy', pp. 111-128, in Anderson and Groove (Ed), *Conservation in Africa : People, Policies and Practise,* Cambridge University Press, Cambridge.

Hardin, G. 1968 'Tragedy of the commons', *Science*. 162, 1243-1248.

Heathcote, R.L. 1983. *The Arid Lands: their Use and Abuse,* Longman, London.

Hitchcock, R. K. 1978. *Kalahari Cattleposts: A Regional Study of Hunter-Gatherers* Pastoralists, and Agriculturalists in the Western Sandveld Region, Central *District, Botswana,* Government Printer, Gaborone.

Hitchcock, 1980. 'Tradition , social justice and land reform', *African Journal Of Law,* 24, 1 1-34.

Hitchcock, R.K. 1987. 'Hunters and herders', *Cultural Survival Quarterly*. 11, 1, 27-30.

Huntley, B. J. 1982. 'Southern African savannas', in B.J. Huntley and B. H. Walker, (Eds), *Ecology of Tropical Savannas,* Springer-Verlag, Berlin.

Huntley, B.J. and Walker, B.H. (Eds), 1982. *Ecology of Tropical Savannas,* Springer-Verlag. Berlin.

Kelly, R.D. and Walker, B.H. 1976. 'The effects of different forms of land use on the ecology of a semi-arid region in south-eastern Rhodesia', *Journal of Ecology* 64, 553-576.

Kgabung, B.B.T. 1999. *The Impact of the Expansion of the Livestock Industry on Subsistence Use of Wildlife and Veld Products in Northern Kgalagadi District, Botswana,* M.Sc. Thesis (Unpublished), Department of Environmental Science, University of Botswana.

Lane, C., Scoones, I. and Toulim, C. 1993. 'Land tenure for pastoral communities', in O. T. Solbring, M.D. and Young, (Eds), *The World's Savannas,* UNESCO and The Parthenon Publishing Group, Paris, New York.

Lamprey, H. F. 1983. 'Pastoralism yesterday and today: the overgrazing problem', in F. Bouliere, (Ed), *Ecosystem of the World, 13: Tropical Savannas.* Elsevier Scientific Publishing Company. Amsterdam , Oxford, New York.

Lewin, R. 1986. 'In ecology, change brings stability', *Science,* 234, 1071-1073.

Machacha, T. 1991. 'The future of the livestock industry in Botswana', in G. Tacheba, (Ed), *Potential for Increasing Livestock Productivity in Botswana,* Proceedings of the Livestock Production Workshop.

Mannathoko, M. 1992. 'The Botswana livestock industry and the environment', *Kalahari Conservation Society Symposium Report,* Kalahari Conservation Society, Gaborone.

Macdonald, I.A.W 1989. Man's role in changing the face of southern Africa: B. J. Huntley, (Ed) *Biotic Diversity in Southern Africa: Concepts and Conservation,* Oxford University Press, Cape Town.

McGowan International and Coopers Lybrand 1988. *National Land Management Project: Incentives/Disincentives Study*, Gaborone, Botswana.

McNaughton, S.J. 1976. 'Serengeti migratory wildebeest facilitation of energy flow by grazing', *Science* 191, 92-94.

McNaughton, S.J. 1985. Ecology of a grazing ecosystem: The Serengeti: Ecological Monographs, 55, 3, 259-294.

Moleele, N.M. 1994. *Ecological Change and Piospheres: Can the Classical Range Succession Model and its Modifications Explain Changes in Vegetation and Soil Around Boreholes in Eastern Botswana?* MSc Thesis. University of Canberra, Belconnen.

Murray, M.L. 1978. *Wildlife Utilization, Investigation and Planning in Western Botswana*, Department of Wildlife, National Parks and Tourism, Gaborone.

Ministry of Agriculture 1993. *Botswana Agricultural Statistics*, Central Statistics Office, Gaborone.

Ministry of Finance & Development Planning, 1991. *National Development Plan 7 (1991-1997)*, Government Printer, Gaborone.

Perkins, J.S. 1991. 'Drought, cattle-keeping, and range degradation in the Kalahari, Botswana' in J.C. Stone, (Ed). *Pastoral Economies in Africa and Long Term Responses to Drought*, Aberdeen University. African Studies Group. Aberdeen.

Perkins, J.S. 1991. *The Impact of Borehole Dependent Cattle Grazing on the Environment and Society of the Eastern Kalahari Sandveld, Central District, Botswana*. Ph.D. Thesis, University of Sheffield, Sheffield.

Perkins, J.S. and Thomas, D.S.G. 1993. 'Environmental responses and sensitivity of permanent cattle ranching, semi-arid western central Botswana', in D.S.G. Thomas, and R.J. Allinson, (Eds) *Landscape Sensitiviyt*, 273-286. Wiley and Sons, Chichester.

Perkins, J.S. 1996. 'Fencing the equity issue: Cattleposts and ranches in the Kalahari Desert, Botswana', *Journal of Arid Environments*, 33, 503-517.

Perkins, J.S. and Ringrose, S. 1996. *Development Cooperation Objectives and the Beef Protocol: The Case of Botswana: A Study of Livestock/ Wildlife/ Tourism/Degradation Linkages*, University of Botswana, Gaborone.

Sandford, S. 1977. *Dealing with Drought and Livestock in Botswana*, Unpublished Report.

Sanford, S. 1983. *Management of Pastoral Development in the Third World*, John Wiley and Sons, Chichester.

Scholes, R. J. and Walker, B. H. 1993. *An African Savanna Synthesis of the Nylsvley Study*, Cambridge University Press, Cambridge.

Sinclair, A.R.E. 1979. 'Dynamics of the Serengeti ecosystem: Process and Pattern', in A.R.E. Sinclair and Norton-Griffiths (Eds) *Serengeti: Dynamics Of An Ecosystem*, The University of Chicago Press, Chicago.

Sinclair, A.R.E. 1983. The adaptations of African ungulates and their effects on community function', In F. Bouliere, (Ed), *Ecosystem of the World, 13: Tropical Savannas*, Elsevier Scientific Publishing Company, Amsterdam.

Solbring, O.T. 1993. 'Ecological Constraints in Savanna Land Use', In O.T. Solbring, and M.D. Young, (Eds), *The World's Savannas*, UNESCO and The Parthenon Publishing Group, Paris.

Skarpe, C. 1986. 'Plant community structure in relation to grazing and environmental changes along a north-south transect in the Western Kalahari', *Vegetatio*. 68, 3-18.

Squires, V. 1981. *Livestock Management in Arid Zone*, Nkata Press, Melbourne.

Sweet, 1986. 'Boreholes for wildlife in the Kalahari – The pros and cons', *Kalahari Conservation Society Newsletter*, 11, 10 -11, Kalahari Conservation Society, Gaborone.

Swift, J. 1982. 'The future of African hunter-gatherers and pastoral peoples', *Development and Change* 13, 159-181.

Tinley, K.L. 1982. 'The influence of large herbivores on savanna structure in Africa', in Huntley and Walker (Eds), *Ecology Of Tropical Savannas*, Springer-Verlag, Berlin.

Tolsma, D.J. Ernst, W.H.O. & Verwey, R.A. 1987. Nutrients in the soil and vegetation around two artificial waterpoints in Eastern Botswana, *Journal of Applied Ecology*, 24, 991-1000.

Tsimako, B. 1991. *The Tribal Grazing Land Policy (TGLP) Ranches Performance to Date*, Ministry of Agriculture, Gaborone.

Tyler, D. 1995. 'Tourism policy and wildlife management policy', *Kalahari Conservation Society Newsletter* 47, 17-18, Kalahari Conservation Society, Gaborone.

Victor, S. 1981. *Livestock Management in the Arid Zone*, Inkata Press, Melbourne.

Veenendaal, E.M. and Opschoor, J.B. 1986. *Botswana's Beef Exports to the European Economic Community: Economic Development at the Expense of a Deteriorating Environment*, Free University, Armsterdam.

Walker, B.H. (Ed), 1979. *Management Of Semi-Arid Ecosystems*, Elsevier Scientific Publishing Company, Amsterdam.

Walker, B.H. Ludwig, D. Holling, C.S. and Peterman, R.M. 1981. Stability of semi-arid savanna grazing systems: *Journal of Ecology*, 69, 473-498.

Walker, B.H. and Noy-Meir, I. 1982. Aspects of stability and resilience of savannas: In Huntley and Walker (Eds), *Ecology of Tropical Savannas*, Springer-Verlag, Berlin.

Warren, A. Sud, Y.C. and Rozanov, B. 1996. 'The future of deserts', *Journal of Arid Environments*, 32, 75-89.

Western, D. and Finch, V. 1986. 'Cattle and pastoralism: Survival and production in arid lands', *Human Ecology*, 4, 1, 77-94.

Westoby, M. Walker, B. and Noy-Meir, I. 1989. 'Range management on the basis of a model which does not seek to establish equilibrium', in *Journal of Arid Environments*, 17, pp. 235-239.

White, R.H. 1993. *Livestock Development And Pastoral Production On Communal Rangelands In Botswana*, The Botswana Society, Gaborone.

White, R.H. 1995. 'Licensing of hunting in Botswana', *Kalahari Conservation Society Newsletter*, 47, 15-17, Kalahari Conservation Society, Gaborone.

Williamson, D.T. and Williamson, J.E. 1983. 'An assessment of the impact of fences on large herbivore biomass in the Kalahari', Botswana *Notes and Records*. 13, pp. 107-110.

World Bank Supervision Mission, 1986. *Current Wildlife Issues in Botswana*, Unpublished Report.

Chapter 15

Globalisation and Wildlife Resource Sustainability in Botswana

J. E. Mbaiwa, M.B.K. Darkoh and J. Nakizito

Introduction

Historically, Botswana's rangelands have supported a variety and abundance of wildlife resources. Large mammals could still be found in remarkable numbers even in the semi-arid environment of the Kalahari up until the 1960s. Campbell (1995) notes that prior to the 1960s, most parts of East and Southern Botswana were occupied by a variety of wildlife species such as elephants, zebras, giraffes and buffaloes. These populations of large herbivores, and indeed, other wildlife species have since disappeared in the south, central, east and western regions of the country. The remaining wildlife resources are currently concentrated in northern Botswana.

A number of studies such as those by Lomba (1991), Campbell (1995), Perkins (1996), Perkins and Ringrose (1996), Albertson (1999) and Mbaiwa (1999) have pointed out that Botswana's wildlife resources are in a state of constant decline. Several factors have been advanced to explain this phenomenon. These include the growth and expansion of human settlements, the development of the livestock industry and its expansion into wildlife areas and ensuing land use conflicts. However, one of the significant factors behind this dramatic decline is the commercialisation and globalisation of the wildlife industry.

Although globalisation, viewed in this chapter broadly as integration into the world market economy, is important in the economic development of a developing country such as Botswana, it has not always had a positive impact on sustainable resource use. As Darkoh (1997) has pointed out, a fundamental problem is that a developing nation such as Botswana has become a part of the global trading system, involving new technologies, new economic demands and new approaches to resource use. The ability or capacity to control overexploitation of natural resources has not balanced these new demands and new capacities to exploit resources. Traditional wildlife management systems have been displaced by European prototypes of management that have been imposed on local people in colonised countries. While the new systems have generally promoted economic development, they have often failed to consider the full social and environmental costs of resource depletion in the developing country.

However, in the last decade, there has been a growing global concern on environmental resource degradation. In 1987, the World Commission on Environment and Development redefined sustainable development as 'a process of change in which the exploitation of resources, the direction of investments, the orientation of technological development, and institutional change are all in harmony and enhance both current and future potential to meet human needs and aspirations' (WCED, 1987: 46). This, therefore, has made economists, sociologists and ecologists to recognise the importance of sustainable development in relation to environmental resource use.

As the concerns of environmental degradation are becoming increasingly global, Botswana has of late responded by introducing policies and measures that are aimed at ensuring the sustainable use of wildlife resources. Examples are the Wildlife Conservation Policy (1986), the Tourism Policy (1990), the demarcation of the country into Wildlife Management Areas and Controlled Hunting Areas (1989) and the Community-Based Natural Resource Projects (1990s). Ironically, the implementation of these policies and measures is mostly to meet the demands of foreign and outside interests. Thus, foreigners largely influence the country's wildlife and tourism industries. As government's stake in wildlife-based tourism is primarily the foreign exchange it can earn from it, and as this industry is mostly dominated by expatriates and foreign investors, the needs of the rural poor communities in wildlife areas are largely ignored. At the same time environmental issues such as wildlife conservation are not given the attention they deserve since foreign investors are much more interested in making quick profits and the government more foreign exchange.

Despite the fact that the globalisation of the wildlife sector in Botswana has wrought serious environmental problems such as the decline of the country's wildlife resource base, it has had a considerable positive effect on the economy of Botswana. This is illustrated by the fact that the country's tourist industry which is mostly wildlife-based contributes significantly to the Gross Domestic Product (about 16 percent). The wildlife industry also remains important to the socio-economic livelihoods of the rural communities living in wildlife areas.

As Botswana's wildlife resources are currently in a state of decline, there is a considerable concern about the prospects of the globalisation process. This chapter, therefore, traces the historical development of the wildlife industry in Botswana. It highlights the impact of commercialisation and globalisation of wildlife resources in the country. The chapter also proposes alternative policies that need to be considered if the sustainability of wildlife resource utilisation is to be achieved in the country.

The Utilisation of Wildlife Resources in Pre-Colonial Botswana

The globalisation of Botswana's wildlife industry in the colonial and post-colonial period cannot be fully understood without an explanation of how the industry was organised prior to European arrival and influence on trade in the country. This section, therefore, makes a brief assessment of the wildlife industry in pre-colonial Botswana.

A common feature of the wildlife industry in pre-colonial Botswana is non-commercialisation and hence non-globalisation, of the industry. History reveals that Botswana was by then characterised by the sustainable use of not only wildlife resources but of most other natural resources in the country. There was knowledge and tradition of natural resource management among the diverse communities or groups of people in the country. The different communities viewed wildlife resources as their property to be utilised for now and the future. Sustainable resource utilisation was inherent within the culture and varying modes of livelihood in the country. The conservation of natural resources was carried out in different regions according to the cultural practices and available resources of the people (Moganane and Walker, 1995).

According to Campbell (1995), until about 1820, wildlife in Botswana was recognised as a single constituent of an environment in which humans formed another related constituent: people and animals co-existed and interrelated. Through this co-existence, humans could maintain environmental stability. In fact, humans were a part of the animal world, and animals were regarded as humans. This could be recognised through totemism, the belief that under certain circumstances, some humans can transpose their spirits into those of animals or take an animal form both before and after death. Animals and birds considered being totems were never killed by their totem holders but respected. In fact, people's relations with the animal world ensured that the latter were not only respected but also protected. The different tribal groups of Botswana varied in totems, ranging from a variety of birds to a variety of animals. For example, the Bangwato and Batawana's totem is the duiker (*phuti*), Bakgatla is monkey (*kgabo*), Bakwena and Bangaketsi is crocodile (kwena), Batlokwa is antbear (*thakadu*), and Barolong is kudu (*tholo*).

The Basarwa (Bushmen) are considered to have been the first people to live in Botswana. Cooke (1985) noted that these hunter-gatherer communities who hunted wild animals and collected wild plants lived in harmony with nature in the Kalahari Desert. Both the Basarwa and all the different tribal groups in Botswana recognised that wild animals were an intimate part of the environment controlled by God and their misuse could bring down God's wrath upon them (Campbell, 1995).

Batswana recognised mutually exclusive hunting groups. There were unwritten laws governing the hunting of wild animals laid down by the chiefs or the community through the *kgotla* system. Everyone had a say in the formulation and implementation of these laws and was as such obliged to observe them. The chiefs actually held wildlife resources in trust for their communities and thus made sure their utilisation on a sustainable basis.

It was forbidden by the chiefs to kill certain wild animals and birds especially those regarded rare and important to society. Such animals were only killed in certain important ceremonies with instructions from the chief. The number to be killed was also given by the chief in order to ensure a continued existence of the species concerned. The leopard is a good example of an animal that was only killed when a new king or chief was installed on the throne. The skin was given to the new king or chief at the installation ceremony as a symbol of honour and respect. The idea of endangered and protected species, therefore, has a long tradition in Tswana custom, it ensured a sustainable wildlife resource utilisation and management.

One aspect common to all Batswana groups is communal hunting. The Tswana speaking groups called it *'letshomo'*. This hunting involved one or more tribal regiments of hunters sent out from the kgotla to an appointed place where they spread out over a large area and slowly close in to trap any game between them. The booty from each game was carefully divided, each participant knowing precisely what portion he should receive. This did not apply to the meat only, but to the skin, and ivory or horns as well.

It can, therefore, be concluded that in pre-colonial Botswana, Batswana placed a highly intrinsic value on wildlife resources. They developed strategies that helped to regulate exploitation of wildlife and ensure that all the people had adequate and readily available natural resources. These strategies were deliberately aimed at the management of living resources for the benefit of the present and future generations, and were deeply enshrined in the culture of the society.

Globalisation of the Wildlife Industry in Botswana from 1850–1885

Although it is generally believed in Botswana that globalisation started in the last three decades of the twentieth century, the process is traceable to the 1850s. This was the period when the country first experienced the influence of European trade that finally contributed to the desire by the British to colonise Botswana in 1885. The traditional wildlife utilisation and management systems in Botswana were severely affected by the arrival of Europeans and their trade expansion in the country (Mbaiwa, 1999). One of the first commodities of trade was ivory. Between 1950 and 1900, ivory from Botswana was being exported via the Cape of Good Hope in South Africa to India and Europe. Campbell (1995) states that this period saw the destruction of vast numbers of animals through hunting and trading activities. According to Campbell (*ibid.*), surplus furs, feathers and ivory were traded north, east and south, across the Okavango, Shashe, Limpopo and Molopo rivers. These activities led to the elimination from the whole of southern and eastern Botswana of species like elephants, rhino, giraffe, zebra, and buffalo. The cheapness of especially ivory and its scarcity in other Southern African states attracted traders into Botswana.

The commercialisation of wildlife especially in the form of trade in ivory and the introduction of the monetary economy changed local communities' attitude towards wildlife resources in general on which they previously depended for their livelihood under a regulated system of usufruct and shared rights access. Local chiefs began to see profit in the commercial activities of hunting by foreigners, turning a blind eye to the subsistence needs and even those values that were so important to their local communities.

White (1995) states that during the 1870s, Francis and Clark's store in Shoshong was exporting annually up to P50,000 (about US$10,417) worth of wildlife trophies from the country. He further states that the actual volume of the trade each year may have involved as much as 5,000 elephants, 3,000 leopards, 3,000 ostrich and 250,000 small fur bearing animals. This information illustrates the fact that European trade expansion in Botswana had tremendous effects on wildlife population.

The expansion of European trade in the country marked the involvement of Batswana in external trade which involved the export of ivory, ostrich feathers, karosses and, to a lesser extent, hippo teeth. Batswana did not regard all these commodities as valuable before the coming of Europeans.

Information from White (1995) indicates that European trade in Botswana introduced and spread the use of guns at an alarming rate in the area. An example is that, by 1874, *Kgosi* Moremi of Ngamiland District personally owned more than 2,000 modern rifles, which he dished out to his people to hunt on his behalf. It is estimated that, there were a total of about 8,000 rifles in Ngamiland District at that time, all these subjected wildlife to terrific pressure. *Kgosi* Moremi relied on his regiments (*mephato*), to hunt for him and provide ivory and ostrich feathers needed for European trade. The Bayei canoeists were expert hippo hunters and were relied upon to produce hippo teeth. White (1995) confirms that the Bayei's traditional hunting system before the introduction of European trade and guns had little effect on wildlife populations. However, after the 1870s, when guns had been introduced into the delta areas, not only hippos suffered from severe hunting but elephants as well. Campbell (1997) notes that rifles were used for hunting larger wildlife species, mainly so that the trophies, ivory, hippo and giraffe hide, furs and ostrich feathers, could be exchanged for more rifles, European-made clothing, ironware, lead, liquor and gunpowder. He further states that Erikson's traders in Ngamiland District had an annual trade of 10,000 rifle barrels and 20,000 pounds of gunpowder during the 1870s.

As a result of the acquisition of guns by Batswana especially by the royal family and the trade in wildlife products with European traders, the tribute system (*sehuba*) became the source of most trade goods used by chiefs in Botswana. Officials or chief representatives travelled throughout their respective territories to collect tribute, and communities paid in whatever commodity they produced. Tlou (1985) states that Bayei paid with hippo teeth, Basarwa with ostrich beadwork, ivory and ostrich feathers. Due to the new European trade in the area, tribute collection became more frequent, systematic and rigorous for the people. Taxation became burdensome. The standing of the chiefs in the provincial administrative system was enhanced because of their role in tribute collection from hunting. They made regular visits to their provinces both to collect tribute, and to check on the movement of traders. Tlou points out that due to trade, the presence of the local chiefs on the people was more felt even in outlying areas than ever before.

The involvement of the people of Botswana in European trade changed the traditional wildlife utilisation aspects in the country. Wildlife species were no longer used only for consumptive and religious purposes, but for commercial purposes as well. The commercialisation of wildlife resources led to the over-harvesting of particular species since the trade was driven by profit making without any consideration for the ecological aspects. The involvement with Europeans also led to chiefs failing to control the use of wildlife as before. European traders became involved in illegal trade and hunting in the country, this contributed to depletion and overharvesting of wildlife resources in Botswana.

Realising the detrimental effects of overharvesting, some local chiefs began to instruct people living near hunting grounds, like the Basarwa in the wildlife areas of the Kalahari and northern Botswana, to drive game away from European poachers or traders who hunted without permission from chiefs. They were also to refuse to guide such hunters to waterholes where game abounded. In northern Botswana, where there are perennial rivers of the Chobe, Zambezi and the Okavango, canoe men were forbidden to transport goods belonging to hunters and traders who violated the Batawana king's orders.

Tlou (1985) states that, while the European traders brought new goods and services which diversified the economy of Botswana, they also caused institutional changes, and increased the degree of political centralisation since Tswana rulers commanded great wealth. It is worth noting that before the British colonial rule was introduced in Botswana, Tswana chiefs directed the socio-economic and political development in their respective territories. With the British assuming political control over the country, there was a shift in the control of natural resource utilisation and exploitation in Botswana. At the turn of the eighteenth century, Batswana hunted mainly using regiments, the chief controlling the trade in valuable trophies. Basarwa continued to hunt with bows and arrows, snares and dogs. During the nineteenth century many changes took place in Botswana, all of which had a great effect on wildlife resources.

Campbell (1995) states that the arrival of Europeans and the spread of the *Mfecane* Wars (these were territorial wars that were fought in the whole of the Southern African region from the 1820s-50s) saw the accelerated use of guns in Tswana societies. At first, guns were used mainly for defence, but soon after the wars, concentrated attention turned to hunting. Regimental hunters of elephants increased, aided by the upsurge of firearms. When the Europeans arrived in the country, at first, chiefs tried to control trade, requiring foreign traders to deal with them through the *kgotla* often held in the capital of each ethnic group, but this soon failed. The arrival of Europeans was responsible for the changing of people's attitudes towards the rights of seniors and collective ownership. The attitudes changed as chiefs lost power and people started to recognise their exclusive property and trading rights. People began to look to wildlife as a commodity for profit and attitudes towards wildlife conservation changed, as commoners could now buy guns and trade with Europeans sometimes directly without going through their chiefs.

In relation to cattle ownership, it is worth noting that prior to European arrival, all cattle belonged to the chief who then gave portions to his people to use as drought power and milk for their respective families. Since cattle belonged to the chief, he could control grazing and watering without much difficulties. However, the European concept of privatisation, and profit and ownership started to make individuals to demand cattle ownership as well.

The chief's strict control over hunting and disposal of the proceeds slackened. Commodification of hunting made people to exploit game mainly for personal gain. They did this to exchange the benefits from wildlife sale for cattle, guns and wagons (Campbell, 1995). The traditional regime all but ceased at this period of time. All who could hunt to make profit did so. The people who suffered were the Basarwa and others belonging to minority groups living in remote areas, which were used and manipulated to produce a constant supply of furs, feathers and skins for their masters (Campbell, 1995). In return, their masters sold these goods to European traders to meet the demands of European markets outside.

In light of the above evidence, it can, therefore, be stated that the process of globalisation and trade in wildlife products affected Botswana on the eve of the British colonisation of the country. With the expansion of European trade into the country, chiefs began to cede and sell their rights to foreigners who traded and hunted on their behalf. As chiefs' central control weakened, individuals in most ethnic groups started having interest in the acquisition of guns mainly for hunting purposes. Since guns provided the means for acquiring new property, commoners also began to grow rich in their own right. Attitudes towards wildlife utilisation changed, people looked on the animals they killed as their personal property, to dispose of as they wished. Hunting for personal gain superseded the in-built social controls which previously ensured utilisation for communal benefit (Campbell, 1995). This marked an end of an era where traditional wildlife management was sustainable in the country. As McNeely (1993) puts it, new and foreign technological innovations tended to favour overexploitation of biological resources and weakened traditional approaches to conservation, especially when a technologically superior group moved into a region occupied by groups with a simpler technology. The collapse of traditional wildlife management systems in the country can, therefore, be explained to be a result of the intrusion of European trade into traditional wildlife utilisation and management systems. With the advent of European firearms and commercial hunting, the previously held attitudes towards sustainable use of wildlife resources changed. People hunted mainly for personal gains buying cattle, guns and wagons with their trophies. Traditional group hunting (*letshomo*) all but ceased, resulting in the deterioration of wildlife resources in the country (Campbell, 1995).

Globalisation of the Wildlife Industry in Colonial Botswana

Botswana came under British Protectorate rule in 1885. The

immediate impact of colonialism was that wildlife resources in Botswana became public property with control vested in the Protectorate Government. Control and access to utilisation values left the hands of the chiefs and their elders and passed onto the colonial Government of Botswana. Realisation of the potential for commercial wildlife use grew much stronger under centralised administration than the territorial use rights for local communities and landholders that have been prevalent in the past in Botswana. Protected wildlife areas especially game parks were created. Under centralised control, *de facto,* open access to the protected areas was limited and new wildlife resource management policies were introduced and promulgated. However, one of the immediate concerns of the new colonial government became the excessive game harvesting due to commercial hunting done by European hunters and traders and the local chiefs. The British colonial administration attempted to control hunting activities by both the Europeans and the local chiefs and their people. Therefore, in response to the issue of overutilisation of wildlife resources, the colonial government introduced laws applying to foreigners, as well as decrees (*melao)* which local chiefs were supposed to apply to their local communities to curb hunting activities. However, these new laws and decrees on traditional wildlife management totally overlooked the traditional customary rites, totems, taboos and the *kgotla* which had existed since time immemorial in these local communities. In certain cases, these regulatory traditional systems were displaced.

As the trade in wildlife products continued, in the latter half of the nineteenth century, one firm in Shoshong exported £50, 000.00 worth of wildlife products annually, at its peak worth some £1.5 million in today's terms (Schapera, 1943, White, 1995). In the late 1880s, there was some decline in trade due to drought and overexploitation of the wildlife resources in the country. This decline is revealed by the combined exports of six stores at Shoshong amounting to an estimated £15,000 per year. Spinage (1991) states that the adoption of sophisticated methods of game destruction, and the changing patterns of land use led to the reduction in game habitat, which, therefore, necessitated game protection.

In 1886, the first statutory game law to curb trade in wildlife products was introduced in colonial Botswana. The law put restrictions on the hunting of certain species of wild animals named and defined as 'game'. It also introduced the idea of suspending hunting in certain months of the year. The law also required people to obtain licences to hunt, capture or sell game (Spinage, 1991).

The introduction of statutory game laws did not solve the problem of unsustainable wildlife utilisation in the country. Hunting unabatedly continued. One notable aspect is that the statutory game laws in colonial Botswana mostly dealt with only one aspect of wildlife conservation, which was the control of hunting. Spinage states that in the colonial period, five major revisions of the principal law (the 1886 law) took place. The revisions introduced little that was new into the basic law of hunting, serving only to make it more detailed.

Probably due to outside influence where protected areas such as Kruger National Park in South Africa, and Albert National Park in the Belgian Congo (now the Democratic Republic of Congo) were established, in 1925, Bechuanaland Protectorate Game Proclamation Act was passed. The proclamation called for the creation of national parks, game reserves and wildlife sanctuaries, whereby wildlife species and areas, or species within defined areas were to be protected. It led to the establishment of Gemsbok National Park (1948), Chobe Game Reserve (1961), Central Kalahari Game Reserve (1961) and Moremi Game Reserve (1965). The British administration further established the Game Control Unit in 1959, to specifically control the harvesting of elephants and the management of protected areas. The law also redefined game into three categories, namely, 'royal game', 'small game', and 'large game'.

The establishment of protected areas over the former hunting and gathering areas of local communities was done without their consent. The site for these protected areas was decided on and chosen by the colonial administration without consultation of the local chiefs and their people. As these areas became declared protected areas, the local communities were removed by force from them, and in some instances, their huts and crops were burnt down, while they were hustled into trucks to give way to wildlife conservation. The new protected area owners denied local people access and benefits from these areas (i.e. wildlife, veld products and visits to their cultural sites). The establishment of the protected areas in Botswana resulted in the emergence of negative attitudes and perceptions of the people towards wildlife conservation. A situation that led to land use conflicts between the traditional socio-economic activities of the local people and the eurocentric wildlife management systems.

In 1961, the Fauna Conservation Proclamation was passed. The law introduced controlled hunting areas, whereby, hunting was restricted by area. The law also declared as unlawful hunting methods such as the use and possession of poisoned bait, poisoned weapons, pitfalls, stakes, nets, gins, traps, set guns, missiles containing explosives, snares, fences or enclosures (Spinage, 1991). It also declared unclaimed parts of the dead animal as government trophies.

White (1995) states that the proclamation provided for the issuing of hunting licences to residents and non-residents who wished to hunt. This proclamation had little impact on subsistence utilisation and hunting of wildlife resources by Batswana residents, but it caused a large increase in sport hunting by non-residents and laid the foundations of the present day wildlife viewing tourism and safari hunting tourism. The law resulted in an influx of wildlife viewers and safari hunters into Botswana from South Africa and Rhodesia (now Zimbabwe) who accelerated the overutilisation of wildlife resources in the country. According to White (1995), the safari hunters hunted game indiscriminately in the hunting areas, resulting in a decline in the use of wildlife resources for subsistence purposes.

With respect to the role played by the chiefs in wildlife management in the colonial era, each local chief was statutorily required by the colonial administration to formulate and pass decrees (*melao*) for their people. The excessive exploitation of game by European commercial hunters made Batswana chiefs to co-operate with the colonial administration in matters of game protection. The chiefs became concerned about the disappearance of game in their own lands due to European trading and hunting activities.

Spinage (1991) states that the chief's decrees probably date well before colonialism was introduced in Botswana, but they intensified after British rule was formally established in Botswana. This was mainly because of the commercial value of wildlife utilisation and possibly missionary influence. However, as already noted, Batswana had traditional wildlife institutions and laws long before the arrival of any European in the country. Although there is little documentation on this fact, unwritten wildlife laws existed in the oral traditions of the people and this knowledge was passed on from one generation to the other by word of mouth.

The decrees passed by the local chiefs were mostly influenced by the colonial administration. As a result, the people of Botswana interpreted them as a step further in denying them rights and access to wildlife resource utilisation in the area. The result was the growing negative attitudes and perceptions towards wildlife conservation in the area. Schapera (1970) states that most of the *melao* were still in force during colonial rule until 1940. Spinage (1991) indicates that the complex game laws of Botswana evolved partly because the ancient customary law of totems and taboos was inadequate to meet the prevailing situation. All these findings show that colonial involvement in wildlife management alienated chiefs and prompted them to fail to enforce the *melao* on their people to effectively ensure sustainable utilisation of wildlife resources in their areas.

Globalisation of the Wildlife Industry in Post-Colonial Botswana

Botswana became an independent country from British rule in September 1966. However, the wildlife policies and institutions of the colonial period were carried out into the new post-colonial era, even though Batswana leaders, rather than the British were now involved in the decision-making process. The wildlife industry continued to be characterised by overharvesting to meet the demands of overseas markets. Von Richter (1969) studied the 1965-67 records of trophy dealers in South-western Botswana. He found that in that period, skins of 40,080 springbok, 155 lions, 131 leopards, 10,445 bat-eared fox, 15,231 other small game furs, 185 fur karosses, together with 6,009 ostrich eggs and 900 kilograms of ostrich features were bought. These purchases reflect legally bought products and do not include the purchases of numerous unlicensed trophy dealers or domestically used products.

The adoption of European ways of wildlife utilisation and management by the new post-colonial government in Botswana is further illustrated by the adoption of the Fauna Conservation Act of 1979. The new law abolished separate regulations for each ethnic group, consolidating them into a single set of regulations applicable throughout Botswana (Spinage, 1991). This law unified all ethnic hunting regulations, making a move to nationalisation of hunting regulations in the country (Thakadu, 1997). The law served mainly to control licensing procedures in the country.

White (1995) states that the impact of these changes on subsistence hunting livelihoods has been damaging, since rural people had to compete for licenses with urban citizens, most of whom are recreational and professional hunters from outside Botswana. As a result, there had been a sharp increase in demand for hunting licenses, out of proportion to wildlife resources. Spinage (1991) states that the centralisation of hunting was a mistake, in that, more hunting licenses had to be sold out to the detriment of wildlife populations. He states that, over 119,000 licenses were being sold each year, with some 33 percent or 40,000 animals being killed annually. The realisation of the potential for commercial wildlife is essential, but current efforts should decentralise control of wildlife resources to local community level.

In modern Botswana, the need to speed up the process of integrating Botswana into the world economy has resulted in wildlife resources playing a major role in the economic development of the country. Botswana's wildlife resources have, therefore, come to be recognised for their consumptive and non-consumptive uses. Globalisation of the wildlife industry has also resulted in the need to develop the tourism sector in Botswana. Actually, the development of Botswana's tourism industry started receiving government attention towards the end of the country's National Development Plan Six in the late 1980s. This led to the formulation and implementation of the Wildlife Conservation Policy of 1986 and the Tourism Policy of 1990.

Prior to the 1980s, the Government of Botswana did not take tourism as an important sector that can speed up the economic development of the country. This is demonstrated by the fact that the Department of Tourism was non-existent. It was only a unit within the Department of Wildlife and National Parks until 1994 when it was elevated to the status of a department. Its main responsibility at that time centred on administrative activities and participation in international fairs that were funded by the European Union. Its growth was hampered by shortage of trained manpower.

As already indicated the need to speed up the process of globalisation of the wildlife industry resulted in the formulation of the Wildlife Conservation Policy of 1986 and the Tourism Policy in 1990. The Wildlife Conservation Policy is the precursor of virtually all the changes that are currently being carried out in as far as new policies and regulations of wildlife conservation are concerned. The Wildlife Conservation Policy was designed to promote economic development of rural areas through the implementation of tourist projects. The policy recognises the potential value of both consumptive and non-consumptive use of Botswana's wildlife resources by the people living in wildlife areas.

The Wildlife Conservation Policy facilitated the formulation and implementation of the Tourism Policy in 1990. The Tourism Policy recognises that revenues generated from tourism should be returned to the rural economy as soon as possible and that wildlife is essential to the tourist attractions of Botswana (Lawson, 1992). The Tourism Policy aims at diversifying the economy of Botswana by outlining initiatives to decentralise control of wildlife resources to districts and local communities and promote rural development in which rural communities utilise wildlife resources for their own benefit.

Apart from the fact that the Wildlife Conservation Policy of 1986 and the Tourism Policy of 1990 were designed to promote Botswana's wildlife and tourist industries to the international world, it is important to note that the two policies were also designed to satisfy the global demands of the concept of sustainable development. Amongst some of the issues, the concept of sustainable development emphasises social equity and that priority should be given to the socio-economic development of the poor. It should be noted that in many parts of the world, the growing numbers of the poor people have inevitably led to the degradation of the environment each day just to make ends meet. The Wildlife Conservation Policy and the Tourism Policy, as a result, encourage the participation of the rural poor communities in wildlife tourist activities through the decentralisation and empowerment of rural communities of Botswana.

The Wildlife Conservation Policy of 1986 also resulted in the country being divided into Wildlife Management Areas (WMAs) and Controlled Hunting Areas (CHAs) in 1989. Mogananne and Walker (1995) state that WMAs, which are proposed to be around protected areas, cover about 22 percent of Botswana's land area, which together with protected areas bring the total wildlife areas to 39 percent of Botswana's land area. The WMA idea encourages community-based wildlife conservation projects and seeks to enhance the conservation of wildlife resources and the biological diversity outside protected areas and private lands, while at the same time affording rural people benefits from wildlife resources in their local areas. The WMAs and CHAs were intended to create a positive attitude and thinking of the local communities towards wildlife resources, thus making them less inclined to poach. The division of Botswana into WMAs and CHAs, therefore, facilitated the implementation of Community-Based Natural Resource Management Projects (CBNRMP). The concept of CBNRMP can be said to be an awakening approach within Southern

African states where the conservation paradigm has been made to shift from a centralised preservationist and protectionist approach to a more integrated approach. This approach recognises the need for the promotion and empowerment of the local communities by linking economic and social development to natural resources management.

The CBNRM concept recognises the incorporation and importance of indigenous knowledge system in natural resources management. The CBNRM proponents regard the local knowledge as the basis for establishing a realistic blend, if not an alternative, to the current inadequate conventional natural resource conservation attempts. The local communities possess a pool of knowledge of the ecosystem in which they live and are involved in adaptive and coping livelihood strategies that ensure that natural resources are used sustainably in their local environment (Darkoh, 1996). It is worth noting that most of the Eastern and Southern African countries that are still endowed with natural resources such as wildlife have since the 1990s adopted and implemented CBNRMP. Mbanefo and de Boerr (1989) say the idea of CBNRMP which was designed to encourage villagers to view wildlife as their own property and utilise and manage them sustainably has been successfully implemented in Zimbabwe since 1986 through the Communal Area Management Programme for Indigenous Resources (CAMPFIRE). Lungu (1990) also claims that CBNRMP is being successful in Zambia through the Luangwa Integrated Rural Development Project (LIRDP) and the Administrative Design for Game Management Areas (ADMADE). Ashley (1995) claims CBNRMP is being implemented through the Living in a Finite Environment (LIFE) programme in Namibia. In Mozambique, CBNRMP is accomplished successfully through the Tachuma Tchato 'Our Wealth' (Cruz, 1995), in Kenya through the conservation of Biodiversity Resource Areas Programme (COBRA) (Masika, 1995) and in Tanzania through the Ujirani Mwena 'Good Neighbourliness'. In giving local communities an active role in the management of wildlife resources, and effectively returning to them some of the control they lost in the past, it is now envisaged that the wildlife resources of Botswana, and those of the rest of the African countries with CBNRMP will become an integral part of the socio-economic development of the rural people, and hence be guaranteed for future generations.

However, the community-based wildlife and tourist projects implemented under CBNRMP in Botswana are currently performing poorly (Mbaiwa, 1999). This is mainly because the development of community-based tourism projects does not match the capability of the local communities. These projects are too elaborate and complicated for the target groups to understand and manage. Since the local communities lack the necessary entrepreneurial skills to run wildlife community-based tourist projects in their area, they are forced to seek the assistance of private sector safari companies who have such expertise. These safari companies are foreign-owned and are able to attract wealthy game hunters from the industrialised states to visit wildlife areas in Botswana. Although the government is expecting that the companies would bring economic benefit and contribute to the rural economic development, little of the money they bring in, is of benefit to the livelihoods of the local people. Safari tours are often paid for outside the country, and the tour operating companies generally use imported food and equipment. The community projects can, therefore, be said to enrich foreign companies at the expense of the local communities. This situation suggests that the CBNRMP in Botswana is not yielding expected benefits to the target groups, instead, it is enriching the few elites and foreign tour operators in the country.

Another important factor that contributes to the poor performance of community-based wildlife tourist projects in Botswana is that decisions are taken based on the demands of the commercial tour operators instead of the local communities who are supposed to be the beneficiaries of the projects. It is important to note that any project that wholly relies on outside influence and support is not likely to sustain itself once foreign aid is withdrawn. The fact that community empowerment especially through training and provision of the necessary skills in wildlife management are left in the hands of the safari operators that happen to win the community hunting tender shows a wrong approach to sustainable community projects. This is an inappropriate strategy since most safari owners lack commitment to develop local skills and knowledge, let alone the sustainable use of the wildlife resources on which their businesses depend.

As already noted, the globalisation of the wildlife industry in Botswana largely depends upon the development of the tourism sector. This is to say Botswana's tourist attraction heavily relies on two products; wildlife resources and the wilderness experience. Although other resources exist such as museums and relics and cultural activities, they have not yet been fully tapped to diversify the country's tourism activities.

Botswana's wildlife-based tourism industry contributes immensely to the economy of Botswana. It contributes about P271.8 million (about US$ 5.6 million) a year in foreign exchange. This, as we have noted, is approximately 16 percent of the country's Gross Domestic Product (GDP). The Bank of Botswana ranks the tourism sector as Botswana's third largest foreign exchange earner after minerals and motor vehicle exports following the slip of the beef industry to the fourth position. Botswana is currently ranking seventh amongst the top 15 African tourist destinations by the World Tourism Organisation, and the third in the Southern African Development Community region after South Africa and Zimbabwe.

The tourist industry pays the following taxes to the Botswana Government and local authorities; corporate tax, airport tax, social tax (paid to the North-West and Chobe Districts), personal or individual tax and sales tax. The tourist industry is also paying the following fees to the Botswana Government and local authorities: park entry and camping fees, vehicle licenses and permits, lease rentals to the Land Boards and local communities and four percent of the audited gross turnover trophy fees and resource royalty to the district councils.

There are many different types of tourist operations, which are mostly located in wildlife areas. These include hoteling, hunting and fishing safaris, photographic safaris and boating. In 1995, the number of tourist visitors to national parks and game reserves stood at 50,000, a two-fold increase over the 1994 figure. Table 15.1 shows that the total number of tourist visits to each park were actually higher, at 76,742. Park and game entrance fees for 1995, which totalled P5.8 million (about US$ 1.2 million), were 34 percent higher than the entrance fees collected in 1994. The marked increase in the number of visitors and revenue collected is attributed to a number of factors, this includes improved management strategies, which among other things, include the introduction of a booking system for national parks.

Table 15.1: Tourist Numbers and Revenue Generated from Protected Areas in 1995

Protected Area	Number of Tourists	Revenue Generated in Pula
Chobe National Park	48,481	3,011,705.00
Moremi Game Reserve	23,504	2,448,316.00
Nxai Pan National Park	915	97,559.00
Makgadikgadi Pans National Park	684	42,504.00
Central Kgalagadi Game Reserve	1,164	110,392.00
Khutse Game Reserve	1,249	71,421.00
Gemsbok National Park	745	53,152.00
Total	76,742	5,835,051.00

Source: Department of Wildlife and National Parks Annual Report of 1996/97.

The Ministry of Commerce and Industry Report of 1997/98 states that direct revenues earned from licences, fees and charges totalled P1,774 million (about US$ 369.6 million), while subsistence hunting licenses contributed P5 million (about US$ 1.04 million). Land Boards and the Department of Lands received P1,705 million (about US$ 223.95 million) for lease fees. Community based tourist organisations together earned more than P1 million (about US$ 0.5 million) through the sustainable utilisation of natural resources. According to the National Development Plan (NDP 8) of 1997/8–2002/3, the number of people employed in tourism related activities such as hotels, airlines, safari companies, handicrafts and transport is estimated at 27,000 people. Mothoagae (1997) notes that there are 47 photographic lodges in the Kwando/Okavango region, excluding seven lodges and hotels in Maun (Maun is the major town that serves as an outlet to the Okavango Delta). Of these lodges 33 are located in Wildlife Management Areas, eight are in Moremi Game Reserve and Chobe National Park and six are in the panhandle and western side of the delta. All these lodges are involved in non-consumptive tourism (mainly photographic) and all use Maun as a base of supply and office administration. About 22 percent of the region, that is, north-western Botswana is zoned as Wildlife Management Areas and of the total, 64 percent is zoned for multi-use which includes hunting, and five percent for exclusive photographic operations. There are, therefore, 16 lodges in the photographic areas (5 per cent), 17 lodges in the multi-use (64 per cent) and ten hunting camps in the multi-use. This land use zonation of the tourist areas was carried out in order to facilitate the process of globalisation of Botswana's wildlife-based tourism.

Globalisation and Botswana's Elephant Problem

The global concern over the degradation of the world's natural resource base has led to the United Nations formulating international conventions and legislation meant to direct the use of natural resources sustainably. One such convention is the Convention of International Trade in Endangered Species of Wild Fauna and Flora (CITES) of 1975. CITES is a United Nations convention designed to control international trade in wildlife products. Emphasis is placed on endangered wildlife species.

In Botswana, CITES has been received with mixed feelings especially because of the trade and utilisation of elephant products. Botswana's adherence to the concept of sustainable development and being a signatory of CITES has led to a trade embargo on Botswana's elephant products at both local and international markets. As already mentioned, CITES regulates trade of species (and items manufactured from their parts) perceived to be in danger of extinction, partially or wholly through international trade. CITES regulates this trade through listing species on three 'appendices' which have restrictions attached to them. Botswana has its elephant resources listed under Appendix I species, which are those threatened with extinction as well as being actually or potentially affected by trade. Export and import of Appendix I species are not allowed by signatory nations except under specialised conditions of non-commercial use, such as scientific.

The international trade embargo on Botswana's elephant products has led to the rapid increase of the country's elephant population. From a population of 45,449 elephants in the 1980s, Botswana's elephants increased to 79,480 by 1995. Botswana's elephant population is concentrated in the north where there is a permanent supply of water. Range ecologists such as Perkins (1996) state that Botswana's elephant population is beyond the range's carrying capacity hence the need to reduce it. Apart from the environmental destruction elephants cause to the northern system especially vegetation, they contribute to land use conflicts experienced in the area. Crop damage by elephants cause land use conflicts between crop farmers on the one hand and wildlife managers and conservationist in Ngamiland District on the other (Mbaiwa, 1999).

However, it is worth noting that the classification of the African elephant population of Botswana, Namibia and Zimbabwe in Appendix 1 species list was temporarily transferred to Appendix II listings of CITES in 1997 during the Harare CITES conference. Appendix II species can be traded only if the exporting nation issues a CITES export permit indicating that the CITES scientific authority considers the trade to be non-detrimental to the species. While the elephant species of Botswana have been temporarily down listed to Appendix II, Botswana is only allowed to trade 25.3 tonnes of ivory with Japan.

The trade is also restricted to stockpiles of ivory acquired from problem animal control and those from natural mortality. This, therefore, means that the live elephants cannot be killed for trade; a situation that is going to make the elephant problem to continue.

Land Use Conflicts in Wildlife Tourist Areas

The issue of the elephant problem in Botswana demonstrates the fact that the process of globalisation and the sustainable use of natural resources fail to harmonise international wildlife conservation legislation with national socio-economic policies and programmes of developing countries. It is common knowledge that the people of Botswana especially those in wildlife areas resist the country's wildlife policies and programmes mainly because of the western approach that is taken in their design and implementation. Local communities happen to experience problems caused by elephants such as crop damage and the destruction of veldproducts (thatching grass, wild fruits and *mophane* worms). Compensation from government happens to be very small and not appreciated by the local communities (Mbaiwa, 1999). Based on this evidence it can be concluded that conservation policies and programmes in Botswana largely ignore the socio-economic activities upon which the livelihood of the rural people depends.

Apart from land use conflicts that result from the elephant

problem, there are several other land use conflicts in Botswana that result from the globalisation of the wildlife industry. Firstly, land use conflicts occur between the livestock sector and the wildlife tourism sector. The expansion of rangelands to accommodate the increasing number of cattle and the need for watering them has resulted in the reduction of areas used by wildlife for grazing. The expansion of the livestock industry into wildlife areas is coupled with the erection of veterinary fences, which are mainly used to control livestock diseases such as foot-and-mouth. These veterinary fences are precautionary measures to control livestock diseases to ensure continuation of Botswana's beef exports to the European Union markets. Veterinary fences introduce an entirely artificial constraint upon wildlife movements between rangelands and water sources. As a result, veterinary fences remain central to any explanation of the dramatic die-off of migratory wildlife species that has occurred in Botswana in the last two decades (Perkins and Ringrose, 1996). The effect of fencing on wildlife areas is illustrated by the example of the Northern Buffalo Fence that has had detrimental effects on wildlife resources in northern Botswana, causing many species to become trapped or die along the fence. Campbell (1973) notes that the visible manifestations of the impact of fences are a build-up of wildlife carcasses along the cordon fence.

Secondly, there is land use conflict between crop production and wildlife management. In East Ngamiland District Mbaiwa (1999) found that the majority (80 per cent) of the local people who were involved in crop cultivation in the last three years experienced crop damage from wild animals. According to Mbaiwa, livestock owners also complained of their livestock being killed or injured by wildlife. Furthermore, according to Mbaiwa, the local communities want to control wildlife resources especially those in protected areas as they consider these to be an integral part of their territorial land resources. The people of Ngamiland District want to take over the control of Moremi Game Reserve and Chobe National Park from the government. These communities see the potential value of the protected areas for settlement, grazing and arable agriculture. The government and the private wildlife tourist sectors see areas occupied by wildlife as a potential source of wealth through hunting and tourism. Conservationists regard the protected areas highly on account of the values of bio-diversity and aesthetics.

At present, local communities do not have any major role in policy making regarding wildlife utilisation and management in the WMAs.

Policies are formulated at the centre and government officials disseminate information to the local people through the *kgotla* about how such wildlife policies and laws should be implemented. The fact that local communities in the country are mostly not involved in wildlife policy formulation means that they have no control over land and wildlife resources. This situation contributes to the unsustainable wildlife utilisation and management in the country. Barnes (1998) states that 'much of the wildlife resources in Botswana are public property and control is vested with central government. Central governments are generally unsuited to any form of active management and are better at passive directing'. As local communities have no control or ownership over wildlife but suffer from livestock and crop damage from the latter, they perceive certain wildlife species as a nuisance. It becomes difficult if not impossible, therefore, for these communities to see any need of involvement in any form of wildlife conservation (Mbaiwa, 1999).

It is important to note that both the colonial and post-colonial wildlife policies in Botswana have advocated for preservationist strategies characterised by top-down approaches, which are divorced from African values and aspirations. It is this strategy that is responsible for brewing conflicts between the Department of Wildlife and National Parks (DWNP), the state office controlling wildlife, and the local communities. The DWNP is viewed with suspicion and mistrust by the local communities in wildlife areas. The harassment, which they face from DWNP when found in protected areas without official permits, increases the hostility between DWNP and the local people. It is important to note that wildlife resources are a resource upon which the survival of local communities is hinged. Cunningham (1985) describes wildlife resources as a buffer to rural communities against poverty, unemployment, health risk and seasonal famine. As local communities fail to obtain direct benefits and have control over wildlife resources in their local environment, they developed an antipathy towards pro-European wildlife conservation attempts which they feel are imposed on them and alienate them from the very source of their subsistence. Adams and McShane (1992) observing this status quo, blame the failure of conservation programmes on their continual reluctance to accept vital links between indigenous economies and the survival of wildlife by building physical barriers between people and wildlife.

Conclusion

It is worth noting that while the wildlife tourism sector generates a lot of revenue for the country, most tourism enterprises remain in foreign hands. Much of the money paid for tours by visitors never arrives in wildlife areas or the country in general. Bookings and payments by tourists for services such as hotel/lodge accommodation are done in either Johannesburg in South Africa or in Europe. It can be argued that even if tourists pay local tour operators in the Okavango Delta area for safari into the wildlife area, a large proportion of this money is used to pay for imported food, equipment and imported expatriate staff. Furthermore, tourists from neighbouring states such as South Africa, Namibia and Zimbabwe usually drive themselves and carry their own food and fuel (petrol and diesel) from these respective countries. Tourism is thus an enclave activity that fails to benefit local communities or Batswana in general except for the country's elite and foreign tour operators.

The fact that the local communities in wildlife areas are getting little or no benefits from wildlife resources suggests that they are most unlikely to appreciate the need to conserve the resource base on which their livelihood depends. This is further complicated by the fact that local people are not included in the decision making process in determining the hunting quota and issuing hunting licences. The implication of this development is that people living in wildlife areas will continue to view regulated hunting through the quota system as foreign to their traditional culture of wildlife utilisation. This may result in illegal hunting by some members of the local communities in wildlife areas.

Since the local communities are not involved in wildlife policy formulation and have no control over land and wildlife resources, it appears difficult for sustainable wildlife management to be achieved in Botswana. For sustainable utilisation of wildlife resources to be possible, control and access to utilisation values should be given to the users, who in this case include the local communities in wildlife areas and other stakeholders in the wildlife industry. Barnes (1998) states that decentralisation of wildlife resources to district and local community levels is vital in that it will empower landholders to take control of the resources and manage them so as to maximise returns. Decentralisation of wildlife resource management to district and local community should also incorporate rights over land use. This has the potential to enhance commitment and ownership of wildlife and land resources by the local people. Once rural communities have access to and benefits from the wildlife resources, they might feel obliged to use them sustainably. The involvement of local communities in wildlife resource utilisation and management is assumed to be an important aspect of sustainable wildlife resource management.

One way in which local communities can be involved in sustainable wildlife utilisation in the country is through the implementation of the Community-Based Natural Resource Management Projects. Although these have been introduced into the country, many are, however, performing poorly and not yielding the expected benefits for the local communities. Continuation of this situation is bound to exacerbate the unsustainable wildlife resource utilisation in the country. The way forward is that means should be found where appropriate and sustainable wildlife and tourist projects are adopted and implemented in wildlife areas. Efforts should be made to give priority to community training as a way of improving rural skills. Small-scale community tourist projects such the establishment of curio gift shops, camping sites, community tour operation, and cultural villages which would provide traditional dishes, accommodation, music and dance (cultural tourism), craft work, walking and boat safaris are a few of the projects that can be encouraged in wildlife areas since they utilise local skills and expertise. These projects appear sustainable and also have the potential to offer needed jobs to impoverished communities in wildlife areas.

The land use conflicts suggest that land use policies and institutions in Botswana are largely reactive to conflict situations instead of being proactive. In the light of all the land use conflicts amongst the various land users and stakeholders in wildlife areas, there is a need for a proactive land use conflict resolution mechanism to be put in place. This can be in the form of an effective institution or policy to specifically deal with land use conflicts in wildlife areas. This initiative needs to be a collective responsibility of all the land users in wildlife areas including especially local communities. Sustainable wildlife utilisation is possible when restrictions agreed upon by all parties are enforced. The local communities should where possible, administer the enforcement of laws. Laws and institutions become sustainable when they come from the local people and other stakeholders and are enforced by them. This reality can, therefore, serve as a basis for the development of a mechanism for land use conflict resolution.

Since livestock production policies in Botswana conflict with wildlife management policies, there should be a review of all current livestock production policies since they do not consider wildlife as a viable land use option. The review should take into account social and environmental implications of agricultural policies in the country. As result, an appropriate Social Impact Assessment (SIA) and Environmental Impact Assessment (EIA) need to be carried out of all existing veterinary cordon fences such as the Buffalo Fence. SIAs and EIAs should precede all future veterinary fences as mitigation measures. Due consideration can be given to remove or realign some of the veterinary fences after a thorough study of the impact has been carried out. However, a lasting solution to land use conflicts can be achieved through the adoption of an integrated land use plan that takes into account the various land users and stakeholders in wildlife areas.

In conclusion, this chapter has revealed that Botswana's wildlife resources are in a state of constant decline. This decline is occurring partly because of the process of globalisation which has promoted the commercialisation of the wildlife resources. Therefore, to stem this problem, it is necessary that wildlife utilisation be governed by the principles and ideals of wildlife and environmental sustainability.

References

Adams, J.S. and McShane, T.O. 1992. *The Myth of Wildlife Africa: Conservation without Illusion,* Norton and Company, New York.

Albertson, A. 1998. *Northern Botswana Veterinary Fences: Critical Ecological Impacts,* Okavango People's Wildlife Trust, Maun.

Ashley, C. 1995. 'Community based tourism as a strategy for CBNRM: options and potentials for achieving conservation and development through non-consumptive tourism in Namibia', in E. Rihoy, (Ed), *The Commons Without Tragedy? Strategies for Community-based Natural Resources Management in Southern Africa,* pp. 56-94, Proceedings of the Regional Natural Resources Management Programme Annual Conference, SADC Wildlife Technical Co-ordinating Unit, Kasane.

Barnes, J. 1998. 'Alternative uses for natural resources in Botswana: wildlife utilisation', in J. Hermans and D. Nteta (Eds), *Botswana in the 21st Century'*, pp. 323-339, Botswana Society, Gaborone.

Barrow, C.J. 1991. *Land Degradation: Development and Breakdown of Terrestrial Environments*, Cambridge University Press, London.

Campbell, A.C., 1973. 'The national park reserve system in Botswana', *Biological Conservation*, 15, 1, 7-14.

Campbell, A.C.1995. 'Utilisation of wildlife in Botswana from earliest times to AD 900', in *The Present Status of Wildlife and its Future in Botswana*, pp. 45-67, Khalahari Conservation Society and Chobe Wildlife Trust Symposium/Workshop, Gaborone.

Campbell, A. 1997. 'A history of wildlife in Botswana', in *Conservstion and Management of Wildlife in Botswana-Strategies for Twenty First Century*, pp. 7-30, Khalahari Conservation Society and Chobe Wildlife Trust Symposium/Workshop, 13-17 October 1997, Gaborone.

Cooke, H.J. 1985. 'The Khalahari today: a case of conflict over resource use', *Geographical Journal*, 151, 75-85.

Cruz, E. 1995. 'Mozambique: community based natural resources management initiatives,' in *The Commons Without Tragedy? Strategies for Community-based Natural Resources Management Programme in Southern Africa* pp. 205-208, Proceedings of the Regional Natural Resources Management Programme Annual Conference: SADC Wildlife Technical Coordinating Unit, Kasane.

Cunningham, A.B. 1985. *The Resource Value of Indigenous Plants to Rural people in Low Agricultural Pontential Areas*, PhD Thesis, University of Cape Town, Cape Town.

Darkoh, M.B.K. 1993. 'Land degradation and soil conservation in Eastern and Southern Africa: a research agenda,' *Desertification Control Bulletin*, 22, 60-68, UNEP Nairobi.

Darkoh, M.B.K. 1996. 'Towards an adaptive and community-based approach to management of natural resources in drylands in Sub-Saharan Africa.' In A. Hjort-af-Ornas (Ed), *Approaching Nature from Local Communities : Security Perceived and Achieved*, pp. 73-99, Motala Grafiska, Motala.

Darkoh, M.B.K. 1997. 'Trade, environment and sustainable development', *Malaysian Journal of Tropical Geography*, 28, 2, 67-82.

Department of Wildlife and National Parks, 1997. *Department of Wildlife and National Parks Annual Report of 1996/97*, DWNP, Gaborone.

Government of Botswana, 1986. *Wildlife Conservation Policy*, Government Paper No. 1 of 1986, Government Printer, Gaborone.

Government of Botswana, 1990. *The Tourism Policy*, Government Paper No. 2 of 1990 Government Printer, Gaborone.

Holmberg, J. and Sandbrook, R. 1997. 'Sustainable development: what is to be done?' in J.Holmberg, (Ed), *Policies for a Small Planet: From the International Institute for Environment and Development*, pp 1938, Earthscan, London.

Lawson, D. 1992. 'The implementation of wildlife utilisation in Botswana: needed administrative changes' in *Which Way Botswana's Environment*, pp. 127-131, Khalahari Conservation Society Symposium/Workshop, Gaborone.

Lomba, R. 1991. *Buffalo Fences: Botswana's Shame. The Death Trap*, Collected Papers on the Buffalo Fence, (Mimeo).

Lungu, F. 1990. 'Luangwa Integrated Rural Development Project (LIRDP) and Administrative Design for Game Management Areas (ADMADE)-Zambia', in A. Kiss (Ed), *Living with Wildlife: Wildlife Resource Management with Local Participation in Africa*, pp. 115-122, The World Bank, Washington, DC.

Masika, A. 1995. 'Kenya; A Review of the Cobra Programme and Potential Linkages with SADC Countries', in E. Rihoy (Ed), *The Commons Without Tragedy? Strategies for Community-based Natural Resources Management in Southern Africa*, pp. 66-89, Proceedings of the Regional Natural Resources Management Programme Annual Conference : SADC Wildlife Technical Co-odinating Unit, Kasane.

Mbaiwa, J.E. 1999. *Prospects for Sustainable Wildlife Resource Utilisation and Management in Botswana: A Case Study of East Ngamiland District*, M.Sc Thesis, Department of Environmental Science, University of Botswana, Gaborone.

Mbanefo, S. and de Boerr, H. 1993. 'CAMPFIRE in Zimbabwe', in E. Kemf, (Ed), *Indigenous Peoples and Protected Areas: The Law of The Mother*, pp. 81-88, Earthscan, London.

McNeely, J.A. 1993. 'People and protected areas: partners in prosperity', in E. Kemf (Ed), *Indigenous Peoples and Protected Areas: The Law of The Mother*, pp. 249-257, Earthscan, London.

Moganane, B.O. and Walker, K.P. 1995. *The Role of Local Knowledge in the Management of Natural Resources with Emphasis on Woodlands, Veld Products and Wildlife: Botswana Case Study*, Forestry Association of Botswana Report, Gaborone.

Mothoagae, M. 1997. 'Wildlife based tourism as the engine of economic growth: policy and institutional issues', in *Conservation and Management of Wildlife in Botswana – Strategies for the Twenty First Century*, pp. 224-231, Khalahari Conservation Society and Chobe Wildlife Trust Symposium/Workshop, 13-17 October 1997, Gaborone.

Perkins, J.S. 1996. 'Botswana: fencing out of the equity issue, cattleposts and cattle ranches in the Khalahari desert', *Journal of Arid Environments*, 33, 503-517.

Perkins, J.S. and Ringrose, S.M. 1996. *Development Cooperations Objectives and the Beef Protocol: The Case of Botswana, a Study of Livestock/Wildlife/Tourism/Degradation Linkages*, University of Botswana, Gaborone.

Schapera, I. 1943. *Native Land Tenure in the Bechaunaland Protectorate*, Lovedale, Alice.

Schapera, I. 1970. *Tribal Innovators: Tswana Chiefs and Social Change 1795-1940*, Athlone Press, London.

Spinage, C. 1991. *History and Evolution of the Fauna Conservation Laws of Botswana*, Botswana Society, Gaborone.

Taylor, R.T., and Martin, R.B. 1987. 'Effects of veterinary fences on wildlife conservation in Zimbabwe', *Environmental Management*, 11, 3, 327-334.

Tlou, T. 1985. *History of Ngamiland: 1750–1906 The Formation of an African State*, Macmillan Publishing Company, Gaborone.

Von Richter, W. 1969. 'Wildlife and rural economy in south west Botswana', *Botswana Notes and Records*, 2, 45-49.

White, R. 1995. 'Licencing, utilisation and public attitudes to wildlife', in *The Present Status of Wildlife and its Future in Botswana*, pp. 79-87, Khalahari Conservation Society and Chobe Wildlife Trust Symposium/Workshop, Gaborone.

World Commission on Environment and Development (WCED), 1987. *Our Common Future*, Oxford University Press, London.

Chapter 16

Community-Based Tourism in Ngamiland District, Botswana: Development Impacts and Challenges

J. E. Mbaiwa

Introduction

The evolution of community-based natural resource management (CBNRM) in Southern Africa from the 1980s was a result of several factors. These include the threat of species extinction due to overutilisation of resources, especially wildlife, through poaching; the inability of the state to protect its declining wildlife resources; land use conflicts; and the need to link conservation and development (Steiner and Rihoy, 1995). In Ngamiland District, located in the north-western part of Botswana, CBNRM implementation has taken the form of community-based tourism. Ngamiland is the major tourist destination in the Botswana mainly because of the abundance of wildlife resources and the wilderness nature of the Okavango Delta which is located within the district. Over 50 000 tourist visit the Okavango Delta every year (Mbaiwa, 2001). Although the majority of these tourists use facilities owned by rich foreign tour operators, a significant number of them make use of facilities owned and operated by the local communities or engage in tourist activities such as hunting and photographic tourism in community hunting areas. Ngamiland covers an area of about 109,130 square kilometers and has a human population of about 110,000 people (Ngamiland District Development Plan, 1997-2003). Community-based tourism in

Ngamiland is carried out through the establishment of Community-Based Organisation (CBO). The establishment of CBOs is a result of the implementation of the community-based management (CBNRM) programme that was proposed by the Wildlife Conservation Policy of 1986. The policy proposed the division of the country into Wildlife Management Areas (WMAs) and Controlled Hunting Areas (CHAs). As local communities obtain direct benefits from tourism, it is hoped that they would be inclined to use natural resources in Ngamiland in a sustainable manner. Through hunting and photographic tourism activities, CBOs in Ngamiland generate substantial sums of money, provided employment opportunities to local people and has resulted in the development of positive attitudes and perceptions of the local people towards natural resource conservation and tourism.

This chapter discusses the development of community-based tourism initiatives in Ngamiland District. It assesses the socio-economic and environmental impacts of community-based tourism and also discusses challenges facing community-based tourism in the district.

Sustainable Development and Tourism

If tourism is to contribute to sustainable development, then it must be economically viable, ecologically sensitive and culturally appropriate (Wall, 1997). The implementation of community-based tourism in Ngamiland District is based on the idea that economic development should conform to the principles of sustainable development. The basic principle of this concept is that of intergenerational equity, which says that our development is sustainable only to the extent that we can meet our needs today without prejudice to those of the future generations. Therefore, the present generation should leave for the next generation a stock of quality of life assets no less than those it has inherited (Pearce et al., 1989). The main thrust of the concept of sustainable development is the utilisation and management of renewable resources for the benefit of today's generations at the same time making the same resources available for future generations (WCED, 1987).

The galvanizing power of the concept of sustainable development lies in the dual benefits it can lead to – it both opens up new opportunities and avoids the trap of trading off environmental goals against economic growth (Darkoh, 1996). If sustainable development is to achieve its potential, it must be integrated into the planning and management systems of individual and corporate business enterprises (Darkoh, 1994, 1999). This suggests that the development of tourism in Ngamiland District should be designed in such a way that it does not lead to an environmental trade off but rather to improvement in environmental condition and human welfare. Sustainable development must therefore give priority to the livelihoods of the poor (WCED, 1987; Redclift, 1987).

Chambers (1986) observes that sustainable development appears to be the terminology of managers, and is not as yet, the terminology of the managed. As a result, in many parts of the world, the growing numbers of poor people have inevitably led to the degradation of the environment each day, just to make ends meet. What this means then is despite the assumptions made about the positive outcomes of the concept of sustainable development, there is need to be aware of and deal with contradictions and other deficiencies of the concept. Scholars such as Redclift (1987), Arnold (1989), Lele (1991) and Warren 1996) argue that the concept of sustainable development has contradictory goals. Nevertheless, it is generally accepted that 'real' development cannot be achieved unless the strategies are sustainable and consistent with social values and institutions. The concepts of sustainable development and ecotourism are interrelated. Apparently, ecotourism is derived from sustainable development. Ceballos-Lascurain (1996: 20) defines ecotourism as 'environmentally responsible travel and visitation to relatively undisturbed natural areas, in order to enjoy, study and appreciate nature (and any accompanying cultural features – both past and present). It is a type of tourism that promotes conservation, has low visitor impact and provides for beneficially active socio-economic involvement of local populations'. The two concepts of sustainable development and eco-tourism advocate for secure livelihoods of the poor in which economic activities like tourism do not degrade the environmental base. Community-based natural resource management projects in Southern Africa, of which Botswana is a part, are based on this broader application of sustainable development, which assumes that economic benefits in the form of community projects and/or household dividends will improve the quality of life of individuals and communities living in natural resource areas. This is expected to make these communities to maintain a sustainable ecological base. The local communities should, therefore, be involved in conserving, protecting and exploiting natural resources in a rationale manner, a goal which sustainable development aims at achieving.

Community-Based Natural Resource Management in Botswana

Botswana's programme of Community-Based Natural Resource Management (CBNRM) is the backbone upon which community-based tourism in Ngamiland District is based. Botswana is not the only country in the region that has so far adopted a CBNRM programme as other Southern and Eastern African states have also done so. The concept of community-based natural resource management can therefore be said to be a relatively recent approach among Eastern and Southern African states, where a general shift is noted in the conservation paradigm from a centralised preservationist and protectionist approach to a more integrated one (Mbaiwa, 1999). This approach recognises the need for the promotion and empowerment of the local communities by linking economic and social development to natural resource management. In Botswana, there are two main overriding policy documents which advocate for the involvement of the rural communities in the sustainable utilization of natural resources. These are the Wildlife Conservation Policy of 1986 and the Tourism Policy of 1990. It is assumed that if the rural communities participate in natural resource utilization and derive economic benefits, this will cultivate the spirit of ownership and will ultimately lead to the sustainable use of natural resources by the rural poor communities.

The Wildlife Conservation Policy of 1986 is generally seen as a blueprint upon which CBNRM in Botswana is based. It proposed the division of all the nine/ten districts in Botswana into Wildlife Management Areas (WMAs) and Controlled Hunting Areas (CHAs). CHAs form smaller management units within WMAs. Botswana is divided into 150 CHAs (Modise, 1996). Most of the CHAs in Ngamiland District are generally adjacent to wildlife protected areas such as Moremi Game Reserve. The idea of Wildlife Management Areas (WMAs) in Botswana arose from a need for conservation and controlled utilisation of wildlife and other natural resources outside national parks and game reserves, along with the desirability of creating buffer zones between parks and reserves and areas of more intensive land use. A Wildlife Management Area (WMA) can thus be defined as an area where wildlife utilisation and management is the primary form of land use. Other types of land use are permitted provided they do not prejudice the wildlife populations and their utilization (Joint Venture Guidelines, 1999). Winer (1995) states that community natural resource management projects in WMAs, as a principal aim, seek to return custodianship of natural resources to the local communities in order for conservation to be linked to rural production systems that generate wealth and not continue to be viewed as being in conflict with them.

The CBNRM concept recognises importance of incorporating indigenous knowledge system in natural resources management. Indigenous knowledge (also referred to as local knowledge, indigenous technical knowledge) denotes a cumulative body of knowledge generated and evolved over time, representing generations of creative thought and actions within individual societies in an ecosystem of continuous residence with an effort of coping with the ever changing agro-ecological and socio-economic environment (Fernandez, 1994; LaDuke, 1994; Laws and Luning, 1996; Serrano, 1996; Warren, 1996; Darkoh, 1996). Proponents of CBNRM regard indigenous technical knowledge as the basis for establishing a realistic blend, if not an alternative, to the current inadequate conventional natural resource conservation attempts. As Darkoh (1996) puts it, the local communities possess a pool of knowledge of the ecosystem in which they live and are involved in adaptive and coping livelihood strategies that ensure that natural resources are used sustainably in their local environment.

Structure and Management of Community-Based Organisations in Ngamiland District

The implementation of Community-Based Natural Resource Management (CBNRM) programme in Botswana has enhanced the participation of rural communities in tourism activities in Ngamiland District. These communities have formed Community-Based Organisations (CBOs) which co-ordinate their involvement in tourism ventures. The Tawana Land Board, the land authority in the district, leases controlled hunting areas (CHAs) to different CBOs for hunting and photographic tourism activities. The establishment of CBOs and the participation of local communities in tourism in Ngamiland District are an indication that conventional methods of natural resource management on their own are not adequate for the sustainable use of natural resources in the area. They are unable to solve the growing problems of natural resource decline such as wildlife. This is partly because conventional methods of resource management are relatively imposed on the local people, and lack appropriateness to the local situation. Hence, they largely remain unacceptable to the majority of the rural people.

The re-introduction of indigenous knowledge into natural resource management through CBOs in Ngamiland District is, therefore, hoped to involve local people in the decision-making process and implementation of resource policies and programmes. In giving local communities an active role in the management of natural resources, and effectively returning to them some of the control they lost in the past, it is now envisaged that the natural resources of Ngamiland District will become an integral part of the socio-economic development of the rural people, thereby guaranteeing that future generations will also benefit from the same resources. This approach is in line with the concept of sustainable development which puts emphasis on intergenerality and social equity. Community-based tourism should therefore demonstrate the extent to which traditional and modern natural resource management practices can be fused to achieve a higher degree of sustainability in Ngamiland.

CBOs are given resource management rights to CHAs leased to them by the Land Board together with security of tenure through fifteen-year leases. The CBOs have three options on which they can operate. These are: managing the CHAs themselves, sub-leasing the resource use rights to a safari company at a fee, and entering into a joint venture partnership with a safari company by holding shares. Lack of expertise and the necessary capital to invest in tourism activities by CBOs have so far favoured the sub-leasing of community CHAs to safari companies by most of the CBOs in Ngamiland except for the Khwai Development Trust. As a way of becoming more transparent and fair, CBOs select lessee or joint venture partners through an invitation to tender. A technical committee comprising of central and local government officials (e.g. from the Departments of Wildlife and National Parks, Tourism and Land Board) assesses and ranks the technical proposals of all tenders received, which are then submitted to the community at the *kgotla* (a traditional public meeting) for the final decision through a public vote on the selection of a lessee or joint venture partner.

In terms of membership, duties and organization, CBOs are guided by their respective constitutions. Generally, all citizens staying in a particular village who are over 18 years of age automatically become general members of their CBO. The Village Trust Committee (VTC) and the Board of Trustees coordinate CBO activities. Members elect Village Trust Committees (VTC) and Board of Trustees at the *kgotla* (public meeting) every two years. CBOs that comprise one village such as those of Khwai, Mababe and Sankuyo do not necessarily have a VTC but a Board of Trustees only, while those that comprise more than one village have both VTCs and Boards of Trustees. The VTC represents the interests of a particular village on the Board of Trustees, which is the supreme board in each CBO. It also reports to the village the deliberations of the CBO and discusses, identifies and manages village tourism projects. VTC also consult members on the usage of the revenues earned by the CBO from tourism activities. The Board of Trustees conducts and manages all the affairs of the CBO on behalf of its members. These affairs can include signing of legal documents such as leases and contracts with safari companies and maintaining close contact with the CBO lawyers. The Board of Trustees also keeps CBO records, financial accounts and reports and present them to members at annual general meetings.

The *kgosi* (chief) and the Village Development Committee (VDC) are ex-officio members of VTCs or Board of Trustees. The k*gosi* is the traditional head of a village and is assisted by the VDC in carrying out development projects in his village. This then explains why, in the absence of the *kgosi,* the VDC chairman acts on his behalf. The inclusion of the *kgosi* and the VDC in the Board of Trustees and VTCs as ex-officio members gives CBOs credibility and support from village authorities. All CBOs operate under a constitution and are registered as trusts according to the laws of Botswana. Therefore, the structure, organization and operations that CBOs demonstrate provide some appropriate atmosphere under which eco-tourism can successfully be implemented in Ngamiland district.

Socio-Economic Benefits of Community-Based Tourism

The following CBOs are discussed briefly to illustrate the socio-economic benefits of community-based tourism in Ngamiland District: Sankuyo Tshwaragano Management Trust (STMT), Okavango Community Trust (OCT) and the Khwai Development

Trust (KDT). Table 16.1 shows seven of the ten CBOs in Ngamiland District, the various village(s) that comprise each CBO, controlled hunting area(s) allocated to each village, types of tourism activities that each CBO is engaged in (mostly cases hunting and photographic tourism activities dominate) and the amount of revenue that each CBO has so far generated.

The Sankuyo Tshwaragano Management Trust (STMT) is one of the first Community-Based Tourism Project (CBTO) to be established in Ngamiland District. It was established in 1996 and it comprises of one village (Sankuyo) which is located in the south-eastern fringes of the Okavango Delta. The village and its trust have been allocated NG 33 and 34 to be used for photographic and hunting purposes, respectively. The STMT has since sub-leased these two community areas to Game Safaris and Crocodile Camp Safaris. Since the trust started operating, it generated P285,000 in 1996 and 1997, P345,000 in 1998, P562,800 in 1999 and P595,400 in 2000. This revenue accrued from lease rentals, selling of animal quota and other community development projects safari operators promise to fund.

Table 16.1: Community Based Tourism Organisations in Ngamiland District

Name of CBO	Village(s) involved	Location	Activity	Revenue Generated Year / Amount	Comments
Sankuyo Tshwaragano Management Trust (STMT)	Sankuyo	NG 33 & 34	Hunting and photo-graphic	1996 285 000 1997 285 000 1998 345 000 1999 562 800 2000 595 000	• Househlds got P200 each in 1997 • built 40 Enviro loo toilets for 40 households (not operating) • purchased a Land Cruiser, built an office and social centre • future plans to build traditional village • Kiosk stopped operating due to losses

Table 16.1: Concluded

Okavango Community Trust (OCT)	Seronga Eretsha Gunotsoga Beetsha Gudingwa	NG 22 & 23	Hunting and photo-graphic	1996 1997 1998 1999 2000	446 000 468 050 625 650 652 340 686 240	• bought a plot at Seronga with shop, bottlestore, guest house and office (shop & bottlestore closed due to mis management of funds • built a shop at Beetsha (shop operating) • Kiosks in Gunotsoga, Eretsha & Gudingwa closed due to mismanagement of funds • purchased 2 Land Cruisers, 1 truck & 1 motor boat (motor boat not operational due to breakdown)
Mababe Zokotsana Community Trust (MZCT)	Mababe	NG 41	Hunting and photo-graphic	2000	675 000	• bought a vehicle
Okavango Kopano Mokoro Community Trust (OKMC)	Ditshiping Boro Xaxaba Daunara Xharaxao Xuxao	NG 32	Hunting and Photo-graphic	1998 1990 2000	680 000 710 000 1 100 000	• bought a Land Cruiser • progress to invest in other activities not forthcoming
Khwai Development Trust(KDT)	Khwai	NG 18 & 19	Hunting and photo-graphic	2000	954 000	• in the process of buying a vehicle
Okavango Jakotsha Community Trust (OJCT)	Etsha 1 Etsha 6 Etsha 13 Jao, Ikoga	NG 24	Photo-graphic only	–	–	• still waiting for head lease from Tawana Land Board hence not operational
Cgaecgae Tlhabologo Trust (CTT)	Cgaecgae	NG 4 & 5	Hunting and photo-graphic	1998 1999 2000 2001 2002	70 750 105 000 315 167 420 336 430 336	• office, campsite, trust hall, craft shop and guest house are operational • bought a Land Cruiser •future plans to operate a bakery and a vegetable garden

Source: Mbaiwa (2000).

The direct household benefits in Sankuyo include the construction of forty Enviro Loo toilets and the distribution of P200 to each household in 1997. Shared community benefits include the purchased Land Cruiser vehicle for community use and the construction of community trust office and a community social centre. Out of a village population of 389, about 55 people in Sankuyo were employed in the community trust in 1998 (Mbaiwa, 1999). However, more employment opportunities and financial gains are expected once the construction of the CBO traditional village is completed. This traditional village includes, among other things, the construction of two safari camps, one in the hunting area and the other is in the photographic area. When completed, the photographic camp is expected to provide services to photographic tourists such as traditional accommodation in traditional huts, traditional dishes, dance and music while the camp in the hunting area is expected to provide accommodation and hunting activities only. The move by the people of Sankuyo means that in the near future, they will not sub-lease their community areas to any foreign hunting and photographic safari company but will run the tourism businesses for themselves. This, to some extent, demonstrates a move towards a culturally and locally controlled tourism as is expected in eco-tourism.

The Okavango Community Trust (OCT) comprises several villages in the panhandale, that is the upper Okavango River area. These villages are Seronga, Eretsha, Gunotsoga and Gudinwa. The OCT was established in 1996 and was allocated NG 22 and NG 23. Like STMT, OCT has sub-leased its CHAs to hunting and photographic safari operators (John Calitz and Machillete Bird Safaris). The sub-leasing of the OCT CHAs generated P 446,000 in 1996, P468,050 in 1997, P625,650 in 1998, P 652,340 in 1999 and P686,240 in 2000 for the trust. Individual benefits to Board members include a sitting allowance of P500.00 for each member per meeting.

In addition to being part of the OCT, the village of Seronga operates its own trust, known as the Okavango Poler's Trust. The Okavango Polers Trust (OPT) was formed with intentions of establishing a viable eco-tourism operation in the panhandle. The trust is involved in canoe (*mekoro)* safaris in the Okavango River and operates a camping site. M*ekoro* safaris are meant to offer tourists an experience of some of the traditional methods of transportation in the Okavango River and Delta. Game viewing and camping in the wilderness are some of the tourist activities offered during *mekoro* safaris. Traditional dishes, dance and music depicting the Wayeyi culture are offered as entertainment to tourists at campsites. From April 1998 to September 2000, the trust received 9586 overseas tourists and generated about P639,118 in the financial year ending in March 1999. Tourists are charged P120 for a one-day *mokoro* safari. The trust also charges P20 per person daily in the community camping site. The trust has so far provided employment to 75 people who work as polers, 20 casual polers, eight boat drivers, six women (who prepare traditional dishes), two office workers, two co-ordinators and one business manager. In addition to these eco-tourism developments, the OPT is currently developing a safari lodge that is also expected to offer traditional services such as music, dance and dishes once completed.

The Khwai Development Trust (KDT) comprises of one village Khwai, which is located on the eastern fringes of the Okavango Delta. The trust started operating in 2000 in NG 18 and 19. The Khwai Development Trust is the only CBO in Ngamiland District that did not sub-lease its community CHAs to any safari operator. Instead, it runs the tourism business own its own. The trust sold its community wildlife quota at an auction sale and made P1, 500, 000.00 in year 2000. Like the STMT, the KDT is also involved in the construction of two safari camps in their CHAs. Employment opportunities have as a result been created for the local people. By April 2001, there were about 78 people, out of a total village population of 429, employed to perform various jobs at these camps. These include 26 people who are employed as builders and 22 employed to construct the roads and tourist trails. In respect to the spot guides, waiters, scullery, cooks and night watchmen, they were four people in each while six were employed as housekeepers. In terms of remuneration, builders were each paid P800 per month, those constructing roads were paid P450 per month, night watchmen P500 and supervisors were paid P750 each.

Socio-Economic Problems of Community-Based Tourist Projects

The major problems faced by community-based tourism activities and CBOs in Ngamiland District is that the concept of Community Based Natural Resource Management (CBNRM) in which Community-Based Organisations (CBOs) are based, is relatively new and generally lacks understanding by the local communities. This is shown by some local communities proposing and engaging in community tourist projects that are too elaborate and complicated for them to understand and manage. For instance, almost all community-based tourist projects have bought Land Cruiser vehicles which have been turned into a free public transport system for the respective villages. The communities have also tried to run businesses such as kiosks, bottle stores and guesthouses but have often failed and closed down due to lack of management and investment skills.

As already noted, lack of entrepreneurial skills to run tourism projects force CBOs to seek the assistance of private sector safari companies who have such knowledge and experience. This problem is exacerbated by the fact that there is a lot of focus on high-income tourist projects requiring high loads of organisation and management that are not easily available in rural areas. Community-based tourist projects would be successful if focus was put on viable and appropriate projects that utilize the locally available skills in rural areas. Although the government is expecting that the safari companies should provide training on how to run the tourist projects and encourage rural development, the multinational nature of safari companies has been of little economic benefit to rural development in Ngamiland District.

Sub-leasing of CHAs by CBOs has resulted in some safari hunting operators exploiting the situation. Their ability to market themselves in Europe where most of them originate makes them obtain huge profits from animal quotas they buy cheaply from the communities in Ngamiland. For example, in 1998, a single elephant that a safari hunting operator bought at P 22,000 (US$ 4,500) from the community was sold at US$ 50,000 (P 250,000) to spot hunters from Europe (Mbaiwa, 1999). This figure respectively rose to P40,000 (US$ 8,000 and US$ 80,000 (P400,000) in 1999 and 2000. Based on these data, it can be seen that community-based tourism can be exploited to enrich foreign companies instead of benefitting the local communities more.

Where local communities have managed to sell their hunting quota or sub-leased their CHAs to safari hunting companies, sums of money have accrued to CBOs as shown in Table 16.1. Lack of entrepreneurial skills by local communities has led to a situation where these funds are either being kept in the bank without being re-invested or have been misused (misappropriated). Failure to re-invest community funds by CBOs in Ngamiland District is largely due to community-based tourism decisions being made as a response to the demands of the commercial tour operators instead of the local communities who are supposed to be the owners of the resources. Local communities lack business negotiating skills when making deals with safari operators and due to this deficiency, unscrupulous safari operators often cheat them. To overcome this problem, local communities rely on outside assistance for advice and for the organisation and management of community tourist projects. For example, local communities mostly depend on the Department of Wildlife and National Parks (a government body) and foreign donor agencies on advice concerning the direction which community-based tourist projects should take. While such advice is necessary, it should be noted that projects, which wholly rely on outside influence and support, are likely to collapse when foreign support is withdrawn. This means that if the capacity of the local communities is not given a priority, community-based tourism is likely to decline once outside assistance is finally withdrawn.

Environmental Impacts of Community-Based Tourism

The implementation of community-based tourism in Ngamiland District has resulted in some positive environmental impact. Earlier studies by Mordi (1991) indicate that Batswana have negative attitudes and perceptions towards wildlife conservation. However, a recent study by Mbaiwa (2001) reveals that there is growing positive attitude and perception towards natural resource conservation and tourism in Ngamiland District. This change of attitude can be attributed to the socio-economic benefits that local communities now derive from tourism in the area. Studies conducted in Zimbabwe (Murphree, 1993; Mwenya *et al.*, 1993) and in Namibia (Ashley, 1995) indicate that people living in wildlife areas tend to put more value and perceive wildlife as a valuable resource when they derive benefits from it; this results in using wildlife sustainably. Findings by Mwenya *et al.*, (*Ibid.*) suggest that people's attitudes are largely based on the personal or community ownership they attach to natural resources such as wildlife found in their area. Rihoy (1995:15) also notes, 'for a community to manage its resource base sustainably it must receive direct benefits arising from its use. These benefits must exceed the perceived costs of managing the resource and must be secure over time'.

The need for community participation in tourism and the growth of tourism in Ngamiland District has generally resulted in the promotion of natural resource conservation strategies and policies being developed by government. Natural resources that are protected include wildlife, veld products and the Okavango River and Delta. Some of the strategies used include formulation of government policies such as the Wildlife Conservation Policy of 1986, the Tourism Policy of 1990, the National Policy on Natural Resource Conservation and Development of 1990, the savingram on Minimum Conditions for Communities seeking Hunting Quotas and Resource Leases of 1995, the Community-Based Strategy for Rural Development of 1997, the Community-Based Natural Resource Management Policy (draft) of 1998, the Joint Venture Guidelines (revised edition) of 1999, the Joint DWNP and ARB CBNRM Policy of 1999, the National Policy on the Use and Management of Natural Resources (draft) of 1999 and the Wetland Policy (draft) of 1999. Most importantly is the conservation of the Okavango Delta, which is the centre of tourist activities in Ngamiland and was in 1997 listed as a wetland of International Importance under the Ramsar Convention. A common feature that exists in all these strategies is that of the promotion of sustainable natural resource use, not only in Ngamiland District but also in the rest of the country. It should of course be observed that it is one thing to formulate and adopt a policy, and another thing to implement it.

Challenges and Prospects of Community-Based Tourism Projects

The development of tourism in Ngamiland District is characterised by two types of tourism that compete with each other for the same resources: community-based and foreign dominated elite. The latter is more organised, developed and obtains higher benefits than the former, a factor which causes resentment and antagonism between the two. While this remains a challenge to both two types of tourism, the challenge becomes more pronounced to the local communities who believe that natural resources found in Ngamiland District are historically theirs and should therefore have a higher share of the benefits accruing than any other new comer to the region. This has as a result been the source of conflict between the two types of tourism in the district.

The predominately foreign elite tourism industry that has so far developed in the Okavango Delta and in Ngamiland District has been referred to in recent literature as 'enclave tourism' (Britton, 1981; Ceballos-Lascurain, 1996). Ceballos-Lascurain (1996) defines enclave tourism as tourism that is concentrated in remote areas in which the types of facilities and their physical location fail to take into consideration the needs and wishes of surrounding communities. The goods and services available are beyond the financial means of the local communities and any foreign currency generated may have only a minimal effect upon the economy of the host location. The majority of tourist facilities such as hotels, lodges and camps in Ngamiland District are foreign-owned and established to meet the needs and interests of foreign tourists. Ndubano (2000) states that about 95 per cent of the accommodation and transport sectors in Maun have foreign involvement, with 60 per cent of them being 100 per cent foreign owned; 35 per cent of them are jointly owned between locals and expatriates, only 1 per cent is 100 per cent locally owned. However, data from the licensing office in the Department of Tourism indicate that out of 103 tourism-related businesses registered in Maun and operational in Maun and in Ngamiland District, 16 (15.5 per cent) are locally owned, 36 (35.0 per cent) are jointly owned (between Batswana and non-citizens) and 51 (49.5 per cent) are non-citizens owned. This suggests that 87 (84.5 per cent) of the tourism-related companies registered in Maun and operating in Ngamiland District have direct foreign involvement. Since tourist facilities are largely foreign owned and prices charged are very high and usually in American dollars, they are unaffordable for the majority of not only the people of Ngamiland District but of Botswana as a whole. The only benefit that local communities derive from these high quality and expensive facilities is that of employment which is characterised by poor quality jobs and low salaries. Mbaiwa (2001) observes that the jobs occupied by local people are of low cadre (mainly manual), where they are employed as drivers, maids, cleaners, night watchmen, gatekeepers and cooks. For example, in Sankuyo Village, local residents are employed in safari companies such as Crocodile Camp safaris and provide manual labour for tasks such as skinning of wild animals in the hunting seasons and tent keeping (house keeping). In Khwai, the local people who work for Tsaro Game Lodge and Khwai River Lodge work as cooks, housekeepers and drivers.

The majority of local people hold poor quality jobs that attract low salaries in the tourism sector while their foreign counterparts, who are few in number, hold managerial positions such as professional guides, managers and accountants; these positions attract higher salaries. A study by Ndubano (2000) in Maun indicates that there are only six (14.3 per cent) local managers out of 42 established posts in the accommodation facilities and 36 (85.7 per cent) of the remaining posts are occupied by expatriates. In respect of salaries, in a sample of 50 local people employed in the tourism sector in Maun, 33 of them earn between P300 and P990 per month. About 62 per cent of the salaries paid to local people employed in tourism-related jobs in Maun are far below P954.78 which is the country's poverty datum level. These findings are consistent with those of BTDP (1999) which state that even though the percentage of foreigners in the tourism employment is small (about 4 per cent in the hotel and lodge sectors), they dominate better paying jobs. The median salaries range from around P500 per month for the lowest paid categories to around P5000 per month for the highest paid. Expatriate salaries are considerably higher than those paid to locals in similar positions, especially for executive managers, general managers and professional guides. The BTDP report further notes that the gap between citizen and expatriate levels of remuneration becomes still wider when benefits and allowances are taken into consideration. Most expatriates qualify for generous tax free gratuities, home leave passages, children's education allowances, furnished housing allowances, and encashment of leave allowances. This trend appears to be consistent in other developing countries such as in the Caribbean. Pantin (1998) states that about nine out of ten management positions in tourism facilities in the Caribbean are occupied by expatriates with average salaries several times higher than those of the unskilled local workers.

The predominantly foreign-owned tourism industry in Maun makes the tourism industry to have a minimal effect on rural development in Ngamiland District and Botswana as much of the tourism-generated revenue by safari companies is not retained within the district or the country. This then means that the dominance of the tourist industry by foreign safari companies makes it possible for a lot of money that is paid for tours by visitors to never arrive in Botswana, since bookings are mostly done outside Botswana (for example, in Johannesburg, America or Europe). The Botswana Tourism Development Programme (1999) indicates that in 1997, tourists who visited Botswana spent an estimated P1.1 billion. Of this gross expenditure, 55 per cent (P605 million) was spent outside Botswana (representing payment to external agents) and a further (16 per cent) P175 million was first-round linkages of receipts due to tourist-related imports (e.g. on food, equipment and expatriate staff). Only 29 per cent (P320 million) was spent in Botswana on local goods, wages, taxes and other activities. These findings are similar to those by Dixon *et al.*, (2000) on the predominantly foreign-owned tourism industry in the Caribbean. They state that in the Caribbean '... investment incentives to foreign firms represent an effective transfer of tax revenue from the domestic economy to the foreign home of the firms in question, since corporations can typically claim foreign tax credits in reporting their income to their home tax authority... the extent that these firms are taxed on worldwide income and do enjoy foreign tax credits, the existence of a tax holiday is of no material to them – they simply end up paying tax to their home country rather than where the investment is located' (Dixon *et al.*, 2000:22).

The dominance of the tourism industry by foreign firms in Ngamiland District is mainly a result of the government's Tourism Policy of 1990 which emphasises the promotion of high-cost, low-volume tourism. The pursuance of this policy is primarily responsible for the development of high quality facilities, mostly owned and operated by foreign investors which then became the foundation of 'enclave tourism'. The exclusive nature of tourism or 'enclave tourism' is also characterised by lack of interaction between foreigners and the local communities, thereby becoming a source of resentment and alienation. As foreign entrepreneurs dominate the tourism industry, citizens view the approach negatively because they perceive the domination by non-citizens as 'selling out' of their resources.

Enclave tourism can be described as the direct opposite of eco-tourism, which is environmentally friendly and takes into account the needs, wishes and participation of the local communities. Eco-tourism is defined by Ceballos-Lascurain (1996:20) as 'environmentally responsible travel and visitation to relatively undisturbed natural areas, in order to enjoy, study and appreciate nature (and any accompanying cultural features – both past and present). It is a type of tourism that promotes conservation, has low visitor impact and provides for beneficially active socio-economic involvement of local populations'. Enclave tourism can also be referred to as internal colonialism. According to Drakakis-Smith and Williams (1983), Dixon and Hefferman (1991), internal colonialism is a phenomenon whereby natural resources in a host region mostly benefit expatriates or outsiders while the majority of the locals are marginalized, either financially or otherwise.

The need to have trained personnel in order to run and manage the community tourism projects efficiently remains one of the major challenges to community-based tourism in Ngamiland District. The problem of lack of trained community tourism business experts will always present room for outside assistance, a factor that is generally not sustainable for any economic project. This issue is related to that of lack of suitable non-governmental organisations in the area that can assist communities in identifying manpower needs and training of local communities in the tourism business. NGOs would also be of assistance in marketing of community products and providing a direct link between communities on the one hand, and government and donor agencies on the other.

Conclusion

Community-based Natural Resource Management (CBNRM) in Eastern and Southern Africa was adopted under a narrow framework of achieving wildlife conservation, rather than the need for social empowerment or economic development in rural settlements. Economic benefits were seen as a means of achieving conservation, especially wildlife resources, as well as being an end in themselves. In Botswana, this fact becomes more obvious when noting that CBNRM was first promoted by the Department of Wildlife and National Parks (DWNP) and later by the Department of Lands. The two departments are respectively responsible for wildlife management and land resources. The Ministry of Finance and Development Planning, which is responsible for finance and economic development, remained on the periphery in promoting CBNRM. Despite this narrow approach enshrined within CBNRM, economic development and financial opportunities in rural areas indicate that CBNRM has grown beyond its original aim. As a result, there is need to shift focus from a purely conservation exercise to include economic development of rural areas as well as the rights of both participating communities and those left outside CBNRM.

Although the implementation of CBNRM has its own limitation, community-based tourism in Ngamiland District has the potential to become a viable and sustainable economic activity that can improve the socio-economic livelihoods of the people living in the area. This can be possible if the challenges that currently face the industry are dealt with. As already noted, the government policy, which emphasises high-cost low-volume tourism, is partly responsible for the development of the foreign-dominated tourism industry in Ngamiland District. The policy encourages the building of capital-intensive infrastructure, which the local communities cannot afford and as a result, they end up sub-leasing their CHAs to commercial safari operators. There is therefore need for government policy to shift towards encouraging local communities to set up reasonably small, manageable operations which require low capital and utilize locally available skills. Mbaiwa (1999) notes that the local communities in Ngamiland District propose projects such as curio gift shops, selling of local souvenirs, traditional campsites, community tour operators and cultural villages which would provide traditional dishes, accommodation, music and dance. Traditional craftwork, walking safaris and canoe (*mekoro*) safaris match the local capabilities and skills and also require little capital outlay to operate. This means that for local communities to benefit from the booming tourist industry in Ngamiland District, they need to engage in small-scale projects which require local skills and knowledge. Such projects do not only require community skills and knowledge but are also compatible with the prevailing environmental aspects and can therefore promote sustainable development of natural resources in Ngamiland District.

While it is correct to state that the original aims of CBNRM were not to give communities living in resource areas ownership over land or resources but an incentive to manage the land resources, the role that CBNRM now plays in Botswana and Ngamiland District has grown beyond this first aim. As a result, community-based tourism in Ngamiland can be more successful and sustainable if full ownership and control of land and all natural resources found in community areas are given to the local people. Barnes (1998: 333) states that 'much of the wildlife resource in Botswana is public property and control is vested with central government. Central governments are generally unsuited to any form of active management and are better at passive directing. For various reasons government in Botswana is small and even less suited than most governments to managing a dispersed, fugitive public resource'. For sustainable utilisation of natural resources in an area to be possible, control and access to these resources needs to be given to the users, who in this case include the local communities in Ngamiland District. Barnes (1998) states that decentralisation of wildlife resources to district and local community levels is vital in that it will empower landholders to take control of the resources and manage them so as to maximise returns, a situation which will oblige them to use wildlife resources sustainably.

Community training and the provision of the necessary skills in the tourism business is left in the hands of the safari operators who happen to win the community hunting and photographic tenders. This is a wrong and inappropriate approach to the sustainability of community-based tourism since most safari owners lack commitment to develop local skills and knowledge, let alone the sustainable use of natural resources such as wildlife on which their business depend. Butler (1980) and Prossor (1994) describe the behaviour of tourism investors as that of making quick profit rather than being concerned with environmental quality and conservation. Butler and Prossor describe tourism as an exploitative industry comparable to the mining of iron ore of which tour operators are always ready to move in to a particular attractive area in times of good business and move out once the environment and tourism resources are degraded. This implies that the government of Botswana should take the lead role to adopt and implement tourism management strategies to ensure the sustainable use of tourism resources in Ngamiland District. Government strategies should also ensure the training of locals in the necessary management and investment fields needed to promote community-based tourism. Once local communities have the necessary investment skills in the tourism sector, it will reduce reliance on foreign safari operators who generally have less commitment to rural development and the sustainable use of natural resources on which their tourism business relies.

References

Arnold, S. 1989. 'Sustainable Development: A Solution to the Development Puzzle?', *Development*, 2, 3, 21-25.

Ashley, C. 1995. 'Community Based Tourism as a Strategy for CBNRM: Options and Potentials for Achieving Conservation and Development Through Non-Consumptive Tourism in Namibia.

In Rihoy, E. (Ed), *The Commons Without Tragedy? Strategies for Community-based Natural Resources Management in Southern Africa.* Proceedings of the Regional Natural Resources Management Programme Annual Conference, SADC Wildlife Technical Co-ordinating Unit, Kasane, pp. 56-94.

Bank of Botswana Annual Report. 1999. *Annual Report and Economic Review 1999,* Bank of Botswana, Gaborone.

Barnes, J. 1998. Alternative Uses for Natural Resources in Botswana: Wildlife Utilisation, in Hermans, J. and Nteta, D. (Eds), *Botswana in the 21st Century,* Botswana Society, Gaborone, pp. 323-339.

Botswana Tourism Development Programme 1999a. *Tourism Economic Impact Assessment,* Department of Tourism, Gaborone.

Botswana Tourism Development Programme 1999b. *Employment Survey Report,* Department of Tourism, Gaborone.

Britton S.G. 1982. 'The Political economy of tourism in the third world', *Annals of Tourism Research,* 9: 331-58

Butler, R. 1980. 'The Concept of a Tourism area Cycle of Evolution', *Canadian Geographer,* 24: 5-12.

Chambers, R. 1986. *Sustainable Livelihoods,* Institute of Development Studies, University of Sussex (mimeo).

Ceballos-Lascurain, H. 1996. *Tourism, Ecotourism and Protected Areas,* IUCN Publication, Gland.

Darkoh, M.B.K. 1994. 'Population, Environment and Sustainable Development: Desertification in Africa', *Desertification Bulletin,* UNEP, 25, 20-26.

Darkoh, M.B.K. 1996. 'Towards an Adaptive and Community-based Approach to Management of Natural Resources in Drylands in Sub-Saharan Africa', In: Hjort-af-Ornäs A. (Ed), *Approaching Nature from Local Communities : Security Perceived and Achieved.* Motala Grafiska, Motala, pp. 73-99.

Darkoh, M.B.K. 1999. 'The Role of Business in Sustainable Development', in *Desertification Control Bulletin,* UNEP. 35, 2-6.

Dixon, C. and Hefferman, M. 1991. *Colonialism and Development in the Contemporary World,* Mansell, London.

Drakakis-Smith, D. and Willams, S. 1983. *Internal Colonialism: Essays around a Theme,* Developing Areas Research Group, Institute of British Geographers, Edinburgh.

Fernandez, M.E. 1994. Gender and Indigenous Knowledge, *Indigenous Knowledge And Development Monitor,* 2, 3, 6-7.

LaDuke, W. 1994. 'Traditional Ecological Knowledge and Environmental Futures', *Colorado Journal of International Law and Policy*. 5, 127-148.

Laws, C.M. and Luning, H.A. 1996. 'Farmers Knowledge and GIS', *Indigenous Knowledge and Development Monitor*. 4, 1, 8-11.

Lele, S. 1991. 'Sustainable Development: A critical Review', *World Development*, 19, 6, 607-621.

Mbaiwa, J.E. 1999. *Prospects for Sustainable Wildlife Resource Utilisation and Management in Botswana: A Case Study of East Ngamiland District*, Upublished M.Sc. Thesis, Department of Environmental Science, University of Botswana, Gaborone.

Mbaiwa, J.E. 2000. 'The Impacts of Tourism in the Okavango Delta in North-west Botswana', a paper presented at a *Workshop on Climate Change, Biodiversity, Multi-Species Production Systems and Sustainable Livihoods in the Kalahari Region*, Maun Lodge, Botswana, 1-13 October 2000.

Mbaiwa, J.E. 2001. *A Survey Report on the Socio-Economic Impacts of Tourism in the Okavango Delta, Botswana*. Harry Oppenheimer Okavango Research Centre, University of Botswana, Maun (*in preparation*).

Mbanefo, S., and de Boerr, H. 1993. 'CAMPFIRE in Zimbabwe', in Kemf, E. (Ed), *Indigenous Peoples and Protected Areas: The Law of The Mother*, Earthscan, London, pp. 81-88.

Modise, S. 1996. 'Botswana: The Role of Wildlife in Developing Sustainable Land Use Management' in Hirschoff, P., Metcalfe, S. and Rihoy, L. (Eds), *Rural Development and Conservation in Africa: Studies in Community Resource Management*, Africa Resource Trust, Proceedings of a Seminar Tour, Harare, pp. 19-25.

Mordi, R. 1991. *Public Attitudes Towards Wildlife in Botswana*, Garland Publishing, New York.

Murphree, M.W. 1993. *Communities as Resource Management Institutions*. Gatekeeper Series No. 36, Sustainable Agricultural Programme of the International Institute for Environment and Development, London, pp. 1-13.

Mwenya, A.N., Lewis, D.M. and Kaweche, G.B. 1990. *Policy, Background and Future: National Parks and Wildlife Services, New Administrative Management Design for Game Management Areas*. USAID, Lusaka.

Ndubano, E. 2000. *The Economic Impacts of Tourism on the Local People: The Case of Maun in the Ngamiland-Sub District, Botswana*. M.Sc Thesis, Department of Environmental Science, University of Botswana, Gaborone.

Pantin, D.A. 1998. *Tourism in St. Lucia*. Sustainable Economic Development Unit for Small and Island Development States, University of the West Indies, St Augustine.

Pearce, D., Markandya, A. and Barbier, E.B. 1989. *Blueprint for a Green Economy,* Earthscan, London.

Prosser, R. 1994. Societal Change and the Growth in Alternative Tourism. In: Cater, E. and Lowman, G. (Eds), *Ecotourism: A Sustainable Option,* John Wiley and Sons Ltd, New York, pp. 19-37.

Redclift, M.R. 1987. *Sustainable Development: Exploring the Contradictions,* Methuen, London.

Rihoy, E. 1995. From State Control of Wildlife to Co-Management of Natural Resources – the Evolution of Community Management in Southern Africa, in Rihoy, E. (Ed), *The Commons Without Tragedy? Strategies for Community-based Natural Resources Management in Southern Africa,* Proceedings of the Regional Natural Resources Management Programme Annual Conference: SADC Wildlife Technical Co-odinating Unit, Kasane, pp. 1-36.

Steiner, A. and Rihoy, E. 1995. *The Commons Without Tragedy? Strategies for Community-based Natural Resources Management in Southern Africa.* Proceedings of the Regional Natural Resources Management Programme Annual Conference: SADC Wildlife Technical Co-odinating Unit, Kasane, pp. 1-36.

Serrano, R.C. 1996. 'Comments on Indigenous and Scientific Knowledge: Some Critical Comments', *Indigenous Knowledge and Development Monitor* 4, 1, 14-15.

Thakadu, O.T. 1997. *Indigenous Wildlife Knowledge Systems and their Role in Facilitating Community-based Wildlife Management Projects in Botswana,* M.Sc Thesis, School of Environment and Development, University of Natal, Pietermaritzburg.

Wall, G. 1997. 'Is Ecotourism Sustainable?', *Environmental Management,* 21, 4,483-491.

Warren, A. 1996. 'Comments on Indigenous and Scientific Knowledge : Some Critical Comments'. *Indigenous Knowledge and Development Monitor,* 4, 1, 13-14.

World Commission on Environment and Development (WCED) 1987. *Our Common Future*. Oxford University Press, London.

Chapter 17

Urban Poverty and Environmental Health: The Case of Kampala City, Uganda

A. Nuwagaba

Introduction

Urbanization in the context of globalization has led to prosperity of some individuals, but in most cases especially in Africa, it has widened social inequalities resulting in poverty among sectors of the urban population (Kalsi 1995). Because the free market economy has spawned the urban system with 'laissez faire' individualism, which is associated with the maldistribution of wealth, the over-arching results have been increased competitiveness leading to isolation and social exclusion of the less competitive individuals.

Urban poverty has its roots in the historically exponential increase of urban population in developing countries. Urban population has been increasingly out of proportion with the growth of the urban economy and employment opportunities (Mabogunje 1994). In the Ugandan case, urban poverty was exacerbated by political strife that characterized the country since the 1970s and 1980s. The inevitable result was the dearth in the institutional capacity of local urban authorities to deal with urban decay characterized by irregularisation of settlements with a significant proportion of the unemployed or/ and under-employed population. If there is any single characteristic that can describe this phenomenon during this period, it is the profusion of poverty in urban areas (Nuwagaba, 1997).

After independence in 1962, the Uganda Government enacted laws for management of urban infrastructure services. The major laws were enshrined in the 1964 Urban Authorities Act that empowered Municipal authorities to effect development control and to provide urban services. The services include water supply, sanitation management, construction and maintenance of access roads in the urban areas. As long as the Ugandan economy was buoyant, there were no problems in the provision of these services. The economic collapse of the 1970s and 1980s precipitated an increase in the poor population who live in substandard settlements characterized by shanty dwelling units and occupation of dilapidated buildings which lack basic infrastructure services, and proper sanitation. The problem emanated from the fact that the decline in economic performance culminated into the dearth of the institutional structure with serious macro-economic disequilibria. Urban population growth however remained on course for example, the population of Kampala City grew by 3.12 per cent per annum between 1970-1980 (MFED 1991). This situation aggravated the sanitation situation, whereby failure of the economy to expand employment opportunities in the urban sector made social services inaccessible to the majority population. Hence forcing people to create 'urban poverty enclaves', with definite sanitation problems.

Poverty in Theoretical Perspective

Poverty, like an elephant, is more easily recognized than defined (Mabogunje, 1994). The word originates from the Latin word 'panpertus' implying restricted means, including *inter alia,* food and financial resources. Misery, an extreme form of poverty originates from a Latin word *miseria,* which means intensive unhappiness, discomfort or suffering.

Poverty Beyond Anonymity

The phenomenon of poverty is a daunting one to operationalise. For many years, the poor were anonymous; they had no identity, no name, and no address (Kok 1994). They were statistically categorized below a computed poverty line. Fortunately, in recent years poverty has been defined with some specificity and the poor have begun to be located in geographical, occupational and social space, in ecosystems and households. To further specify the poverty problem and to determine how susceptible it is to macro-policy and development interventions, there is need to desegregate the poor in terms of how immutable the household's state of poverty is. Kok (1994) has examined varying degrees of immutability including:

> Stage in the family life-cycle poverty meaning poverty stricken family background where the only way out is through self-made opportunities by family members and acquisition of gainful employment.
>
> Life time poverty which connotes a vicious circle of poverty synonymous with Chamber's deprivation trap cum cluster of disadvantages. People are found entangled in this web.
>
> Intergenerational poverty refers to poverty mostly among *'new comers'*, for example, recent migrants. This type of poverty is however relative, the intensity of poverty increasingly diminishes across generations.
>
> Acquired poverty means poverty that to some degree is self-inflicted. For example, poverty that afflicts drug addicts.

Such an analysis will enable us to deal with poverty in a more realistic and differentiated fashion. The implication here is that poverty emanates from different situations. It manifests itself differently and hence approaches to alleviate it ought to take this important aspect into consideration. Many efforts in poverty alleviation have not been appropriately targeted due to a lack of clear understanding of the different ramifications of poverty. As Meerboer (1984) clearly indicated, the description of poverty in terms of income, consumption or expenditure although useful and practical, can be misleading, especially in the context of policy discussions. To use a medical term, poverty is a 'syndrome', consisting of interrelated features. Its exact configuration varies from individual to individual and across countries. Poverty has different faces. Inadequate income and lack of purchasing power are only one part of the picture (*Ibid.*, 1994).

Perceptions on Why the Poor are Poor

Two rather contrasting perspectives, each of which leads to an alternative set of policy prescriptions, have dominated the analysis of poverty. On the one hand, poverty is attributed to the personal failings of the individuals concerned which leads to self-perpetuating cycles of social pathology (Wrratten, 1995). On the other hand, it is viewed as the inevitable outcome of an unfairly structured political and economic system that discriminates against powerless and poverty stricken groups. The former perspective is intellectually rooted in *laissez faire* individualism and racial discrimination (Pryer, 1993).

The alternative, radical view draws from Marxist theory and is closely associated with the development of basic needs tradition. Policy prescriptions involve a more interventionist role for the state in promoting equity, analysis of poverty as a social construct rather than an individual problem, and the redistribution of assets and decision making power at international, national, regional, community, household and inter-household levels (Wratten, 1995).

It should be reiterated that, urban areas in themselves do not precipitate poverty; poverty can emerge if dysfunctions emerge in the urban system. It is the process of urbanization with increased competitiveness that creates widespread social inequalities. In large measure, this has resulted in low income and vulnerable populations which go by the brand *'urban poor'*. Studies conducted by Nuwagaba (1997) and Onibokun (1995), among others, have shown that poverty enclaves are characterized by poor and further deteriorating environment with poor drainage systems, poor waste disposal methods and grossly inadequate infrastructure services in the midst of a proliferation of socio-economic activities and unplanned dilapidated housing structures. In Uganda's urban areas, there is pervasive poverty, widespread unemployment and underemployment, which have resulted in high dependency ratio and predominance of informal activities.

Another indication of poverty is the emergence of a significant number of socially marginal people such as peddlers, hawkers and market vendors including evening market vending. There is no aggregate figure that can be used as an indicator of the magnitude of poverty in urban settlements in Uganda. Nevertheless, the political fine print is that the economic activities and the level of irregularization characterized by 74 per cent of the urban population putting up in dilapidated dwelling units which make up the urban settlements manifests the prevalence of poverty (Nuwagaba, 1997; Mwesigwa, 1987).

The implication of such pervasive poverty is that the capacity of the population in peri-urban settlements to afford infrastructure services such as water and electricity supply is limited. It should however be reiterated that all population sub-sectors in an urban area possess functions which are very crucial for the overall functioning of the urban system. According to Oscar Lewis, the poor are assumed to be marginal to urban development due to innate and culturally determined personal characteristics and associated deviant behavior (Wratten, 1994). On the other hand, other scholars for example Perlman J. *et al.,* (1976) propose the alternative marginalisation thesis, which ascribes more positive role to the activities of poor urban communities and emphasizes the structural barriers which exclude their participation in formal economic, political and social institutions. Oscar Lewis's (1972) anthropological studies of poor families in Mexico City and San Juan, Puerto Rico, led him to the conclusion that those trapped in poverty had characteristically different behavioural patterns and values from the dominant society and culture. He argued that:

> The culture of poverty is both an adaptation and reaction of the poor to their marginal position in a class-stratified, highly individualistic society. It represents an effort cope with feelings of hopelessness and despair which develop from the realization of the improbability of achieving success in terms of the values and goals of the larger society (Wratten, 1994: 21).

The hypothesis of the separate sub-culture was used to explain the perpetuation of poverty from generation to generation. Children born into poor families grow up with a weak family structure, ineffective interpersonal relations, a present time orientation and unrestrained spending patterns. This environment engenders values such as helplessness, dependence, a sense of inferiority, resignation and fatalism. Such children would be less interested in education, the value of work or self-improvement. They would have a low ability to plan ahead or to identify and react to changing opportunities, (Wratten 1994). These adaptive responses to economic deprivation and social marginality would, in turn, make the disadvantaged position of the poor yet more entrenched. Characteristics such as male machismo, sexual promiscuity, teenage pregnancy, illegitimacy and female-headed households, a rigid division of family responsibilities between men and women, alcohol and drug abuse, and poor education performance, serve as 'own goals', preventing the poor from rising out of their situation and precluding economic, social and political integration into the mainstream culture (Emory, 1992).

Lewis' work has provoked widespread debate and criticism. Concern has been expressed regarding the research methods used, the lack of representativeness of the families studied, and the ethnocentricity with which Lewis judges other peoples' lives. Wratten argues strongly that the lifestyle of the poor is not based on deviant values but is instead a rational response to having insufficient money. Poverty results from discriminatory structures which denies the non privileged the means to realize their aspirations. Mwesigwa (1987) reinforce this view pointing out that although squatter settlements are built in peripheral areas, where there is less competition for land from high income groups, squatters are usually established urban residents who are well integrated into the city economy. It appears that squatter settlements are highly heterogeneous and contain middle-income as well as low-income residents. The dispute over the validity of the culture of poverty hypothesis, and the shift toward a structural explanation of poverty reflects wider shifts in development thinking. In the 1970s modernization theory prescribed a blue print development strategy based on rapid economic growth and accorded a residual role for social policy. This perspective was challenged by dependency theorists who claimed that urban and rural poverty in the Third World was intrinsically linked to the process of capital accumulation in the North. On the other hand, the proponents of basic needs strategies called for a redefinition of the goals of development to emphasize equity and democracy as well as growth (Streeten 1978).

Sen's Thesis of Poverty, Entitlement and Deprivation

Sen's theory (1987) is premised on the concepts of entitlement and deprivation. After his study of poverty and famine in India, Bangladesh and Ethiopia, Sen made a starvation analysis and concluded that poverty is a function of the structure of production and exchange in the economy. This theory presupposes that entitlement relations determined one's access to resources depending on one's ownership bundle. According to Sen (1987: 4), the ownership bundle includes;

- whether one can find employment and if so, for how long and at what rate;

- what one can earn by selling his non-labor assets and how much it will cost one to buy whatever one may wish to buy;
- what one can produce with one's on labor power and resources and what one can manage;
- the cost of purchasing resources and the value of products one can sell; and
- the social security benefits one is entitled to and the taxes one is liable to pay.

It is clear from Sen's thesis that poverty cannot be relegated to one's inherent abilities. Poverty cannot be explained as an individual responsibility but a structural problem that relates to the labour market, production relations within an economy, the fiscal regimes as well as the social safety nets that governments have put in place for the vulnerable population. Individual access to resources and services depend more on the socioeconomic configuration of the economy rather than on individual capacity alone. What seems to emerge is that urban poverty is a complex phenomenon. It is influenced by various factors including institutional contexts, governance structures as well as the distribution of socioeconomic opportunities.

The Concept of Environmental Health

In this chapter, environmental health refers to sanitation management including solid waste disposal, human excreta disposal and waste water management. The concept also refers to water supply, household drainage and general household aesthetics.

Sanitation in Urban Poverty Enclaves

In cities around the world, the living conditions of hundreds of millions of people threaten their health, impose misery and have potentially catastrophic social consequences. Conditions are worse for poor people, particularly in developing countries. Urban growth has out-stripped society's capacity to meet human needs, leaving millions of people with inadequate incomes, inadequate diet and inadequate housing and other social services.

The Uganda Case

In Ugandan urban areas, the problems of environmental sanitation seem to have resulted from rapid population growth, overcrowding and high density of human settlements with congestion amidst poor social infrastructure services. This has culminated in improper disposal of commercial and domestic waste. Most of the urban poor enclaves have poor drainage system because this is where land is cheap and does not have value and relevance for the affluent. Persistent stagnant storm water and silage around homes characterize housing accommodation, a condition that provides excellent habitat for disease vectors. This has provoked a pattern of health problems for both children and adults that include diseases typical of underdevelopment. It is against this background that, the plight of the urban peripheral populations must be addressed as an integral part of the overall development because they affect the overall urban environment.

There seems to be a significant relationship between sanitation and spread of disease. In Kampala City, common diseases include malaria, typhoid, diarrhea, worms, respiratory infections and cholera epidemics, all which to a significant extent, relate to the unsanitary conditions in the City. Table 17.1 below shows the distribution of morbidity among children in Kampala City.

Table 17.1: Distribution of Child Morbidity and Mortality in Bwaise Zone – Kampala District

Perspective **Child morbidity**	**Frequency**	**Percentage**
Less than Monthly	4	6.7
Monthly quarterly	34	56.7
Greater than Quarterly Annually	14	23.3
More than Annually	2	3.3
N/A	6	10.0
Total	60	100.0

Table 17.1 concluded

Perspective	Frequency	Percentage
Ailments		
Malaria	58	96.7
Diarrhea	49	81.7
Measles	24	40.0
Worms	29	48.3
Typhoid	16	26.7
Scabies	12	20.0
Respiratory infections	13	21.7
Child mortality in the area		
Very high	12	20.0
High	35	58.3
Moderate	17	11.7
Low	6	10.0
Total	60	100.0

Source: Nuwagaba (1994: 32).

The preceding analysis has indicated that most of the peri-urban settlements are in a poor state. The perpetually increasing population in peri-urban settlements results in overcrowding of dwelling units. The settlements have been affected by waste from various activities, which is not well disposed of. Poor sanitation has also negatively affected the environment. This has mainly resulted from lack of 'private toilet facilities' for a significant proportion of households, which has resulted in the disposing of human excreta in the open drains in Kampala City. The major health problem here has been the outbreaks of cholera that has hit Kampala City with calamitous consequences (*New Vision Newspaper,* March 1998 p.11).

The sanitation problem has been exacerbated by poor drainage. Most of the peri-urban settlements are located in flood plains with big parts of the area under swamps while most of the built up tracts are soggy. The implication from such an environment coupled with poorly maintained drains due to poor methods of waste disposal, has been the creation of breeding ground for mosquitoes with high risk of infectious diseases such as malaria and typhoid.

Water Supply

There is an impressive distribution of water supply in major urban areas in Uganda (Mwesigwa 1987). The only problem is affordability of the water tariff, which sometimes compromises on the amounts used. This sometimes forces people to resort to springs and wells which constitute high risk sources of contamination. The corollary problem is the attitude of the people towards water supply, communities feel that 'water is water'; which literally means that all water is the same and therefore why pay a high tariff for such a 'free' commodity. This implies an awareness gap on sanitation.

Another aspect aggravated by the housing pattern and congestion is the storage of water. In Kampala City, most households store water in the bedroom or sitting room (Kampala City Council, 1999). Considering the poor nature of their houses, their small size and the big number of people that share a room, the underlying health hazards cannot be over-emphasized. Table 17.2 provides the status of water supply in Kampala district.

As can be viewed from the table, there is an impressive distribution of piped water supply in Kampala central division. Out of the total number of households considered, 86 per cent uses tap water.

What the table does not indicate however is that most households fall far short below average water consumption per capita. Because the water tariff depends on the amount of water consumption, households have attempted to trade off the potential high cost by consuming less water. The average water consumption per capita per day (wc/capita) is 20 litres which falls far too below the standard requirement of 50 litres per capita per day. Table 17.3 indicates the existing gap regarding water supply in the city.

Table 17.2: Current Status of Household Water Supply in Kampala Central Division

Variable	Number	%
Source of Domestic water		
1. Tap	309	86
2. Well	5	1
3. Spring	31	9
4. Vendors	14	4
Total	359	100.0

Table 17.2 concluded

Variable	Number	%
Availability of water in the area		
1. Yes	293	82
2. No	66	18
Total	359	100.0
Monthly expenditure on water		
1. < 5,000	212	59
2. 5,000 < 10,000	50	14
3. < 10,000	57	27
Total	359	100.0
Water Sufficiency		
1. Sufficient	287	7
2. Not sufficient	72	93
Total	359	100.0
Storage of drinking water		
1. Fridge	25	7
2. Jerrycan	334	93
Total	359	100.0

Source: Public Health Department, Kampala City Council (April 1999: 28).

Table 17.3: The Gap Regarding Water Supply in Kampala

Variable	Expected standard score	Actual standard score	Sanitation Gap score	Sanitation Gap %
1. Water present at time of visit	3,610	3,410	200	5.54
2. Litres consumed per person per day	3,610	1,122	2,488	69.39
3. Drinking water	3,610	2,335	1,275	35.32
Total	10,830	6,867	3,963	36.59

Source: Sanitation Quality Gap for Kampala Central Division, Kampala City Council (1999: 31).

The composite sanitation gap of 36.59 per cent exists between the expected standards and the current reality. The implication is that there is a short fall of 69.39 per cent to the expected level of water consumption. However, there is a discernible gap of 35 per cent regarding drinking water. The majority of household members tend to regard tap water as safe enough for drinking and therefore do not see the need of boiling the water.
Cost effective intervention measures for closing the gap include:

- The need to extend the distribution of water to close the existing gap from a significant proportion of the households who either use water obtained from springs or wells.
- Given the dense settlements with poor drainage and poor excreta disposal, it is necessary that households increase their access to piped water.
- Local urban authorities could achieve improved access to safe water through provision of stand pipes and water kiosks managed by community based water committees. These committees have been tested in Wobulenzi in Luwero district near Kampala City and they seem to be reasonably successful.
- It is envisaged that households will increase their water consumption as a result of increased standpipes and more competition in water business. Most of the water currently used in households is purchased at high costs from vendors. A one twenty-litre jerrycan is purchased from vendors at Ug. Shillings100/= as compared to Ug. Shillings 12/= per twenty litres purchased from National Water and Sewerage Corporation (Nuwagaba, 1999).

Excreta Disposal

The standard of hygiene and excreta disposal is an influential aspect in the prevalence of disease and their spread. In the entire country, national latrine ownership for both urban and rural areas is estimated at 65 per cent with wide variation in quality. Out of these, 22 per cent are unsanitary open pits (Ministry of Health, 1998).

While the standards differ from area to area and even within an area, peri-urban environments are especially known for generally low standards with differences mainly being in the degree. Kalanzi

(1987) found that 32 per cent of the households in Mulago slum area in Kampala City did not have a private pit latrine. In Kifumbira Slum also located north of Kampala City, another study (Kamya, 1992) found that some people ease themselves in nearby bushes or banana plantations. The same study also identified the rampant sharing of toilets as leading to difficulty in keeping them aesthetically clean. The pit latrines are mainly of the raised category and shallow due to water logging. Human waste litter is a problem to reckon with in these areas and many households lack a urinal, bathroom and/or kitchen. Though most if not all households (claim to) have toilets, a single toilet can be shared by up to four households. These could be referred to as 'public toilets'!

Urinals are generally rare in Bwaise parish located north of Kampala City with 65 per cent of the households without urinals (Nuwagaba 1994). Most of these residents do not have access to a bathroom and many households mainly share the few available bathrooms, thus their standards of cleanliness are very poor. The issue of hygiene is further exacerbated by the fact that the bathrooms also serve as urinals.

In general standards of waste disposal and hygiene in those settlements is deplorable. This problem is exarcebated by congestion and high density of dwelling units (Nuwagaba, 1997). The table below shows the sanitation quality gap for excreta disposal in Kampala City.

It is extremely ironic that a city like Kampala which is believed by many Ugandans to be a place of 'Milk and Honey' experiences such a sanitation gap regarding access to toilet facilities. As Nuwagaba (1987) has indicated, the toilet ownership in Kampala constitutes only 83 per cent of the total city population. One wonders what the remaining population of people use for relief! It is puzzling why people flock to such environmentally hazardous urban areas. As Todaro (1999) puts it, it is not the current opportunities that sustains the urbanization spiral but the perceived opportunities that influences people to migrate to cities. Recommended cost effective interventions for excreta disposal include:

- The long-term provision of flush toilets in Kampala Central division. The reason being that most settlements are located in low land and soggy areas that cannot allow deep pit latrine construction.

- In the medium term however, there is need for ventilated inspection pipe latrines which could be constructed by the city council and tendered to the private sector operators.
- In the short term, there is need to demolish illegal structures that lack toilet facilities. This however requires tremendous political will since such slum areas constitute densely built areas with high population density. If unsatisfactorily handled, such breakage could result in serious social and economic disruptions with far reaching political implications.

Table. 17.4: Sanitation Quality Gap for Excreta Disposal for Selected Households in Kampala Central Division

Variable	Expected standard score	Actual standard score	Sanitation Gap score	Sanitation Gap %
1. Presence	3,610	3,220	390	10.80
2. Accessibility: Distance	3,610	2,742	868	24.04
3. Accessibility: Cost	3,610	3,330	280	7.76
4. Accessibility: Privacy	3,610	2,845	765	21.19
5. Technology	3,610	1,835	1,775	49.17
6. Accessibility: Locking	3,610	2,545	1,065	29.50
7. Level of filling	3,610	1,829	1,781	49.34
8. Accessibility: Soiling outside (around/approach)	3,610	2,040	1,570	43.49
9. Smell	3,610	1,895	1,715	47.51
10. Aesthetics	3,610	2,630	980	27.15
11. Support facilities:				
Toilet paper	3,610	1,145	2,465	68.28
Hand washing/ water	3,610	1,855	1,755	48.61
12. Toilet (Stance) Population Ratio	3,610	2,010	1,600	44.32
13. Toilet ownership	3,610	2,075	1,535	42,52
14. Insect Infestation	3,610	2,034	1,576	43.66
15. Children's stool	3,610	1,290	2,320	64.27
16. Evidence of Buvera (plastic bag) Toilet	3,610	3,010	600	16.62
Total	61,370	38,330	23,040	37.54

Source: Ministry of Health, Disease Surveillance Centre, Kampala (1999: 84).

Refuse Disposal

Poor refuse disposal definitely goes in tandem with slums. Approximately 38 per cent of the households in Kampala City dump household refuse in the open space 'around the house' (Nuwagaba 1997). In many areas of the country, the standards of refuse disposal are generally low. According to UNICEF (1989), poor rubbish disposal is a national problem. Less than 5 per cent of the households in Uganda have proper rubbish pits and approximately 91 per cent of the families just scatter their refuse within or outside their compounds. The poor disposal of refuse is evident from studies done, (Mwesigwa 1987) which indicate that the majority just dump litter anyhow 'around the house'. 'Around the house' may mean as near as 3 metres from the house. Indeed, some houses are just located amidst a continuous sea of refuse litter. Table 17.5 shows the gap between the expected standards of refuse disposal and the reality.

Table: 17.5: Sanitation Quality Gap for Refuse Disposal for Selected Households in Kampala Central Division

Parameter	Expected standard score	Actual standard score	Sanitation Gap score	Sanitation Gap %
1. Refuse container: size	3,610	1,500	2,110	58.45
2. Refuse container: material	3,610	1,955	1,655	45.89
3. Refuse container: cover (isolating vermins)	3,610	1,385	2,225	61.63
4. Evidence of littering	3,610	2,442	1,168	32.35
5. Presence of vermin/pests	3,610	2,075	1,535	42.52
Total	18,050	9,357	8,693	48.16

Source: Nuwagaba (1997: 41).

The sanitation gap is mainly a culmination of both attitudinal as well as technical problems. It was evident from Nuwagaba's (1997) study that even areas which were well served with refuse skips did not utilise them despite heaps of garbage around houses. On the other hand, there were areas which lacked garbage skips and hence had no alternative but to throw garbage in the open.

The implication here is the need for community education and participation in ensuring a sanitary and healthy environment. More investment in skips and acquisition of more landfills by municipal authorities alone may not solve the problem. The population should be sensitized about the concept of environmental health.

Drainage

The problem of poor drainage-cum flood plain nature of most of the settlements in Kampala City has made it difficult to construct appropriate pit latrines. Because the water table is very low, the few pit latrines available have been made shallow resulting into quick filling. Overcrowded dwelling units and the fact that in many instances up to six households share one toilet facility compounds the problem. Similarly, because of the low-income nature of the inhabitants in the settlements, there are very few cesspool services in the area. Therefore, the major option available to filled up pit latrines has been the opening of the filled pits and letting the sewage gush out into the open drains.

Table 17.6: Sanitation Quality Gap for Drainage for Selected Households in Kampala Central Division

Parameter	Expected standard score	Actual standard score	Sanitation Gap score	Sanitation Gap %
1. Utensils washing/ drying	3,610	2,207	1,403	38.86
2. Bathing	1,025	2,520	1,090	30.19
3. Drainage: waste water sullage/grey	1,530	1,025	2,585	71.61
4. Drainage: Storm water	3,610	1,530	2,080	57.62
5. Refuse/Kaveera choking of drains	2,115	2,115	1,495	41.41
6. Night soil dumping	2,755	2,755	855	23.68
Total	21,660	12,152	9,508	43.90

Source: Sanitation Quality Gap Study for Kampala Central Division, Kampala City Council (March 1999: 48).

People indiscriminately dump garbage and dirty water in the drains, further complicating the drainage problem. This has led to pooling of water in many parts of the area even during the dry season. The practice is unsanitary and provides a haven for disease vectors like mosquitoes, flies and rodents. The gap regarding the expected drainage standards in Kampala City and the current reality is indicated in Table 17.6. which shows that refuse disposal, especially where Buvera (polythene bags), are used is a major problem in the city. This has resulted in choking of water drains resulting in blockage of channels culminating in flooding during the rainy season and dusty conditions during the dry season. This problem further complicates the flow of storm water as indicated by the 58 per cent gap resulting in breeding grounds for mosquitoes, a condition that increases morbidity in the area.

It should be reiterated that the high density of irregular settlements has been greatly facilitated by the existence of these flood plains that are less demanded by the affluent in the city. (The flood plains are mainly water logged. These areas are less attractive to the affluent individuals who prefer to purchase land and reside in well-drained areas of the city).

Aesthetics and Housing

Housing is a major parameter of the sanitation phenomenon. The nature of a dwelling unit, structure, spacing size and ownership of the houses, affects the frequency of sickness among household members. Most of the slum houses are predominantly single-room shelters commonly known as 'mizigo', a local description of a tenement. These structures are built in such a way that there is virtually no space between them. Clusters of shelters are just separated by a corridor or verandah. In this type of housing, a single room acts as a bedroom, sitting room, store and so on. Pathetically four people share a single room, and in some circumstances more than four people may share a single room as well!

Where some houses have some space, which passes for a compound, in the strict sense of the term, they are actually more of public paths and mini playgrounds than compounds. The general characteristic of these dwelling units in urban enclaves is poverty. The foundation of the dwelling units is poor leading to many houses assuming slanting postures, without ventilators and characterized by evidence of breaking walls and wearing away. But perhaps the most detrimental aspect of housing in this group is the 'clustering syndrome' where either separate houses are built too close to each other or in most instances where a building is structured in such away as to constitute numerous single room units (mizigo), each accommodating a big household.

Policy and Legal Instruments

Government Strategy on Urban Poverty

The growth of urban poverty and deteriorating urban environment in Uganda has necessitated government interventions. The major thrust of government policy had since 1960s been a presumption that urban areas are better off in terms of socioeconomic infrastructure (Stren and White 1989; Mabogunje 1994). Notwithstanding, there is increasing government concern about urban unemployment, low incomes, poor infrastructure facilities, appalling sanitation and lack of access to housing and increasing social exclusion. These ills led to a number of policy programmers and strategies.

Statutory Instruments

After independence in 1962, the Uganda Government enacted laws for management of urban infrastructure services. These were enshrined in the Public Health Act 1969 recently repealed by the Local Government Act 1997, which empowered Municipal Authorities to effect development control and provide urban services. The Act further empowered Municipal Authorities to demolish illegal structures such as dwelling units without toilets (Public Health Act 1969: Section 37).

The services provided include water supply, sanitation management, construction and maintenance of access roads in the urban areas. As long as the Ugandan economy was buoyant, there were no problems in the provision of these services. However, from early 1970s to late 1980s, the Ugandan economy virtually atrophied. The period was characterized by political turmoil resulting in the collapse of the industrial sector. This led to open unemployment and widespread poverty among the urban population (Nuwagaba, 1997).

By 1979 when Idi Amin's military junta was overthrown, the economy was in total doldrums with income per capita of US $120 as compared to over US$450 in the 1960s. (Background to Budget 1987).

The drastic decline of the industrial sector and the collapse of the economy had adverse impact on the institutional capacity of the local urban authorities. There was widespread corruption, characterized by misallocation and embezzlement of public funds including those meant for infrastructure services (Mwesigwa 1987). However, population growth in urban centres remained on course. For example, between 1969 and 1980, Kampala city grew by 3.14 per cent (MFED 1991). The population growth rate was much higher than the growth in the economy including the industrial sector. This resulted in a lack of capacity for the economy to absorb the growing population. The major result was the decline in institutional performance leading to poor infrastructure service delivery by the local urban authorities.

Constitutional Provisions

Until 1975, most of the land (60 per cent) of Kampala City was managed under a system of land tenure known as Mailo tenure, a form of free hold where individuals acquired huge chunks of land under the British colonial administration. This is a system of land tenure where individuals possess certificates of the title certifying the *bona fide* rights over the land resource facilities. This system applies, for instance, to the major slum settlements in Kampala such as Bwaise, Katwe, Kibuli and Mulago, which stand on land held under mailo tenure. Most dwelling units in these areas are illegal and some are devoid of such crucial facilities as toilets. The Mailo land is held under private certification. The owner possesses a land title and holds land in perpetuity. This form of land tenure prevailed until 1975 when President Amin declared all land in Uganda under leasehold and vested in to be held by the state. However, due to highly charged political volatility of land ownership, this decree was never implemented. The land office continued to operate illegally both the mailo tenure title and the leasehold tilte-land held by the municipal authorities.

The 1995 Constitution which was promulgated by the constituent Assembly during Yoweri Museveni's rule, repealed the 1975 land Reform Decree and stipulated that:

all land in Uganda becomes freehold and vests the holding and user rights in private citizens [Article 237 Cap (8)].

This has imposed tremendous implications and consequences on urban resource use management and development. The implication for mailo and free holding is that most of the population has to acquire land through purchase which requires enough financial resources. However, the difficulty with mailo and freehold tenure system is that the owners of titles to particular land parcels may not have enough resources to develop their land resources. Because the landowner has security of tenure and there is no effective design, zoning, land use or infrastructure requirement legislation, the urban authorities seem to be unable to enforce planned development control in these areas. This has had far-reaching environmental consequences especially the emergence of squatter settlements without appropriate infrastructure.

Attitudinal Issues in Urban Environmental Management

Household Perception of Sanitation

Perception of sanitation refers to the community's views of the methods of environmental management. These include individual and community initiatives-cum-activities directed at improving urban environmental quality.

According to Nuwagaba (1997), there is a high level of awareness of environmental problems in Kampala among the population in peri-urban settlements. However, there is an insignificant proportion of the population which contribute to mitigatory actions against environmental degradation.

The explanation for this anomalous relationship between knowledge of the environment and mitigating the adverse effects of environmental degradation is that people may be conditioned to ignore their values due to lack of means to promote good sanitary environment. The issue is affordability rather than carelessness.

A deeper analysis however indicates that some clusters of the urban population seem to regard the concept of environmental sanitation as alien *(ibid.)*. In most areas, household members actually believe that their sanitation standards are not as low as an outsider might think.

The Way Forward

Poverty is a structural crisis requiring structural solutions. Governments have the responsibility to reduce household poverty but the people also have an obligation to take up measures directed at poverty alleviation. There is need to develop positive attitudes that support poverty alleviation initiatives with the support of both government and non-governmental organizations. The complexity of the poverty question, requires policy instruments. However ingenious, poverty alleviation programmes will be effectively operated (and targeted on the poor) only if the poor can exercise politically 'effective demands on civil society. Donors alone cannot of course substantially influence the civil societies of other peoples countries, the input of the indigenous people and their governments'.

New Paradigms for Intervention

Participation is a crucial factor in bridging environmental health gaps, because of the increasing role of the civil society in socio-economic development. Since governments in developing countries, Uganda inclusive, are increasingly divesting themselves out of social service provision, it is in this view that local participation in environmental management activities is premised. There is need to change the orientation of physical planning structure. There is need for demystification of the existing as well as new structural plans for the local community. This is called 'participatory planning', a system of planning where the target beneficiaries are involved in the planning and implementation and reviews of various strategies and activities under implementation.

Supply forces households to resort to other less healthy sources of water such as wells. Reasonable regularization should be effected and affordable services provided. These include clean water supply using standpipes and construction of toilets such as VIPs which can be contracted to private operators. This could be achieved through collaboration with international as well as indigenous NGOs. With the current Local Council (LC) political organizational structure, this is not a far-fetched proposition.

The Concept of Minimum Service

Although there is an impressive distribution of infrastructure services especially water and electricity supply in the peri-urban areas, an anomalous relationship exists between availability of services and actual use of such services. The problem is that the bills for electricity are astronomical, which often results in massive power supply cutoffs.

Lack of electricity means that most households use charcoal for both cooking and lighting. The environmental hazard associated with using charcoal stoves and water disconnection in a heavily congested settlement without proper kitchens cannot be underestimated.

The Role of Civil Society

Despite the little awareness of environment health in urban settlements, the population lacks alternatives to improve their environment due to their socioeconomic context. They reside in peri-urban localities with high and fast growing population, which is partly associated with growth of the informal sector. This sector facilitates easy access to economic opportunities. The concentration of high population densities in these settlements, and the proliferation of socioeconomic activities are conducive to inadequate infrastructure services and environmental degradation.

In the light of the foregoing, there is a dilemma of how to assist this population which is aware of the environmental problem facing them, yet has been forced to suffer poor environmental conditions due to the dynamics of urbanization and the inevitable socioeconomic differentiation of the whole urban population superstructure. Endeavors to improve the environment have been initiated by different stakeholders. Local communities NGOs, women and youth organizations have initiated some efforts and yet others locate their genesis from international organizations such as UNDP and UNICEF, which collaborate with city municipal authorities. Major weaknesses in the latter efforts are that they seem not to be coordinated, they lack sustainability and most of them though useful, are 'foreign' to communities they are intended to serve, having originated from 'outside' and without consultation with peri-urban communities.

A case in point is the drainage project initiated by UNDP in Bwaise, a peri-urban area in Kampala. The project was successful in implementing the construction of the drainage channel but the very residents who should be maintaining the channel currently block the channel with solid waste.

There are two major emerging issues here:

- The first one is the need to inquire into how these community-based efforts are initiated, their objectives, management and problems arising from there with the aim of coming up with sustainable strategies on how to improve their performance.
- Secondly, there is need to initiate and articulate a coordinating system for the various on-going efforts to enhance their efficiency and efficacy. This effort should further involve finding ways through which to encourage the population to initiate more appropriate environmental management practices on their own.

PLA and Adaptive Planning

The proliferation of slums and emergence of slum subculture in Uganda's urban landscape is unfortunate. But even more unfortunate is that tangible intervention initiatives are yet to be put in place to halt the onslaught of slums. By virtue of the urban monetary economy, the urban poor have no access to even the basic human resources such as shelter and enough food.

This obviously makes them the most vulnerable group. Their situation is exacerbated by the affluence of their rich neighbours who look on as the poor suffer poverty. This means that any intervention must adopt a learning approach that involves the local beneficiaries not as recipients but as planners, implementers as well as participants in evaluation of their activities. In substandard settlements, abject poverty, improper attitudes and perception of the communities toward environmental health, nurture a relationship between disease and poor sanitation. Any effort that is directed at addressing the sanitation problem must focus on the people, the main beneficiaries.

Conclusion

What emerges from the above analysis is that the peri-urban environment problem is multi-dimensional and must be addressed as such. Sustainable urban environmental health and welfare will require building partnerships. It calls for a multi-dimensional approach founded on simultaneous and mutually reinforcing initiatives. To ensure success and sustainability, however, community participation should be the guiding principle. All stakeholders including government, civil society NGOs and CBOs, as well as multilateral agencies must work together to ensure sustainable urban growth with improvement in people's welfare amidst a healthy environment. As Mougeot (1993) asserts, efficient and sustainable management approaches must address linkages between urban waste, poverty and people's perception of environmental degradation and their degree of concern must be clarified, and action plans must be drawn.

References

Emory, C 1992. *The Poverty Debate: Politics and the Poor in America,* Westport, Greenwood Press, Connecticut.

Halfani, M. 1993. *Urban Poverty in Anglophone East Africa,* University of Dar-es-Salaam, Dar-es-Salaam.

Kalsi, A. 1995. *Urban Poverty Reduction Programmes in Developing Countries,* World Bank, London.

Kalanzi, J. 1987. *A Survey Report on the Health Problems among Under-Five in Mulago Slum Area, Kampala City Council, Kampala.*

Kampala City Council,1999. *Sanitation Quality Gap Study in Central Division Kampala, Kampala City Council, Kampala.*

Kamya, S. 1992. *Morbidity in Children Under-five in Bwaise,* Makerere, Kampala University.

Kok, W 1994. 'Setting The World To Work'; in *Poverty and Development,* Harvester Wheatsheaf, London.

Lewis, O. 1972. *Life in a Mexican Village Tepoztlan Restudied,* University of Illinois Press, Urbana.

Mabogunje, A. 1994. 'Overview of Research Priorities in Africa', *Urban Research in the Developing World,* 2, 26-47.

Maxwell, D. and Zziwa Z. 1993. *Urban Farming in Africa: The Case of Kampala, Uganda,* ACTS Press, Nairobi.

Meereboer, T. 1994. *The Anti-Poverty Agenda,* Hamondsworth Viking, The Hague.

Ministry of Health, 1998. *Annual Report.* MOH, Entebbe, Uganda.

Ministry of Finance and Economic Development, 1987. *Background to the Budget,* MOFED, Kampala.

Ministry of Finance and Economic Development, 1991. *Population and Housing Census,* MOFED, Kampala.

Mougeot, L. 1993. *Urban Environment Management,* IDRC, Ottawa, Canada.

Mwesigwa, D. 1987. *The Structural and Spatial Significance of the Small Scale Economic Sector in Kampala City, Uganda,* University of Dar-es-Salaam Press, Dar es Salaam.

Nuwagaba, A. and Kisamba-Mugerwa, W. 1993. *Comparative and multidimensional analysis of common and private resource land tenure and development in Uganda,* Makerere Institute of Social Research (MISR), Kampala.

Nuwagaba , A. 1994. *Impact of Macro Economic Policy on Housing Investment in Uganda,* Makerere University, Kampala.

Nuwagaba, 1997. *Sustainable Human Settlements in an urban world,* Makerere University Press, Kampala.

Nuwagaba, A. 1999. *Sanitation Quality Gap,* Kampala City Council, Kampala.

Onibokun, A. 1995. *Poverty in Nigeria.* CASSAD, Ibadan.

Perlman, J.E. 1976. *The Myth of Marginality: Urban Poverty and Politics in Rio de Janeiro,* University of California Press, Berkeley.

Pryer, J. 1993. 'The impact of adult ill health on household income and institution in Khulna, Bangladesh', in *Environment and Urbanization,* 5, 2, 35-49.

Sen, A. 1981. *Poverty and Famines,* Oxford University Press, Oxford.

Sengendo, H. 1994. *Urban Poverty, Shelter and Urban Governanance,* Makerere University Press, Kampala.

Simpson, S. 1999. *Nationalising the Anti-Poverty Agenda?* IDS Bulletin, University of Sussex.

Streeten, P. 1978. (Ed) Special Issue on Poverty and Inequality, *World Development Report,* 6, 3, 60-78.

Stren, R. and White, 1989. *African Cities in Crisis; Managing Rapid Urban Growth,* Boulder, Westview.

The New Vision (Newspaper) March, (1998), p. 8, Kampala, Uganda.

The Republic of Uganda, 1967. *Constitution of the Republic of Uganda,* Government Press, Kampala.

The Republic of Uganda, 1995. *Constitution of the Republic of Uganda,* Uganda Government Press, Kampala.

Todaro, M. 1999. *Economics for the Developing World,* Macmillan, New York.

Townsend, P. 1979. *Poverty in the United Kingdom,* Penguin, London.

Wratten, C 1995. 'Conceptualising Urban Poverty', *Environment and Urbanization,* 7, 1, 26-34.

Chapter 18

Government Policy and Environmental Waste Management in Maiduguri Urban Area, Nigeria

S. O. Okafor

Introduction

This chapter focuses on solid waste, which is defined as all the wastes originating from human and animal activities that are normally solid and relinquished as useless or unwanted. It embraces the heterogeneous mass of discarded material from households and commercial activities as well as the more homogeneous accumulations of a single agricultural or industrial activity (Heeramun 1995; Odeyemi 1995).

The insightful studies of Nze (1977), Roberts (1993), and Heeramun (1995), among others, have addressed environmental sanitation problems of developing countries. They all contend that with rapid population growth and urbanization, urban areas in these countries, Nigeria inclusive, inevitably generate enormous filth and waste. But the authorities assigned to control or clean up urban environmental waste have, unfortunately, been unable to carry out this important responsibility even in the face of the massive build-ups of the wastes.

One could identify four factors responsible for these phenomena in Nigeria:

- there are faulty and deficient infrastructures, manifested in poor storage and disposal facilities. Added to this, is the use of foreign methods and equipment which the country cannot afford to sustain and maintain (Nze, 1977; Roberts, 1993; Oyesiku, 1995;

Abiodun, 1995);

- there is inadequate governance, which has encouraged the sharing of power among various levels of government responsible for urban waste management. This has lent itself to overlapping of responsibilities in some urban areas and in others, centralization of control of activities (Roberts, 1993; Abiodun, 1995);
- financing is very poor, as collection, treatment, and disposal of wastes are capital-intensive. With poor funding, environmental sanitation authorities find it difficult to meet the cost of running these activities (Abiodun, 1995); and
- there is also the issue of human factor. This pertains to the poor attitude of the Nigerian public towards environmental hygiene. This explains the littering of wastes at points in the urban areas where they should not be (Nze, 1977; Sani, 1977; Lekwa, 1995).

These factors can be attributed to the government policy directives. There is, therefore, a need to address the effects of government policy on the management of waste in an urban area. Maiduguri is chosen as an appropriate case study because of its rapid urbanization. Its location on the border between Nigeria, Chad and Cameroon encourages cross-national migrations.

The primary objective of this study is to identify the impact of government policy on the management of environmental waste in Maiduguri urban area. Specifically, the writer has examined:

- government declared policy directives for the control of environmental waste in Nigeria;
- the implementation of these directives in Borno State generally, and Maiduguri urban area, in particular;
- the effects of the policy directives on the management of environmental waste in the urban area; and
- the people's reaction to the policy directives, including the mechanism for waste control in the urban area.

A systems approach to the study of administration referred to as the policy management has been utilized for this paper. Policy management can be defined as the evaluation of the process through which government purposes are implemented. Policy is seen as inputs or demands brought to bear on government which get articulated into a preferred choice position or defined objective. Management is the conversion of the policy choice into programmes to produce the output of the administrative or bureaucratic system. The impact of programmes on people turns out as part of their environmental satisfaction or dissatisfaction.

As a systems approach to the study of administration, it puts the political and evaluative aspects of the governmental context together with the administrative and implementation dimensions of bureaucratic behaviour. Policy management perspectives focus on three processes - administrative problem (operating decisions), management response (implementation), evaluation (assessment/ management of impact) (Howard and Okafor, 1983; Okafor, 1991).

The data were collected from both primary and secondary sources. For purposes of data adequacy, the writer resorted to applying methodological triangulation, thus enriching his data by relying on survey and available data methods.

The primary sources consist of focused group interviews. The Sole Administrator in-charge of Borno State Environmental Sanitation Board and his team, and the Head of Administration and his assistants in Maiduguri Metropolitan Council were interviewed. Two separate interview guides were used for the interviews. Fifty-one street traders and one hundred and eighty-seven residents in the various wards in the urban area were conveniently selected and interviewed. The major sources of secondary materials are government publications and records, academic journals and books. These documents are useful in throwing some light on government policy perspectives, problems of environmental sanitation in urban areas in Nigeria, and other developing countries, the nature of wastes and the setting of the study.

Both qualitative and quantitative data analyses were used. For quantitative analysis, simple percentages and frequency distribution of the variables were used to see the basic distribution of the samples. Borno State is made up of twenty-one local government areas, of which Maiduguri urban area is one. Maiduguri came into prominence initially as the seat of Government of Borno Province. In 1968, following the introduction of twelve-State Structure in Nigerian Federalism, it became the administrative headquarters of the newly created North-Eastern State. In 1976 the twelve-state structure was changed to nineteen. North-Eastern State was split into three States - Bauchi, Borno and Gongola. In 1991, Borno State was divided into two, Borno and Yobe states; and Gongola State was broken into two states - Adamawa and Taraba. Maiduguri continued as the seat of Government for Borno State. Apart from serving as administrative headquarters initially for Borno Province, then for the North-Eastern State, and later for Borno State, it has been the seat of the Shehu of Borno. The Shehu is the most prominent traditional ruler in the State.

With these features, Maiduguri urban area has been the largest centre for migration, commercial and industrial enterprises in Borno State. Since 1968, its growth has been accelerated, attracting people from all over the federation. The urban area is cosmopolitan in nature and its population heterogeneous both ethnically and in terms of occupation. Its estimated population in 1991 was placed at 629,496 persons (Population Commission, 1991). Projecting for 1996 using a yearly growth rate of 2½ per cent, the current population of Maiduguri is estimated at 712,217 persons. The number of wards in Maiduguri has recently been increased from twenty-one to thirty-four.

It has one large market, popularly known as 'The Mondaymarket' and seven satellite markets. The Monday-market is a state government parastatal arrangement, managed as a company. Besides the markets, the urban area has six motor parks. Other commercial enterprises and manufacturing industries in the urban area include metal fabrication, food and allied industries; printing, furniture, block making, leather shoe, and general industries; services provided by insurance and financial institutions, hospital and medical clinics, pharmaceutical and chemists, hotel and restaurants, legal, transport, petroleum filling stations, departmental stores and supermarkets, bookshops, motor and vehicle sales, and general trading.

The informal sector is dominated by some of the above-named service enterprises, the self-employed, street and petty-traders, road-side motor vehicle mechanics, welders, panel beaters, meat roasting (Suya) kiosks, tailors, shoe-repairers, electrical and electronic equipment repairers. These commercial enterprises and manufacturing industries are potential sources of wastes. These sources are further increased by the presence of street traders. This is a feature which, as Roberts (1993) has observed, is peculiar to the informal sector of the economy of urban areas in Africa.

Government Policy Directives on Environmental Waste in Nigeria

Conscious of the need to maintain a clear and habitable environment in its human settlements, the Federal Government of Nigeria has for quite sometime made policy statements, aimed at coming to grips with the problems of environmental wastes. For our discussion of these statements, we depend on five major sources – National Development Plan reports, the 1979 and 1989 Constitutions of Nigeria, War Against Indiscipline Programme (Phase Five), Federal Environmental Protection Agency (FEPA) and Government National Policy On Environment, and National Rolling Plan reports.

National Development Plan Reports

The first two national development plans covering the periods 1962-68 and 1970-74 offered nothing realistic in regard to the sewerage, drainage and refuse disposal sector. In the *Third National Development Plan 1975-80 report* government made the following observations on the sector:

> In many urban areas outmoded methods are still being employed in disposing of human wastes; streets are frequently littered with domestic garbage because of the inability of township authorities to organize regular collection and disposal, and extensive flooding is often experienced especially after heavy rains because of the inadequacy of existing storm water drainage system. Some modern sewerage treatment plants can be found in a few large establishments and institutions, but there is not a single Nigerian town with a modern or central sewerage system at present (cited in Sani, 1977: 106).

The Government did not find the environmental situation in the urban areas acceptable and was therefore determined to improve on it. It declared that during the plan period 1975-80, feasibility studies for central sewerage and integrated drainage systems should be carried out for all the state capitals in the early years of the plan period, and implementation should start towards the end of the period.

During this period, the State Governments were expected to spend a total sum of N273.9 million in the sewerage and drainage sub-sector. Half of this amount was for feasibility studies and actual construction whilst the remaining half was earmarked for the purchase of vehicle for refuse disposal (Sani, 1977).

It would appear that by the end of the *Third National Development Plan* period in 1980, progress made on environment sanitation in urban areas was not impressive. Hence the government came out with some pronouncements to that effect in the *Fourth National Development Plan, 1981-85.* Examining past efforts to make the environment of human settlement in Nigeria clean and tidy, it observed, amongst other things, that:

- in spite of the efforts of the government in the sewerage, drainage and refuse disposal sector, there was a wide gap between the present condition and the standards required for a healthy environment;
- refuse disposal was poorly organised in the urban area;
- because of lack of adequate finance and executive capacity, local government councils have not been able to carry out refuse disposal efficiently; and
- performance in the sector has not been encouraging during the plan period 1975-80. In terms of expenditure, fifteen states spent less than 50 per cent of their financial allocation for the whole plan period while six states utilized less than 10 per cent. All in all, the states in the federation utilized only 14 per cent of their total allocation. When considered by the index of physical performance, no state government achieved more than 25 per cent during the plan period; and
- with regard to refuse disposal, it was observed that only little progress was made in the purchase of refuse disposal vehicles, and the provision of dustbins and incinerators, and very few states have made some progress in the provision of public conveniences.

The major problems experienced by the sector were listed as low priority and inadequate attention was given to them. These problems were manifested in several ways, including inadequate funding, poor institutional arrangements for service delivery, and failure to take full account of all the requirements in the physical planning and design of built-up areas, particularly in the area of storm water drainage. The problem experienced with refuse disposal in the urban areas was with refuse collection and disposal. In many cases, refuse was dumped at random, and where bins were provided, they were hardly emptied with the result that the bins overflowed with refuse and formed unsightly heaps on the streets. One important factor held responsible for this unsatisfactory situation was the lack of effective institutional machinery for refuse collection and disposal.

The government therefore reiterated that its major objective during the plan period, 1981-85, was to maintain a clean and habitable environment in Nigeria's human settlements. In order to meet this objective, it directed that the following policy directives be implemented:

- Local governments should establish institutions such as waste disposal boards, which would be charged with the responsibility for environmental sanitation in their localities. These boards would be empowered to undertake public works relating to sewerage, drainage and refuse disposal systems, and to make adequate provision for their maintenance. Efforts would be made to ensure that the local governments were adequately staffed and equipped for these very important functions.
- With regard to refuse disposal, the policy of the government would be directed towards organizing efficient and regular collection and disposal of refuse starting from household bins to local collection and disposal centres. Appropriate machinery and equipment would be procured and maintained for the purpose.
- As part of the effort to provide local governments with adequate staff, Sanitary Inspectors Department, fully supported with Health Inspectors, would be established in each local government area to enforce local government laws relating to environmental sanitation.
- To ensure a sustained source of funds for services to be rendered and ensure their reliability, a reasonable cost recovery system would be introduced in the form of licence fees for industrial waste disposal and service charges for residential consumers.

To encourage the involvement of the three tiers of government in the federation in this sector, the government stipulated thus:

- The Federal Government should be concerned with the building of central sewerage and drainage systems at Abuja – the Federal Capital.
- The State Governments should be involved in the areas of sewerage, storm water drainage and refuse disposal projects in the State. With regard to refuse disposal, the main thrust of the project includes the purchase of refuse collection bins, construction of incinerators, or refuse collecting centres and the purchase of refuse disposal trucks, loading tractors and bulldozers.
- The major involvement of local government in this sector should be the provision of public toilets and disposal of refuse.

The Constitution of Nigeria

The 1979 Constitution, which was promulgated towards the end of the Third National Development Plan period, assigned, *inter alia*, the following sanitation functions to local government councils:

- Establishment, maintenance and regulation of markets, motor parks, and public conveniences.
- Construction and maintenance of roads, streets, drains and other public highways, parks, open spaces, or such public functions as may be prescribed from time to time by State House of Assembly.
- Provision and maintenance of public convenience and refuse disposal.
- Control and regulation of outdoor advertising and hoarding, shops, kiosks, restaurants and other places for sale of food to the public.

The 1989 Nigerian Constitution assigned the same functions to the local government councils, thus putting the issue of environmental sanitation within the competence of the councils.

One may observe some inconsistency in the policy directives of the government. The 1979 Constitution assigned certain environmental sanitation functions to the local government. The same government in the *Fourth National Development Plan* argued that local government councils were unable to perform these functions satisfactorily because of lack of funds and executive capacity. It, therefore, suggested the establishment of a separate body, Waste Disposal Board, by Local Government Councils to take charge of these functions. Despite this move, the 1989 Nigerian Constitution reassigned the same function to the Councils.

War Against Indiscipline (WAI) Programme

By 1984, the Federal Government had started its crusade on environmental sanitation through its War Against Indiscipline (WAI) programme (Phase Five). According to this programme, every last Saturday of the month was proclaimed national environmental sanitation day.

Human and vehicle movements were restricted between 7 a.m.

and 10 a.m., and people were expected to stay at home and clean their environment. State Governments joined in this crusade on environmental sanitation, some specifying an additional day in the month, frequently, the middle of the month, for the exercise.

Federal Environmental Protection Agency (FEPA)

The Federal Government, in its effort to realise an environmentally wholesome society, established the Federal Environmental Protection Agency by Decree 58 of 1988. The Agency was established to monitor environmental changes and set up strategies aimed at preventing or reducing environmental abuse by individuals or corporate entities (Eshett, 1995, Roberts, 1993).

In November 1989, the Federal Government declared its national policy on the environment. The goal of this policy was to achieve sustainable development in Nigeria in general. Specifically, it was designed to secure for all Nigerians an environment adequate for their health and well-being, raise public awareness and promote understanding of essential linkages between environment and development, and to encourage individual and community participation in environmental improvement efforts (FEPA, 1989).

Strategies set out for implementing the national policy on sanitation and waste management sector include:

- Specification of waste disposal sites that guarantee the safety of surface and underground water system.
- Originating and enforcing standards for adequate sanitary facilities for the disposal of human and other solid wastes in dwellings, housing estates, and public facilities in both urban and rural areas.
- Establishment of monitoring programmes, including periodic surveillance of approved waste disposal sites and their surroundings and wastewater treatment systems.
- Provision of information on the appropriate methods and technologies for the treatment and disposal of wastes.
- Regulation, registration, and licensing of all major land-based waste disposal sites and system (FEPA, 1989: 12).

Currently there is a strong move to institute FEPA offices in all the States of the Federation.

The National Rolling Plan

In spite of all efforts so far to secure a clean and healthy environment for Nigerians, the government, in the First National Rolling Plan report 1990-1992 described the prevailing situation with regard to sewerage, drainage and refuse disposal sector as unsatisfactory.

Government's concern about the magnitude of the environmental problems called for further intervention to reduce the problem. The intervention included the establishment of Task Force on Environmental Sanitation and Protection.

Though some measure of success was recorded following the intervention, there problems were still experienced in this sector, namely: the absence of effective institutional machinery for refuse collection and disposal, and lack of funds. The former was manifested by refuse being randomly dropped, refuse bins irregularly emptied when they were provided, thus over flowing and forming ugly heaps along the streets. Lack of finance was caused partly by budget cutbacks and partly the inability of the governments to formulate appropriate cost recovery measures on services rendered. The budget cutbacks were reflected in the low level of government funding.

The major objects of government policy in this sector still remained the maintenance of clean and habitable environment in Nigeria's human settlement. In order to attain these policy objectives, the government directed the implementation of the following measures:

- Strengthening the existing Disposal Boards to enable them to effectively execute the environmental sanitation programme in their localities through the formulation and implementation of adequate legislation affecting the dumping of refuse in places and manner that would facilitate their easy removal.
- The policy of providing public conveniences in all strategic places in rural and urban settlements should be more seriously pursued.
- Government would encourage studies on possible economic uses of the household and human wastes and improve on the existing refuse disposal system.

With regard to the projects which the three tiers of government in the federation were directed to execute, the Federal Government was to continue and complete the Abuja sewerage network and sewerage treatment plant during the plan periods. State Governments were to deal with storm water drainage schemes and sewerage treatment plants. Local governments programmes were to be concerned with improving refuse collection and disposal systems as well as purchasing refuse disposal vehicles, and constructing public toilets and conveniences.

Government policy directives declared in the *National Rolling Plan Report 1991-93* were similar to those enunciated in the *Rolling Plan Report 1990-92.* For the implementation of the policy directives, albeit lofty and fraught with handicaps, we turn to the next section to view their implementation in Borno State generally and in Maiduguri urban area in particular.

The Implementation of Government Policy

Directives on Environmental Waste in Borno State

The implementation of Federal Government policy directives has led to (i) the establishment of alternative and supportive institutional mechanisms for the management of environment waste in the State, and (ii) the shift of greater power and role in sanitation matters from the local government to the State Government. The implementation of the policy directives would be better discussed in their chronological order.

The North-Eastern State Government, following the Federal Government policy statements during the Third National Development Plan period, provided for five specific environmental sanitation programmes and the expansion of refuse collection systems in Maiduguri urban area and four other towns in the State. The thrust of the policy was more on feasibility studies.

The Borno State Government Edict 1977, referred to as Edict Number 6 of 1977, assigned most of the functions relevant to environmental sanitation exclusively to its local government councils, and, in the case of our study, to Maiduguri Metropolitan Council. Other functions are to be handled concurrently by the local government councils and the State Government. Examples of those functions reserved for the local government councils were markets and motor parks; sanitary inspection, refuse and night soil disposal; public convenience; regulation of hawking of wares or the erection of stores on or near any street. Functions assigned concurrently to the local government councils and the State Government include meat inspection and abattoirs, control of water and atmospheric pollution, and of piped sewerage systems.

The 1979 Nigerian constitution and later that of 1989 allocated, as we have shown earlier, the same functions to local government councils, and thus tended to support the Edict. This action by the governments accords with the popular belief that the management of environmental waste has traditionally been the function of local government authorities.

The indication given by the Federal government in the Fourth National Development Plan report, in its review of past governmental efforts to secure a clean and healthy environment for human settlements in Nigeria, was that the local government councils have been unable to manage environmental waste because of lack of finance and executive capacity. Consequently, the Federal Government recommended certain measures, which we have discussed previously. The implementation of these measures tended to shift the management of waste gradually away from the local government to the state government in the first instance. Later, state institutions were established to take full charge of the management of environmental waste.

The Borno State government enacted the Borno State Environmental Sanitation Edict, referred to as Borno State Edict No. 3 of 1986. The purpose of the Edict was to make comprehensive provisions for keeping Borno State environment always clean and free of disease. The onus of keeping Maiduguri urban area clean and healthy was to fall on the State Ministry of Health, owners or occupiers of premises and Maiduguri Metropolitan Council. The State Commissioner of Health was empowered by the Edict to assign a medical officer to take charge of any area of the State for purpose of environmental sanitation control.

It became mandatory for every owner or occupier of any premises to keep his premises and surrounding clean and tidy. He was also to provide adequate receptacle for storage and transferring of waste to authorized collection centres. As provided by the Federal Government in the *Fourth National Development Plan*, Health Officers were empowered to enforce environmental sanitation laws, and to impose fines not exceeding N100 on first offenders. Subsequent offences could earn fines up to N200.

Maiduguri Metropolitan Council or authority responsible for the collection of refuse was assigned the duty of providing adequately public dust bins for the purpose of receiving and disposing of wastes. Every local government council in the State was directed, in line with

Federal Government policy directive, to make proper provision of public convenience at all strategic places.

The shift of greater power and role from the local government to the State Government in sanitation matters started to manifest itself with Borno State Government, enacting *Environmental Sanitation Edict No. 3 of 1986.* This piece of legislation empowered every court in the State to try and impose punishment for environmental sanitation offences under the Edict. In the same year, the *Environmental Sanitation Edict No 4,* established Environmental Sanitation Court for the State, thus transferring the power of trying environmental offences and imposing punishment from every court in the state to Environmental Sanitation Court only. The implication is that these courts have dispossessed both the medical officers of the State Ministry of Health and the Health Officers of Maiduguri Metropolitan Council of their powers to enforce environmental sanitation laws in the urban area.

The policy directives of the Federal Government had allowed local government councils in the States to establish Waste Disposal Boards to take over management of environmental waste from them. In spite of this provision, the *Borno State Environmental Sanitation Edict No. 5 of 1986* established State Environmental Sanitation Board to perform the functions to the Disposal Board. Local Governments in Lagos State had similar experience with Lagos State Government, creating Waste Disposal Board instead of allowing them to do it (Roberts, 1993). Functions allocated to the Sanitation Board by the Edict were extensive. They include collection and disposal of both wet and dry refuse, including human wastes, street cleaning and maintenance, control of air and water pollution, market sanitation, installation of refuse collection devices at appropriate points in the urban area.

The Board was also entrusted with some legislative powers to make by-laws, and regulations concerning environmental sanitation matters like sanitary inspection, public conveniences, removal and disposal of all types of refuse, street cleaning and maintenance, control of industrial waste, air, and water pollution.

The Federal Government suggested, as we have stated earlier, the introduction of cost recovery system as a means of providing adequate funds for waste management. In the Edict, the State Government in line with this suggestion made provisions for the sanitation board to secure adequate funds with which to discharge its functions or meet with its obligations. Thus, the Board was empowered to borrow with the prior approval of State Executive Council such funds as needed from within Nigeria. Apart from money provided by the Government, there were other avenues set aside by the Edict for the Board to source for finance.

On May 10, 1989, the *Borno State Environmental Sanitation (Amendment) Edict,* referred to as Edict No. 1 of 1990, was enacted. This Edict repealed Edict No. 4 of 1986, divided the State into five zones, and provided for an environmental court in each zone. These courts were given extensive powers including the power to fine on the spot any person or body corporate that committed any offence under the Edict.

Borno State Environmental Sanitation (Amendment) Edict 1994 increased the punishment for environmental sanitation offences to a fine, not exceeding N200 or three months jail sentence or both for first offenders. Subsequent offenders were liable to a fine not exceeding N500 or six months or both. For corporate bodies, the punishment to be imposed was a fine not exceeding N2000 in the first instance and N5000 for subsequent offences.

Borno State Government joined the Federal Government in its crusade on environmental sanitation through WAI Programme (Phase Five), by declaring the second Saturday of each month the State Environmental Sanitation Day. No human nor vehicle movement is allowed in the state streets between 7 a.m. – 10 a.m. People are expected to keep their environment clean during that period.

In line with Federal Government declaration in the *First National Rolling Plan 1990-92,* the Environmental Sanitation Task Force was established by Borno State Government in 1993. The Task Force was established to supplement the Environmental Sanitation Board in the State, but it has no work force of its own, and depends on that of the Board.

The Effects of Policy Directives on Waste Management in Maiduguri Urban Area

We have observed in the preceding section that the implementation of Federal Government policy directives has led to three major developments:

- the establishment of Environmental Sanitation Board with a supporting institution, Environmental Task Force, to take over the

management of environmental waste in Borno State generally and particularly in Maiduguri urban area,
- the creation of Environmental Sanitation Court to enforce sanitation laws, and
- the gradual shift of power and role from the local government to the State Sanitation Board on sanitation matters.

To examine the effects of these policy directives on the management of environmental waste in Maiduguri urban area will imply considering:

- how the Sanitation Board has coped with waste management in the urban area,
- how the Sanitation Court has played its role in enforcing sanitation laws, and
- how the Sanitation Board - local government relations have affected certain specific environmental sanitation matters in the urban area. These issues will be examined in this section.

There are six functional elements associated with waste management, namely:

- waste generation,
- on-site handling and storage,
- collection,
- transfer and transport,
- processing and recovery, and
- disposal.

They run the gamut from the waste generation to its final disposal (Heeramum 1995). These elements are discussed here as they relate to Maiduguri urban area, before we come to the conclusion on the effectiveness of the Sanitation Board in respect of its waste management, in the last section of this chapter.

Water Generation and Volume

Solid waste generation in the household is influenced by a number of factors such as family income, eating habits, and life styles; climatic factors, including seasonal variation. The effect of the last factor is that certain food items are not available all the year around. However, it is estimated that the average household generation per day in Maiduguri urban area ranges between 1.54 and 2.28 kg, and per capita generation fall between 0.2 and .18 kg (Kagu, 1996). The writer has to add here that the collected waste represents only a part of the total generated waste. In an interview conducted with 185 residents in various wards in the urban area, the question, 'where are wastes from your house/compound stored?' - was asked. Eighty one (44 per cent) of the residents stored theirs in a receptacle to be deposited at the collection centres, 54 (29 per cent) held that the wastes were kept in the compound to be burnt, and 41 (22 per cent) dumped theirs on roadsides, while 9 (5 per cent) gave irrelevant responses.

Wastes generated in Maiduguri urban area are of two types - commercial and industrial, and urban wastes. The main sources of commercial and industrial wastes are business (Informal and Formal), markets, motor parks, repairs, maintenance, manufacturing and construction premises, used wrapping materials including plastic polythene bags, cartons, empty bottles etc. Urban wastes are mostly derived from households/compounds, institutions, and street cleaning. Street wastes such as leaves, papers, sand, polythene bags, wrapping materials and carcasses (Dead animals) are forms of wastes generated from street cleaning. Kagu (1996) mentioned the following as the sources of wastes in households/compounds in the urban area - cooking (54 per cent), farm products (6 per cent) and others (40 per cent) which include wastes from gardens or the like, such as oranges and bananas, or residues of farm products. Cooking tends to be the major source of household wastes. The major components of waste are ashes, garbage, rubbish, and garden waste. The components of household refuse, according to Kagu, are leaves/wood (27 per cent), papers (16 per cent), food (7 per cent), metal (4 per cent), polythene bags (23 per cent), and unclassified refuse (13 per cent).

Waste Storage

The State Environmental Sanitation Board provides no household/compound with dustbin for waste storage. Nor does the Board provide household with a regular daily waste collection service. An interview which the writer had with the Chairman of the Board and his team revealed that an attempt was made in 1987 to provide households with dustbins but the exercise was stopped after that year. The dustbins were not effectively used, there were no funds to provide them because of their costs, and there was some difficulty in emptying the dustbins when they were full. Domestic waste is therefore stored in various ways, as revealed by our interviews, before they are collected.

185 of the various wards in the urban area were asked the question: 'Where are wastes from your house/compound stored?' 55 of the respondents (30 per cent) said they used receptacle without cover for storage, 41 (22 per cent) dumped them on the road-side, 28 (15 per cent) stored them in the back-yard of the compound, 26 (14 per cent) dumped their waste in an enclosure built with brick/block/ stone within the compound, 13 (7 per cent) stored them in a pit dug in the compound, and 9 (5 per cent) used other means.

They were then asked to state who provided the receptacles used for storage. Of the 101 respondents, 44 (44 per cent) indicated that the receptacles were provided by the owners of the house/compound, 35 (35 per cent) mentioned tenants, 13 (13 per cent) cited Maiduguri Metropolitan Council, and 9 (9 per cent) mentioned the Environmental Sanitation Board. This tends to support the view expressed by the Chairman of Sanitation Board. Those receptacles reportedly provided either by the Maiduguri Metropolitan Council or Environmental Sanitation Board might have been a carry over from the days when Metropolitan Council or the Sanitation Board provided them.

Waste Collection

In general, designated collection centres receive wastes from households/compounds. It is from these centres that they are transported to their ultimate destinations - the disposal sites. Either owners of the houses or tenants in a compound undertake the burden of depositing household waste at these centres. This view is confirmed by responses received from the sample that were asked the question, 'Who disposes of the wastes gathered or stored in your compound?' Out of 176 residents who answered the question, 76 (43 per cent) affirmed that owners of compounds undertook the burden, 63 (36 per cent) tenants, 27 (15 per cent) the Sanitation Board, while only 10 (6 per cent) indicated that the Maiduguri Metropolitan Council did the job. Thus, Sanitation Board's involvement in this regard seems to be minimal.

There are only two legal collection centres. One is the refuse bunker and the other is the legal dumping site. There are also illegal dumping sites. A legal dumping site is a place officially set aside for dumping wastes. An illegal dumping site is an unauthorised place used for dumping wastes. The Chairman of Sanitation Board and his team informed the writer in an interview that there are twenty-three refuse bunkers and one hundred legal dumping sites in Maiduguri urban area. The latter are scattered over the whole wards. Out of the twenty-three bunkers, only eight are currently functional. Many wards do not have refuse bunkers and therefore depend on legal and illegal dumping sites. As the illegal sites are difficult to police, illegal dumping itself is also difficult to stop.

This observation tends to accord with the responses received from our sample. The question asked is 'Does your ward have refuse bunkers?' Out of 178 residents who provided answers to the question, 126 (71 per cent) affirmed that their wards do not have them, whilst 51 (29 per cent) held that they have them. One person did not give any relevant answer. The Chairman of Sanitation Board revealed that the Board cleared the collection centres at least twice weekly. Domestic collection of wastes from the centres is slow and labour-intensive. The collection involves using brooms, rakes, forks, wheelbarrows, and shovels.

The Sanitation Board has six zone offices in the urban area. Each zone has between ninety and one hundred staff members. At the head of each area office is a zone officer who is of the rank of a Health Superintendent. It is the function of the Zone Officers to supervise and co-ordinate sanitation activities of their respective areas including streets, open spaces, markets and motor-parks. Educational and medical institutions manage the waste generated from their environment.

The writer was also told during the interview with the Chairman of Sanitation Board that the Board's street cleaners keep the streets clean on daily basis. On sanitation days, non-Board workers assist the cleaners. The major waste generated from the street is sand. The sand is generally deposited on the roadside. Another type of waste generated is leaves. When the leaves are small in quantity and dry they are burnt on the roadside. But when they are in large quantity and wet, being difficult to burn, the Board is invited to clear and dispose of them.

The total workforce employed by Maiduguri Metropolitan Council for environmental sanitation when the management of wastes in the urban area fell under its competence was 1,130. The distribution of the workforce into categories was labourers 575, drivers 77, supervisors 8, and health inspectors 370. When the sanitation board took over sanitation function from the Council, 666 workers were deployed to the Board. The number and categories of staff deployed were labourers 380, drivers 74, health inspectors 206, supervisors 6 (Interview with the Head of Administration, Maiduguri Metropolitan Council). The current workforce size is difficult to ascertain from the Sanitation Board.

Transport and Transfer

Transfer of wastes is generally from the collection centres into vehicles and subsequent movement to disposal sites. The Chairman of Environmental Sanitation Board revealed in an interview that when the Sanitation Board took over sanitation function from Maiduguri Metropolitan Council, the Board had the following vehicles - 4 pail loaders, 15 tippers, 2 tractors, 1 caterpillar bulldozer, 1 gulley emptier. By 1996, the Board had only five functional vehicles - namely 1 pail loader and 4 tippers. The main form of transport used in transferring wastes to disposal sites consists of tippers, pail loaders. The Chairman estimated that when vehicles are functional, they could make between 90 to 100 trips daily from collection centres to disposal sites.

One major problem experienced by the Board in the transfer of wastes from collection centres to disposal sites is the open roof nature of vehicles used. Because their roofs are open, wind tends to litter the streets with the wastes in the vehicles while they are moving. This littering is what Oyesiku (1995) has described as a significant contributory factor that further degrades Nigerian cities and rural environments.

Processing and Recovery

These embody all the techniques, equipment and capabilities used to increase the efficiency of solid waste management (Heeramun 1995). This is an area very much neglected in the management of waste by the Sanitation Board. There is only one incinerator serving the whole urban area. It is situated close to the Abattoir serving Maiduguri urban area. The writer is treating incineration as a processing method following Heeramun's stand that incineration is a processing and not a disposal method. There is no work force attached to this functional element. The processing and recovery of items from the wastes are carried out privately by beggars who search for items like plastics, used bottles, mutilated metals, zinc, materials, bowls and papers for sale in order to earn a living. These items when recovered are sold to companies who recycle them. Stray animals such as goats, sheep, and cows feed on the dumps. The action by beggars and animals are carried out at the disposal sites and collection centres.

Waste Disposal

This is another neglected area in the management of waste in Maiduguri urban area. Three methods of waste disposal are identified - dumping, burning and tipping. The most common method is the uncontrolled crude dumping at illegal dumping sites. The Sanitation Board and certain voluntary bodies ultimately clear the wastes so dumped. Burning of wastes, as we stated earlier, are mostly done in compounds, particularly for household wastes. Wastes generated in open spaces, markets, motor-parks, and streets inclusive are either burnt on the spot or carted away to riverside or other remote spots for burning. At present, there are four disposal sites. These sites are not landfills. In fact, the Chairman of State Environmental Sanitation Board called them dumping pits. These pits are not dug for dumping wastes. Rather, they emerge as a result of construction work. Road construction companies dig the pits to obtain sand/gravel that are used for building roads. By the time the roads are completed, the pits are left. The Sanitation Board looks for these pits and uses them as disposal sites. The present sites are located at the Bama Road off University of Maiduguri Campus, off Polo Road near Old Government Residential area, Damboa Road, and Bulumkutu Road. These pits are generally far from human settlements. It was revealed in an interview with the Chairman of the Sanitation Board that the pits are covered with soil by the landowners when they are full. At times, wastes collected from the urban area are used for land reclamation purposes.

Given the foregoing, it is not absolutely clear whether the Borno State Sanitation wastes disposal methods do really come under the surveillance of FEPA. Disposal sites used are not approved sites and neither FEPA nor any body has raised any alarm. Though Borno State Environmental Sanitation Board is the official body entrusted with the responsibility of waste management, the degree to which it copes with this function has been discussed. The Board is supported unofficially by non-governmental groups like individuals, families, waste disposal contractors, communities and nature (rain and wind) (Kagu 1996).

The Role of the Sanitation Court in Enforcing Environmental Sanitation Laws

Table 18.1 shows the number of sanitation offenders tried by the court, number of people jailed, and the amount of revenue realised from 1991 to 1995. Just as the number of people tried tended to have risen sharply in 1994, so did the amount of the revenue realised. As a supporting institution, the court deserves a more than average pass mark in the performance of its role. People's reaction to the court appears mixed. We asked the residents the question, 'Should the Environment Sanitation Court have a branch in the ward?' Out of 180 respondents, 89 (49 per cent) responded positively, and 88 (49 per cent) answered negatively, 3 (2 per cent) gave responses which are not relevant. The reasons given for their responses were also mixed - it will make sanitation more effective, sanitation lawbreakers will be controlled by it, the court is corrupt, establishing it in the ward is unnecessary.

Table 18.1: Showing Number of People Tried / Jailed for Environmental Sanitation Offences and Money Raised: 1991 - 1993

Year	Number of People Tried	Number of People Jailed	Amount Realised x (Naira)
1991	579	385	19,130.00
1992	436	293	20,625.00
1993	497	353	25,345.00
1994	1,384	862	103,200.00
1995	1,562	847	166,949.00

Source: State Environment Sanitation Court, Maiduguri.

With the establishment of the Sanitation Board, Maiduguri Metropolitan Council became absolved from Sanitation matters. This change has tended to affect its attitude towards sanitation issues. I will use the control of public conveniences and street traders to illustrate this point. Maiduguri Metropolitan Council perceived these two cases as revenue generating ventures, the Sanitation Board, while keeping mute on public conveniences, sees street traders as nuisances. Following Federal Government directives, 60 public conveniences were built in Maiduguri urban area. 51 out of the 60 are located in the markets and motor-parks, and eight in the wards. Currently, for those located in the markets and motor-parks, three are not functioning, two have been contracted out and have been converted to public baths. For those in the wards, two are not functioning, two have been converted to shopping facilities, and three are now used as car wash facilities. With these changes, only one public convenience is available for use in the urban area. The major reason for diverting these public conveniences to other uses is the bid by Maiduguri Metropolitan Council to generate local revenue to supplement the grants received form Federal and State Governments.

The question is: 'How could Maiduguri Metropolitan Council convert public convenience in the wards from their normal use to other uses without intervention from the Sanitation Board?' One plausible explanation is that the Sanitation Board has been unable to maintain them because of its poor resources. This must have given the Maiduguri Metropolitan Council a free hand to divert them to other uses, particularly now that sanitation matters are no longer reflected in its goals.

Street trading is a common feature of the informal sector of the economy of most African towns, Nigerian towns inclusive. In Maiduguri urban area, street traders are found everywhere but they tend to concentrate more outside the premises of major markets and motor parks and the main post office. These street traders are either married men or women with families. They hail from Borno and neighbouring states like Yobe, Adamawa, Taraba, Kano and have all received informal education.

The Task Force has done everything possible, including the use of Environmental Sanitation Court, seizure of goods displayed for sale to stop street trading. In spite of the draconian Task Force action, the street traders still operate. No sooner does the confrontation by the Task Force end, no matter how serious the scuffle, the investment lost in the seizure of goods, the fine imposed by the court, than they return to their illegal bases as members of the Task Force turn their backs. The Task Force has even arranged to resettle them in some of the satellite markets. But this has not provided the solution.

The writer observes that the Task Force perceives these street traders differently from Maiduguri Metropolitan Council. To the Task Force, they constitute a nuisance. But to the latter, they are sources of revenue generation.

We asked 52 street traders the question: 'Do you pay fees for trading on the street?' 42 (82 per cent) admitted that they paid fees. Others stated that they did not. Of the 42 who admitted they paid fees, we posed the question: 'To whom do you pay the fees?' 21 (50 per cent) affirmed they paid to Maiduguri Metropolitan Council, 13 (31 per cent) held that they paid to Monday Market Company management. 7 (17 per cent) and 1 (2 per cent) stated that they paid to State Government and Environmental Sanitation Board respectively.

To find out what the street traders feel the government should do to stop street trading, we asked fifty-one respondents, 'What in your views should the Government do to stop street trading?' All of them responded that the government should build more market stalls.

Environmental sanitation days are schemes imposed by the governments on the people. It is the only means through which the people themselves are involved in waste management matters. The people themselves, from our observation, tend to accept this view.

The question, 'Do you observe sanitation days?', was put to 183 respondents. 175 (96 per cent) affirmed that they did, 8 (4 per cent) admitted that they did not. We went further to ask, 'What do you do during that time (7 a.m. –10 a.m.)?' One hundred and 61 (88 per cent) respondents stated that they used it to keep their premises and their surroundings clean, 12 (7 per cent) confessed that they used it to rest, 10 (5 per cent) held that they used the time to do their personal or family work.

To find out the popularity of sanitation days amongst people in the urban area, we asked the question, 'Will you support any move to scrap sanitation days?' Out of 183 respondents, 145 (79 per cent) replied that they would not support such move, and 38 (21 per cent) stated they would. These responses are indications that sanitation days are acceptable to people in Maiduguri urban area.

People tend to co-operate and work with the Environmental

Sanitation Board and the Task Force. This is manifested in the way individuals, communities and corporate bodies in Maiduguri urban area invite the Board to assist in demolishing illegal structures. Pressure groups, like Gamboru Progressive Union or Market Associations, report cases of street trading in their areas and seek the intervention of the Task Force.

Conclusion

We have mentioned some major Federal Government policy directives on management of environmental waste in Nigeria. Those aspects which have been implemented in Borno State have led to the establishment of certain institutions required for the management of waste in the State and in Maiduguri urban area, in particular. The effects of such policy directives on the management of waste in Maiduguri can be determined by the contributions of these institutions to waste management. Their contributions and the reaction of people in the urban area to them have been examined. One conclusion reached is that though officially the management of environmental waste is entrusted to the Sanitation Board, the Board does not carry out the six functional elements involved alone. This is because some non-governmental groups make their voluntary contributions to waste disposal. There are definitely some functional gaps, which should be filled by the Board. It has also been established that people in the urban area tend to accept the institutions.

One important aspect of the policy statements, which has not implemented, is what the Federal Government has described as cost recovery scheme. The implementation of the scheme would have assisted in raising funds to supplement government subventions for the management of waste. The Borno State Government, in line with Federal Government policy directive provided, in the *State Environmental Sanitation Edict No. 5 of 1986*, avenues for the Sanitation Board to source for funds from which to Supplement State Government subvention for waste management. These avenues have not been explored and that has resulted in the lack of finance for the management of waste. Poor funding is a major problem in that sector.

Another effect of Federal Government policy directives on waste management is manifested in the organisational structure of the Environmental Sanitation Board. This, one feels, requires review for enhanced management performance. The Board, as it is currently organised, is centralised. The State Government controls funding and management. The only contribution from the majority of people who are waste generators and beneficiaries of the effort of the Sanitation Board is the Sanitation day's contribution imposed by the Federal and State Governments. This top-down, welfare state approach is not adequate.

Waste management in Maiduguri requires a system that encourages synergy. The people who are the waste generators are encouraged to participate in the management of waste, and the government, whose concern it is to secure clean environment for its people is also involved. Each side should make its contributions according to its ability. The system introduced by *Borno State Edict No.3 of 1986* was very close to the arrangement because it encouraged synergy. The onus of keeping Maiduguri clean, according to the Edict, was left in the hands of the State Ministry of Health, respective owners or occupiers of premises, and the Maiduguri Metropolitan Council, which, if well utilised, will bring synergy. Each entity had its own assigned functions. This arrangement should have been allowed to grow to maturity, but, unfortunately, its life was suddenly terminated.

Decentralisation, is capable of bringing about synergy and improved waste management performance. The management of waste should not be concentrated at the State level. This should involve establishing at each ward, an environmental sanitation committee, which should be entrusted with the control and monitoring of the following functional elements associated with the management of waste: handling, storage, and transfer of waste from compounds/households and streets to legal collection centres. Membership of the Committee should be open to ward heads and residents who are to be elected into the committee. The monitoring, overseeing of the activities of ward Sanitation Committee and the control of processing, recovery and transfer of wastes from the collection centre to disposal sites should be left to the Sanitation Board.

Based on these observations, one could recommend the decentralisation of waste management in Maiduguri urban area. Such decentralisation will involve getting the active participation of communities at the grass root not only in the control but also in the management of solid waste.

References

Abiodun, Y. 1986. 'Urban waste management in Nigeria - the way forward', *Vanguard,* Lagos, January 9, p. 17.

Borno State of Nigeria, 1977. *Local Government Edict* (Edict No. 6), The Government Printer, Maiduguri.

Borno State of Nigeria, 1986. *The Borno State Environmental Sanitation Edict* (Edict No. 3), The Government Printer, Maiduguri.

Borno State of Nigeria, 1986. *The Borno State Enviromental Sanitation (Amendment) Edict* (Edict No. 4), The Government Printer, Maiduguri.

Borno State of Nigeria, 1986. *The Borno State Environmental Sanitation Board Edict* (Edict No. 5), The Government Printer, Maiduguri.

Borno State of Nigeria, 1989. *The Borno State Environmental Sanitation (Amendment) Edict* (Edict No.1 1990), The Government Printer, Maiduguri.

Borno State of Nigeria, 1994 *The Borno State Environmental Sanitation* (Amendment) *Edict,* The Government Printer, Maiduguri.

Egunjobi, L. 1995. 'Urban Governance and the Quality of Urban Physical Environment in Nigeria,' in Onibokun, A. and Faniran, A. (Eds), *Governance and Urban Poverty in Anglophone West Africa.* Fairchild Design and Printers, Ibadan, pp. 127-143.

Eshett, E.T. 1995 'Urban Governance and Environmental Quality in Nigeria,' in Onibokun, A. and Faniran, A. (Eds), *Governance and Urban Poverty in Anglophone West Africa.* Fairchild Designs and Prints, Ibadan, pp. 154-166.

Federal Republic of Nigeria, 1970. *Second National Development Plan 1970-74, First Progress Report,* The Central Planning Office, Federal Ministry of Economic Development and Reconstruction, Lagos.

Federal Republic of Nigeria, 1975. *Third National Development Plan 1975-80* (Revised), The Central Planning Office, Federal Ministry of Economic Development and Reconstruction, Lagos.

Federal Republic of Nigeria, 1981. *Fourth National Development Plan 1981-80* (Revised), The Central Planning Office, Federal Ministry of Economic Development and Reconstruction, Lagos.

Federal Republic of Nigeria, 1989. *National Policy on the Environment: Federal Environmental Protection Agency,* Jeromelaiho and Associates Limited, Lagos.

Federal Republic of Nigeria, 1990. *First National Rolling Plan 1990-1992, Volume I,* The Planning Office, Federal Ministry of Budget and Planning, Lagos.

Federal Republic of Nigeria, 1991. *National Rolling Plan 1991-1993, Volume I,* Federal Government Printer, Lagos.

General Secretary (Umara Baba Rabi), 1994. Letter titled 'Congratulation/Appreciation Letter' addressed to the Chairman, Task Force on Environmental Sanitation, Borno State in *Borno State of Nigeria: Report of the Special Committee on Environmental Sanitation,* Annex 111, p. 7 (Unpublished).

Heeramun, K.K. 1995. 'Solid Waste Management In Mauritius - An Alternative to Sanitary Landfill,' *Waste Management* 2: 80-127.

Howard, L.C. And Okafor S.O. 1983. 'Centre-periphery relations: the influences of the past on the present in Borno State', *Annals of Borno* 1: 57-74.

Iyioke, A. 1993. 'When buckpassing overtakes good governance', *The Guardian,* Lagos July 1, p. 6.

Kagu, A. 1995) *Urban Refuse Generation and Disposal: A Case Study of Maiduguri Metropolis,* M.Sc Dissertation, University of Maiduguri.

Lekwa, A. 1995. 'Towards efficient waste disposal', *Daily Champion,* Lagos January 15, p. 4.

Modupe, O. 1995. 'Urban governance and the quality of urban environment in Nigeria', in Onibokun, A. and Faniran, A. (Eds), *Governance and Urban Poverty in Anglophone West Africa.* Fairchild Designs and Prints. Ibadan, pp. 173-186.

North-Eastern State of Nigeria, 1975. *Third National Development Plan 1975-80, North Eastern State Programme,* Ministry of Finance and Economic Planning, Economic Planning Division, Maiduguri.

Nze, F.C. 1977. 'Managing urban wastes in Nigeria for social and economic development'. *The Nigerian Journal of Public Affairs,* 7: pp. 27-39.

Oderinde, J.A. 1995. 'Improving the quality of urban environment in Nigeria - appraisal of past roles of government', In Onibokun, A. and Faniran, A. (Eds), *Governance And Urban Poverty in Anglophone West Africa.* Fairchild Designs and Prints, Ibadan, pp. 144-153.

Odeyemi, O. 1995. 'Environmental pollution by solid and liquid wastes in Nigeria.' Lecture delivered at Obafemi Awolowo University, Ile-Ife, Nigeria.

Omorogbe, S.K. 1995. 'Human waste as a social problem: a sociological analysis of waste generation and management in Benin city', Paper presented at Obafemi Awolowo University, Ile-Ife, Nigeria.

Okafor, S.O. 1991. 'Cooperatives and rural development in Borno State, Nigeria', *CAFRAD, African Administrative Studies,* 36:35-52.

Okwu, M. 1996. 'UNDP harps on proper waste disposal, management in developing countries', *Vanguard* (Lagos), January 16, p. 16.

Oyesiku, K. 1995. 'A review of approaches to refuse transportation and disposal by the public authorities in Nigeria', Paper presented at Obafemi Awolowo University, Ile-Ife, Nigeria.

Parker, A.N. 1995. *Decentralization: The Way Forward for Rural Development?* (Policy Research Working Paper), World Bank, Washington, DC.

Principal Accountant (Mohammed Abba), 1994. Letter titled 'Re: Authority of Loading and Off-loading of Goods at The Vegetable Market, Maiduguri', dated March 23, 1994 addressed to the Chairman, Task Force Committee on Sanitation *In Borno State of Nigeria,* Report of the Special Committee on Environment Sanitation, Annex 111, p. 32 (Unpublished).

Roberts, F.O.N. 1993. 'Garbage crisis and federalism in Nigeria: the political economy of environmental sanitation in Lagos State', *Annals of the Social Science Council of Nigeria,* 5: 85-109.

Sani, H.A. 1977. 'Fiscal and institutional arrangement for environmental development in Nigeria', *The Nigerian Journal of Public Affairs,* 8: 99-123.

Secretary (Shettima Alkali), 1994. Letter titled 'Re: Vegetable Market' addressed to the Chairman, Task Force on Sanitation, dated January 3, 1994, In *Borno State of Nigeria, Report of the Special Committee on Environmental Sanitation,* Annex III, p. 3 (Unpublished).

Chapter 19

Adaptive Strategies of Small Fishermen to Environmental Degradation Induced by Oil Pollution in the Niger Delta: Implications for Rural Planning

A. Avwunudiogba

Introduction

Fishing is a major occupation among the various communities in the Niger Delta (Nigeria) and along the offshore areas where intensive crude oil production activities has been carried out since 1952 (Shell, 1995a, 1995b). Recently, the tempo of various activities associated with crude oil industry, including prospecting, exploration, production and refining has been on the increase with far reaching impacts on the environment of the Niger Delta. One of such impacts has been the increasing incidence of oil pollution with disruptive effects on the ecology of the fragile aquatic ecosystem, its productivity and the economic base of the rural fishermen who dwell in the Niger Delta. This chapter examines the strategies adopted by small-scale rural fishermen in the Niger Delta to deal with the disruptive impacts of oil production activities on their economic base.

Conceptual Considerations

According to Okafor (1988), developing economies face a dilemma. Although they are expected to meet the material needs of the population by exploiting the available resources, they simultaneously must ensure that national resources are not squandered through corruption and mismanagement. In other words, there is the need to strike a comfortable balance between the process of development on the one hand and ecological conservation on the other hand. And this quest has given rise to a variety of 'development concepts' among which the concept of 'eco-development' is particularly relevant to this discussion.

Eco-development refers to a development strategy that is ecologically sensitive and incorporates economic, social and environmental concerns in the process of development. In a way, the concept of eco-development is related to the more recent fanciful concept of 'sustainable development' which has been defined as 'development which meets today's aspiration without compromising the needs of future generations' (Aina et al, 1992).

In line with the concept of eco-development, Okafor (1988) argues that every rural environment is made up of eco-units, eco-groups and ecosystems delicately balanced and interrelated to form a dynamic system. This system performs optimally in a relatively undisturbed environment.

Accordingly, any change brought by human activity in the quest for development usually has these components disturbed and the balance dislocated with the degree, intensity, direction and severity of dislocation depending on how much of the natural environment is still preserved in the process of resource exploitation and development. Furthermore, dislocation of the ecological system usually results not only from over exploitation of the environmental natural resource base but also from complete disregard for environmental conservation considerations in the bid to achieve economic growth in the shortest possible time. This shortsightedness often leads to environmental overload and consequent disruptive ramifications. There must, therefore, be a conscious effort to maintain environmental harmony and integrity in the face of unrelenting demand for more material goods (Moran et al 1980). This inevitably calls for a pattern of resources exploitation and development which can be accommodated by the maximum carrying capacity of the environment. It also involves the consideration of the environmental base of the population along with development (Okafor 1988) in order to avoid ecological shock (Craik, 1971). Against this framework of eco-development, it is important to pursue human-centred intervention strategies in the development process in order to ensure a constructive relationship between man and his environment.

Unfortunately, the practicability of the concept and principle of eco-development in rural areas of most developing countries seems to be an impossible task. This is because the exploitation of natural resources for the benefit of all appears to be a moral imperative while at the same time the resources are dwindling quite fast. With such a moral imperative, it is little wonder that the exploitation of natural resources over much of our rural areas is not within the limit that preserves the concept of eco-development.

The Study Area

The study is part of a wider survey of the impacts of the petroleum industry on the Niger Delta environment. The study was conducted in two communities in parts of the Niger Delta area of Nigeria (Fig 19.1). The two communities studied namely Opukushi and Tunu, are made up of several villages that trace their origin to a common ancestry.

Fishing is the main occupation of the people and is conducted mainly along the creeks that form the blackish water of the Niger Delta distributories. Employing fishing techniques that are relatively indigenous with little capital investment is in harmony with the environment. Fishing is done mainly with hand propelled canoes and with simple equipment such as line hooks, gill nets, fish trap and the like.

Apart from dug up fishponds where fishes are held in captivity and reared, the fish industry in solely dependent on the productivity of the natural aquatic ecosystem. The regenerative capacity of the aquatic ecosystem is therefore very fundamental to the sustainability of the fishery industry.

Under the natural balance, the fishing industry which employs over 80 per cent of approximately six million people in the Niger Delta generates enough income for large households and in the process serves as a major source of protein for the local people. However, based on oral interviews, informal discussion and some studies (e.g., Adeniyi *et al.*, 1983), it would appear that the disruptive ecological effects of crude oil pollution in the Niger Delta have combined to reduce the carrying capacity and ecological productivity of the aquatic environment on which the fishing industry has been thriving.

Figure 19.1: The Niger Delta showing the Study Area

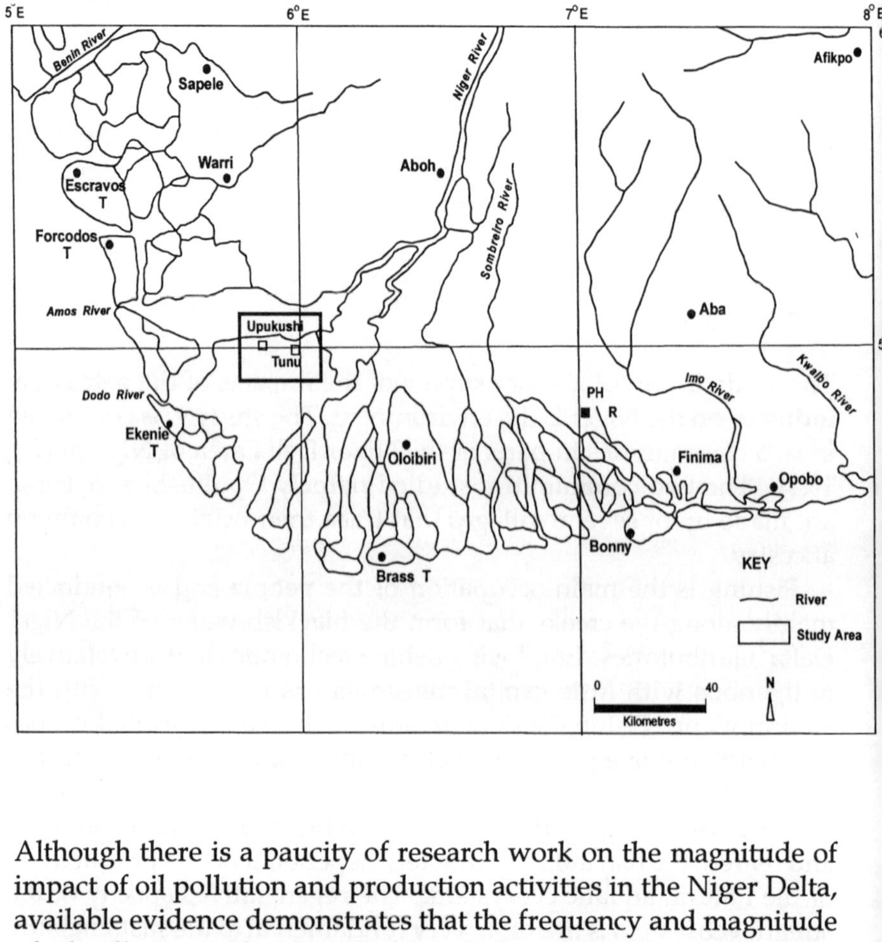

Although there is a paucity of research work on the magnitude of impact of oil pollution and production activities in the Niger Delta, available evidence demonstrates that the frequency and magnitude of oil pollution in the Niger Delta including the study area is on the increase (Ikporukpo, 1988).

Furthermore, it has been well documented that petroleum hydrocarbon deposits in streams and creeks can generate areation problems by decreasing biological oxygen demand (BOD), and chemical oxygen demand (COD) on which the productivity of any aquatic ecosystem depends (Blumer et al, 1971). The deposit of crude oil film can negatively alter the food chain of the aquatic ecosystem by preventing, for example, the productivity of phytoplakton and zooplankton, which are the primary producing components of the aquatic ecosystem.

Changes in streams' water chemistry induced by pollution may also lead to incidence of thermal pollution and the non-biodegradable heavy metals (lead, zinc) associated with crude oil often settle down among the sediments at the stream bottom where they again have negative effects on the productivity of benthic organisms.

The overall impact of these processes is the negative disruptive effect on the ecosystem. Granted that the incidence of oil pollution has been on the increase in the Niger Delta and the study area, the reports of the locals regarding the decreasing quantity of fish catch may be credible. Indeed, the few studies that have been conducted (Adeniyi *et al.*, 1983, Nwakwo and Ifeadi, 1988) have shown that the incidence of oil pollution in the Niger Delta has greatly reduced the productivity of the fishing base. Clearly, fishing is a major occupation and by implication a major source of income for the majority of the people living in the Niger Delta. Consequently, diminishing fish resources is likely to result in a decline in income. Hence, the people must develop some adjustment mechanisms to cope with the effect of declining fish resources and at least maintain their normal standard of living. This study identifies these adaptive strategies and examines their implications for rural development.

Methodological Considerations

The following analysis is based on the findings of a questionnaire survey and in-depth group discussions that were conducted to ascertain the adaptive strategies of fishermen to decreasing fish resources. This methodology was employed to obtain data on the impact of crude oil spillage, and the increasing pollution on the aquatic resources. It was also used to establish the trend in the fish catch over the years; the impact of the declining fish productivity on household income; and the adaptive strategies and actions taken to curtail the declining trend in the productivity of the aquatic ecosystem. In all, three hundred and fifty households were interviewed while five in-depth group discussion sessions were held with community leaders and other representatives of the sampled communities.

Perceived Impact of Pollution on the Aquatic Ecosystem

The general consensus among the communities (Table 19.1) is that oil pollution and associated exploration activities have been largely responsible for the declining trend in the fishery resources of their localities over the years. There is however, some spatial variation in the perception of the rate of decline among the communities, which probably reflects the duration of oil exploration activities, and more importantly, the intensity, magnitude and frequency of oil spillage in the various localities.

Table 19.1: Perceived Impacts of Oil Pollution on Fishery Resources

Perceived level Impacts	Opukushi	Tunu
Decreasing	26.6%	64.5%
Decreasing Abruptly	73.4%	35.5%

Source: Fieldwork, 1994–1995.

For example, while majority of the people in Opukushi perceived the rate as being abrupt, in Tunu it was believed that the rate was gradual. It is interesting to note that Opukushi has a longer history of crude oil production with facilities such as flow stations, flow lines, and more active oil wells than in Tunu. As a result, the degree of change in the ecosystem is more pronounced in Opukushi than Tunu.

Occupational Combination

This represents the first step in the adaptive strategies. The strategy involves the combination of varied occupations in order to augment income for the household. Table 19.2 illustrates the various levels of occupational combinations and the income derived from them.

Table 19.2 does not show the apparent sex bias in terms of the occupational combinations. For example, men generally combine fishing with logging or fishing with gin brewing or local transportation with outboard engine boat while women mostly combine fishing with petty trading or with minor farming and fuelwood gathering. The level and type of occupational combination is strongly tied to the limitation of the immediate environment.

Table 19.2: Occupational Combination in the Communities

	* % involved now	* % willing to remain next year	Estimated income without combination Naira	Estimated income with combination Naira	Difference of income in Naira
1. Fishing & Lumbering	35	50	28,000	32,000	+ 4,000
2. Fishing & Trading	20	90	29,000	14,000	-15,000
3. Fishing & Farming	15	85	28,000	10,000	-18,000
4. Fishing, Lumbering & Casual jobs	5	20	28,000	31,000	+3,900
5. Lumbering& Transportation	4	64	28,500	24,000	-4,500
6. Fishing& Hunting	6	34	29,300	12,000	-13,300
7. Fishing & gin brewing	15	78	28,000	18,000	-10,000

Source: Fieldwork, 1994-1995.
* % of the number involved in particular occupation combination.

The physical and ecological limitations of the environment coupled with smallness of the settlements and the general low level of tempo of economic activities have actually limited the scope and number of occupational combinations available to the community. For example, the inability of the environment (physical and social) to sustain the various occupational combinations for a long time is reflected in the number of people who plan to remain in their combined occupations for the subsequent years.

In this regard, 50 per cent of men who generally combine fishing with lumbering do not intend to remain in it in the subsequent years because the stock of available matured mangroves trees is fast diminishing as a result of increasing number of people going into the lumbering industry. It has become increasing difficult for them to find mature trees. Similarly, those who combine fishing with hunting do not plan to remain in this activity because of fast disappearing game. Thus the observation by Aina and Salau (1992, and Mayor, 1988) that environmental degradation is often associated with high level of poverty is glaring in the Niger Delta. As the people's means of sustenance is eroded by the effect of oil pollution activities, the people resort to the exploitation of other marginal aspects of the environment which induce further environmental degradation.

However, in terms of total income earned, there seems to be a little difference for those who combine fishing with lumbering while there is none for other categories of combinations. Rather, income figures are lower when they do not combine occupations. Indeed the differences in income status are statistically significant (chi-square: = 0.5).

The overall implication is that even with occupational combination, the income status of many families/communities has not changed very much. Thus, there must be some other adjustment strategies that are adopted by the communities. This is the subject of the next section of this discussion.

Migration from the Communities

Migration from the communities represents the most far-reaching adaptive measure that has been taken by the people to counteract their dwindling source of livelihood. There are two dimensions to the outward migration process. This involves those categories of people who migrate to the neighboring rural communities in search of better fish catch and those who migrate to far away urban centers in search for white-collar jobs. A few people have migrated to the neighboring countries adjoining the coastline of Nigeria. Table 19.3 shows the destination of out migration and potential out-migrants while Table 19.4 shows the characteristics of the out-migrants. In aggregate terms, the direction of movement by the respondents one year prior to the study (Table 19.3) indicates that 20 per cent moved to nearby rural communities within the Niger Delta. Most of the people who have moved to other rural areas were attracted by the prospects of better fish catch, occasioned probably by the lesser amount of crude oil pollution in those areas. Table 19.3 also shows the numbers of those who plan to migrate within the next one year in search for better economic prospects as the principal motivating factor.

On average, based on Table 19.4, old people moved to nearby rural areas, especially women. The men folk moved farther away, especially to urban centres.

Overall, those who migrate within the rural areas have little education while those who are younger with a bit of education moved far away to urban centers such as Warri and Port Harcourt where they expected to find some employment.

Table 19.3: Destination of Out-migrants and Intending Migrants

Destination	Number in the past one year*	Number in the Next one Year**
Near by rural areas	20.4%	11.0%
Farther rural areas	40.6%	50.7%
Urban centres	35.0%	33.3%
International	4.0%	5.0%

Source: Field work, 1994-95.
* % of the number of out migrants.
** % of the total number of intending out migrants.

Table 19.4: Characteristics of the Out-migrants

Destination	Average Age	Sex Ratio Male:Female*	Education Attainment
Nearby rural areas	54.6	1 : 5	No formal education
Farther rural areas	28.9	3 : 2	No formal education
Urban Centers	27.5	5 : 0	Primary & Secondary
International	26.6	5 :0	Primary & Secondary

Source: Fieldwork, 1994-95.
*For every 5 persons that move.

Apart from the general implications of outward migration on the communities, there are two important peculiar implications of the migration that are of practical importance to rural planning and development. The first is the increasing number and rate of settlement disaggregation, dualization and abandoment. The second is the issue of increasing number of women-headed households.

The process of settlement disaggregation begins with the establishment of an outpost settlement initially designed as a working post or fishing post. Often, the head of the household moves and stays in these working stations (camps) for up to two weeks before returning to their mother settlements. With time other kith and kin as well as other members of the community are attracted by better fish catch opportunities from the distant camps and are 'pushed' to join the first set of migrants. In due course, the settlements become stabilized and permanent. However, the number of housing units in such area does not on average exceed 200. Some may have as small as 50 housing units and households. As soon as fish catch starts to show signs of decline, people are compelled to move further away. These adjustment mechanisms, as reflected in Table 19.5, explain the high incidence of settlement fragmentation, disaggregation and abandonment.

Table 19.5: Number of Settlements that have Broken off from the Mother Settlements and Number of Abandoned Settlements in One Year

Mother settlements	No of daughter settlements abandoned	No of abandoned settlements
Opukushi	54	29
Tunu	41	12

Source: Fieldwork, 1994-95.

One obvious implication of the settlement disintegration and disaggregation is in the area of provision of rural infrastructure, and amenities for the oil producing areas. The issue of neglect of those rural communities, which produce the wealth of the nation, has become increasingly topical in recent times. As a way of addressing this imbalance, the Federal Government of Nigeria set up the Oil Mineral Producing Area and Development Commission (OMPADEC) with the important mandate to provide social amenities and improve the quality of life of the oil communities in general. With increasing disaggregation of settlements, the Commission may find it difficult and more expensive to provide these amenities, as the minimum required for sustaining efficient utilization may not exist in one particular location. At another level, it may be argued that the provision of the social amenities in the mother settlements may help to pull back the migrants. To some extent, this argument appears plausible. However, it is doubtful if provision of social amenities like schools, pipe-borne water and health services without addressing the fundamental issue of the eroded economic base of the people will be enough to attract back the out-migrants. Perhaps, a more interesting development is that women now head the majority of the households. This is closely associated with the migration of men to far away urban centres in search of white-collar jobs. According to this study's findings, women now head over 50per cent of the households in the community. Furthermore, the women complained that their husbands rarely remit money back home for their upkeep. Very often, some males migrate to urban areas with their youngest wives while others marry new wives when they reach their destination. As a result, the burden of caring for the family in the communities now rests squarely on the women folk. This important changing role of women in the development of these communities has far reaching implications for rural development. It also calls for a review of the status of women in traditional African societies.

In most African societies, women are generally not empowered. They are usually given less important duties and are relegated to the background when it comes to the issue of material possession. For example, women do not hold title to landed properties as this is the exclusive reserve of men. The result is that whenever the need for payment of compensation for ecological damages and land acquisition for oil exploration activities arises, government and multinational companies involved in the oil industry prefer to deal directly with the men. Since most of the men live outside their original communities, the income earned from such compensation is mostly spent outside the recipients' communities and hence the multiplier effect is not felt in the local economy of these rural settlements. This development further diminishes the already dwindling economic base of these people. This may explain why the remaining 'stranded' men folk resort to sabotaging the oil installations resulting in non-accidental oil spills.

Sabotage of Oil Installations

A dangerous development in the oil producing communities is the increase in the reported cases of sabotage of oil installations leading to incidence of oil spillage. For example, between 1989 and 1995 over 221 reported cases of oil spillage were recorded and of this amount, 50 per cent of the volume was mainly attributed to corrosion of aging facilities especially flow lines. Twenty two percent of the spillage has been linked to the oil producing processes including engineering and drilling activities while the remaining 28 per cent were due to willful sabotage of oil installations. Ikporukpo (1988) reported the growing number of reports of sabotage and oil spillage in the Niger Delta. He argued that the sabotage is done deliberately to induce monetary compensations from oil producing companies operating in the various communities. In 1991 for example, more than $5.5 million US dollars was paid out to communities in compensation for oil spills (Shell, 1995b).

However, the issue of whether the motivation behind sabotage activities is primarily to induce compensatory income is highly debatable. This is because, according to the Nigerian Law, compensation is not paid to communities where proven cases of sabotage are established. During this investigation, it became apparent that the issue of sabotage of oil installation generated lengthy debates without a consensus opinion as to the reason behind the act. The communities tended to deny that deliberate sabotage was behind oil spillage in the Niger Delta.

When presented with photographic evidence of the occurrence of sabotage in their community in the past, the majority of the respondents claimed that it was the handiwork of thieves who cut oil pipelines for sale. As such, the phenomenon of the growing incidence of sabotage of oil installations in the Niger Delta is a complex one involving social, economic and political issues. According to the law, these communities cannot be rewarded for momentary damages to their properties due to oil spillage involving the sabotage of oil installations. Therefore, it can be hypothesized that the principal motivation behind the sabotage is to attract public sympathy, through political blackmail, so as to attract attention to their plight. While this issue requires further research, the implications for the sustainability of the oil industry and its contribution to the further degradation of the rural environment demand swift and urgent attention. This is because whatever the motivation behind environmental damage, it is certainly a negative adaptive strategy.

Conclusion

This study has attempted to demonstrate the extent to which human disruption of the delicate balance of the ecosystem in the quest for development through exploitation of mineral resources in the Niger Delta has destabilized the social and economic life of the people and in the process induced several adjustment mechanisms. In a disturbed physical system, for example, a river system, several adjustment mechanisms are automatically put in place to restore the system to a new equilibrium. Under such a scenario, the time lag required for the attainment of this new equilibrium may be immaterial. However, where such a system affects humans, the social, economic and political consequences of this adjustment time lag and in particular the adjustment mechanisms cannot be overlooked because of their far-reaching implications for development.

The study has identified certain adaptive strategies commonly adopted by the crude oil producing communities of the Niger Delta to curtail the increasing disruption of their environment by oil producing activities. The implication is that any policy or programme adopted to address the issue of the environmental deterioration within the context of planning must take cognizance of related adaptive tendencies. It is important to adequately incorporate the adjustment mechanisms as part of the data base for the formulation and development of policies and programmes that will enhance rural development in Nigeria in general and the oil producing communities in particular.

References

Adeniyi, E.O., Sule, R. and Agaye, G. 1983. 'Environmental and Socio-Economic Impacts of Oil Spillages in the Petroleum Producing Riverine Areas of Nigeria', In *The Petroleum Industry and the Nigerian Environment,* Lagos: Nigerian National Petroleum Corporation, pp. 233-248.

Aina A. T. and Salau A.T. 1992. *The Challenge of Sustainable Development in Nigeria,* Nigeria Environmental Study/Action Team (NESF), Ibadan.

Blumer, M. *et al.,* 1971. 'A small oil spill', *Environment,* 13, 2, 2-12.

Chokor, B.A. 1987. 'Environment-Behavior - Design Research Techniques: An appraisal and Review of the Literature with Special Reference to Environmental and Planning-Related Information Needs in the Third World', *Environmental and Planning,* 19, 7-37.

Craik, K. H. 1972. 'An Ecological Perspective on Environmental Decision - Making', *Human Ecology,* 1, 1, 69-80.

Ikporukpo, C.O. 1988. 'Managing Oil Pollution in Nigeria: Towards an Interactive Approach'. In Sada P.O. and Odemerho F. O. (Eds), *Environmetal Issues and Management in Nigerian Development.* Evans Brothers (Nigeria Publishers), Ibadan, pp. 224-299.

Mayor F. C. 1988. 'Geographic Information System for Managing Sustainable Development: The UNESCO Approach', *ITC Journal,* 4, 350-353.

Moran J.M. *et al.,* 1980. *Introduction to Environmental Science.* W.H. Freeman & Co, San Francisco.

Nwakwo, N. and Ifeadi, C.N. 1988. 'Case Studies of the Environmental Impact of Oil Production and Marketing in Nigeria'. In Sada, P.O. and Odemerho, F. O. (Eds), *Environmental Issues and Management in Nigerian Development.* Evans Brothers (Nigerian Publishers), Ibadan, pp. 208-223.

Okafor F. C. 1988. 'Rural Development and the Environment: Degradation Versus Protecton'. In Sada, P.O. and Odemerho, F.O. (Eds), *Environmental Issues and Management Issues: and Management In Nigerian Development,*Evans Brothers, Ibadan, pp. 208-223.

Shell Petroleum Nigeria Company Limited, 1995a. *Nigeria Brief: The Ogoni Issue,* Academy Press, Lagos.

Shell Petroleum Nigeria Company Limited, 1995b. *Nigeria Brief: The Environment,* Jeromelaiho and Associates Ltd, Lagos.

Chapter 20

Context, Principles and Practices of Conservation Strategies and Environmental Action Plans

M.A. Mohamed Salih

Introduction

The context of national environmental policies, plans and strategies has been influenced by public concerns about symptoms of growing environmental crises. In the developing countries, national environmental strategies have been influenced by their colonial past and current position in the global economic system. By and large, their environmental policies are poverty-driven and are related to deforestation, land degradation, over-crowded housing, ill health, inadequate drinkable water supply, food shortages and air pollution in environmentally poorly planned industrial and mega cities. However, in both developing and developed economies, environmental policies have some identical objectives, that is, to govern, regulate and control human activities and their interactions with the environment in order to optimise the use of the natural capital which the environment represents as well as to mitigate, reduce or eliminate environmental hazards and pollutants.

In the North, where some charge that the dominant global environmental crises have originated, people began to experience the consequences of these crises (air pollution and water contamination, the incidence of nuclear fallout, oil spills, nuclear, chemical and toxic production and disposal) on their health (Carson, 1962; Capra, 1985; Tolba and El Kholy, 1992). Second, they expressed fears of economic losses due to exceeding the *Limits to Growth*

(Meadows, Meadows and Randers, 1995) and not observing the *Limits to Competition* (The Group of Lisbon, 1995) and their consequences on environmental degradation as a result of over exploitation of natural resources. Third, ethical, moral and in some instances post-modern sentiments and concerns with the well being of the human and non-human world have been raised. Reference has also been made to post-modern sentiments enhanced by a degree of economic influence among a certain social strata in the industrialised world.

In 1972, the Stockholm Conference on the Human Environment further exposed the global nature of resource dynamics as a process which should be dealt with in an interdependent world. The objective of the North in holding a conference on human environment was to stage an international event focused on the problems which concerned their publics such as pollution, over-consumption of natural resources and global population growth. The South insisted that the agenda should be widened to include issues of global poverty and development aid levels. The South also insisted that the solution to rampant poverty in the developing countries is found in economic development and the provision of environmentally safe technologies to use its resources for its own benefits.

The shift in development policy was expressed in the Action Plan for the Human Environment (APHE) which recommended, among other things, that governmental and intergovernmental organisations should adopt environmental policies aiming at species conservation, forests and atmospheric marine pollution to development policy, technology transfer and the impact of environment on trade. The message from Stockholm Conference and its implications for global resource dynamics is that there is need for economic growth but with proper safeguards to reduce its costs on humans, natural resources and environmental life support system. This common understanding has also shifted form indefinite growth to the realisation that economic growth means an increased level of consumption of food, consumer goods, energy, minerals, forests, etc.

National conservation strategies (NCSs) are means of providing a comprehensive, cross-sectoral analysis of conservation and resource management issues, to integrate environmental concerns into the development process. They are aimed at identifying the country's most urgent environmental needs, stimulate national debate and raise public consciousness, help decision-makers set priorities and allocate human and financial resources, and build institutional capacity to handle complex environmental issues. NCSs have been strongly process-oriented. Information has been obtained, and analysis undertaken by cross-sectoral groups. NCSs have sought to develop political consensus around issues identified through such group interaction. Their results include policy documents approved at high level, action plans and specific programmes and projects (Reid *et al.*, 1994: 36).

As indicated in the definition of NCSs, they result in national environmental action plans (NEAPs) which are defined as, 'plans undertaken by the host country organisation (usually coordinated by a ministry) with technical and financial assistance form the World Bank, various international organisations, NGOs and other donors. NEAPs are designed to provide a framework for integrating environmental considerations into a nation's overall economic and social development programmes, sometimes in response to structural adjustment imperatives. They also may make recommendations for specific actions, outlining the environmental policies, legislation, institutional arrangements, and investment strategies required' (Reid *et al.*, 1994: 37).

While the objectives of national conservation strategies and national environmental plans overlap considerably, they are both implemented with reference to specific aims and targets. The principles of environmental policies could be very clear and highly desirable. The challenge often is how to implement them. This chapter examines the global and national context within which national conservation strategies and national environmental action plans are undertaken and formulated, the conditions necessary for their success and the constraints which might derail the objectives.

From World Conservation Strategy to National Conservation Strategies

The dominant thinking during the 1970s was to prepare a World Conservation Strategy (WCS) in order to show how conservation can contribute to development. In 1977, the United Nations Environment Programme (UNEP) commissioned the International Union for the Conservation of Nature to prepare a document on global perspective on conservation. After two decades of collaborative efforts by conservation science institutions, mainly IUCN, the World Conservation Strategy was developed by funds provided by UNEP,

UNESCO, FAO and World Wildlife Fund (WWF). In 1980, the IUCN published *World Conservation Strategy: Living Resource Conservation for Sustainable Development,* which emphasised the earth's limited supply of natural capital. WCS also articulated the notion that degradation of nature is a threat to present and future survival of humans and non-humans alike. Table 20.1 shows the priority requirements and objectives of WCS. These include: (a) the maintenance of essential ecological processes, i.e. environmental life systems, (b) the preservation of genetic diversity, both the genetic material in different varieties of locally adapted crop plants or livestock and in wild species, and (c) the sustainable development of species and ecosystems, particularly fisheries, wild species which are cropped, forests and timber resources and grazing lands.

Although WCS was hailed as an important global environmental conservation strategy in highlighting for the first time the concept of sustainable development, Adams (1990) reported six major problems:

- relative weakness of conservation in the planning policy-making, combined with the sectoral nature of planning;
- environmental planning rarely allocates land uses rationally and at times marred with preference for natural capital over human livelihood;
- lack of adequate legislation, and weak and overlapping natural resource management agencies;
- shortage of trained human resources, inadequate educational facilities, and absence of basic environmental data;
- lack of adequate financial support for conservation policies, lack of awareness of the benefits of conservation and its relevance to every day concern; and
- lack of enthusiasm for conservation-based rural development, i.e. rural development that combines short-term measures to ensure human survival with long-term measures to safeguard the resource base and improve the quality of life.

Table 20.1: Priority Requirements of the World Conservation Strategy

A. *Ecological processes*

- Reserve good crop land for crops (par 5.1).
- Manage cropland to high, ecologically sound standards.
- Ensure that the principal management goal for watershed forests and pastures is protection of the watershed (para 5.6).
- Ensure that the principal management goal for coastal wetlands is the maintenance of the processes on which fisheries depend (para 5.8).
- Control discharge of pollutants (para 5.8).

B. *Genetic diversity*

- Prevent the extinction of species (para 6.1).
- Preserve as many kinds as possible of crop plants, forage plants, timber trees, livestock, animals for aquaculture, microbes and other domestic organisms and their wild relatives (para 6.4).
- Ensure on-site preservation programmes to protect:
  - The wild relatives of economically valuable and other useful plants and animals and their habitats;
  - The habitats of threatened and unique species;
  - Unique ecosystems; and
  - Representative samples of ecosystem types (para 6.8).
- Determine the size, distribution and management of protected areas on the basis of the needs of the ecosystems and the plant and animal communities they are intended to protect (para 6.10).
- Coordinate national and international protected programmes (para 6.12).

C. *Sustainable utilization*

- Determine the productive capacities of exploited species and ecosystem and ensure that utilization does not exceed those capacities (para 7.4)
- Adopt conservation management objectives for the

Table 20.1 concluded

utilization of species and ecosystems (para 7.2) • Ensure that access to a resource does not exceed the resource's capacity to sustain exploitation (7.3) • Reduce yields to sustainable levels (para 7.4) • Reduce incidental take as much as possible (para 7.6) • Equip subsistence communities to utilize resources sustainably (para 7.6) • Maintain the habitats of resource species (para 7.7) • Regulate international trade in wild animals and plants (para 7.8) • Allocate timber concessions with care and manage them to high standards (para 7.9) • Limit firewood consumption to sustainable levels (para 7.10) • Regulate the stocking of grazing lands to maintain the long-term productivity of plants and animals (para 7.11) • Utilisation of indigenous wild herbivores, alone or with livestock, where domestic stock alone would degrade the environment.

Source: IUCN, 1998.

By and large, some of the above problems have persisted, while others have taken new dimensions and magnitude. In the African context, these problems have been compounded by the depressing economic decline of the 1980s which, in some countries, continued for almost a decade. Issues of direct nature to the debate on environment and (under) development will be discussed later. The initial policy response aimed at reversing these alarming trends in resource depletion was by launching the World Commission on the Environment and Development (WCED) armed with the World Conservation Strategy (WCS) and other multilateral environmental initiatives. One of the main points which WCED borrowed from WCS and successfully advanced to become a major global ideal is the concept of sustainable ecosystem and its translation into sustainable development.

The global popularisation of sustainable development was achieved through the publication of two independent reports: *Brundtland Report* (1987) and *Environmental Perspective to the Year 2000*. These were a result of deliberations of the World Commission on Environment and Development (WCED). According to the official definition of sustainable development, it is routed in multilateralism and the interdependence of nations. For sustainable development to be achieved, it requires the restructuring of the international economic system of cooperation. Hence, sustainable development is development that 'meets the needs of the present without compromising the ability of future generations to meet their own needs'.

According to *Our Common Future* (1987), sustainable development involves progressive transformation of the economy and society, which involves changes in the environment and therefore ecosystems cannot be kept intact with the onset of development. The use of non-renewable resources such as minerals and fossil energy reduces their stockpile for future generations. This means that sustainable development is a process of change in which the exploitation of resources, the direction of investment, the orientation of technological development, and institutional change are all in harmony and enhance both current and future potential to meet human needs and aspirations.

The main shift in development paradigm in relation to sustainable resource use and management as embodied in sustainable development requires that all nations of the world would adopt national conservation plans within more encompassing national conservation strategies. The main objective of these strategies and plans is to operationalise the concept of sustainable development and its objectives with the possibility of:

- reviving growth;
- changing the quality of growth;
- meeting essential needs for jobs, food, energy, water and sanitation;
- ensuring a sustainable level of population;
- conserving and enhancing the resource base;
- reorienting technology and managing risk; and
- merging environment and economic decision-making.

Brundtland Report (1987) and *Environmental Perspective to the Year 2000* were used as basis for the United Nations Conference on Environment and Development (UNCED) process and the Earth Summit in 1992, which adopted three major documents:

- Rio Declaration on Environment and Development;
- Agenda 21, a long-term action programme for the environment and development to the year 2000; and
- a non-legally binding authoritative statement of principles for global consensus on the management, conservation and sustainable development for all types of forests.

Carley (quoted in Reid, 1995: 11) provided a synopsis of the extent of the environmental crisis and range of the problematic policy issues which national plans and strategies should take into account:

- Land degradation, deforestation, biodiversity loss and pollution from the use of excessive fertilizers and pesticides and soil salinatization from improper irrigation.
- Inter-generational flows of natural and man-made capital in a 'full world' where natural capital is the limiting factor of production.
- Growing inequality between world's rich and poor, and the need to address poverty and basic needs and food security in an increasingly degraded environment.
- Industrialization, urbanization, capital and population mobility, and growing aspiration for the environmentally degrading lifestyles.
- Complicated and an increasingly difficult equation linking issues of governance, planning, market, democratised decision making and their varying definitions of what sustainable development strategies entail.

As have been defined in the introduction to this chapter, NCSs and NEAPs are designed to overcome such complex and inter-sectoral problems that expand almost all human activities. For these strategies and plans to succeed, Reid *et al.*, (1994: 43-45) outlined the necessary conditions for successful implementation of national environmental strategies. Considering the significance of these conditions or principles, they are quoted at length as follows: (a) a defined need and purpose, (b) strong organisational location, (c) high level political support (parliamentary, cabinet or head of state) and financial support to ensure government backing and hence the integration of environmental concerns into socio-economic development planning, (d) commitment of key participants, with a wide public support, and (e) conducive political and social climate. Other conditions include (f) wide understanding, (g) clear objectives, (h) adequate resources and (i) effective communication.

The question to ask here is whether these conditions were in place when African NCSs and NEAPs were conceived. It should also be made clear from the outset that these conditions have to be supported by experienced, skilled human resources and institutions, capable of delivering results that match the objectives and purpose of NEAPs.

From National Environmental Plans to National Environmental Policies

From the 1970s, environmental policies were subsumed under a host of public policies dealing with specific sectors (agriculture, forestry, fishery, water management etc), including measures to regulate industry and transport, to control pollution, manage and dispose waste and hazardous chemicals etc. Today, environmental policies have become increasingly cross-sectoral and integrated economic, social and environmental policies (such as national conservation strategies, national environmental action plans) built into sustainable development strategies (Tolba and El Kholy, 1992). Environmental institutions are guided by environmental action plans (NEAP) and national conservation strategies (NCS).

Industries are increasingly requested by national environmental institutions to develop self-regulating mechanisms inspired by environmentally compatible technologies. Not all environmental quality control and environmental regulation institutions are governmental. National and transnational non-governmental organizations (NGOs), environmental pressure groups, environmental social movements, grassroots and citizen organisations provide effective environmental institutions. Non-governmental organisations, citizen and grassroots organisations lobby, monitor, alert and pressure governments, business and industry to improve on their environmental practices or abandon environmentally damaging policies.

It is also increasingly evident that anticipatory/preventive environmental policies are more common policy instruments than reactive policies (Utton and Daniel, 1973; Baumol and Oats, 1975; Goldin and Winters, 1995; Opschoor and Turner, 1994; Hanley, Shogren and White, 1997). Environmental policies are often built around five main principles (Turner and Opschoor 1994:4):

- the polluter pays principle i.e. the polluter pay the cost of meeting socially acceptable environmental quality standards;
- the prevention or precautionary principle, which explicitly recognises the existence of uncertainty (environmental and social) and seeks to avoid irreversible damages *vis-à-vis* the imposition of a safety margin into policy; it also seeks to prevent waste generation at source, as well as retaining some end-of pipe measures;
- the economic efficiency/cost effectiveness principle, applying to both the setting of standards and the design of the policy instruments for attaining them; the subsiderity principle;
- to assign environmental decisions and enforcement to the lowest level of government capable of handling it without significant residual externalities; and
- the legal efficiency principle, to preclude the passage of regulations that cannot be realistically enforced.

The principles of environmental policies are operationalized through: (a) laws (legal) instruments, (b) fiscal instruments and (c) institutions. Strategically, environmental policy laws, instruments and institutions aim at sustainable development and resource management for better environmental quality control through policy windows pertaining to regulation, incentives and internalisation of environmental costs. First, environmental laws are concerned with regulating activities that have potential to cause environmental hazards, including the authorization of discharges to the environment, containment of toxic substances, setting standards and licensing of emission levels and manufactured products etc (for more details refer to Mohamed Salih, 1997a and b).

Environmental policy instruments consist of financial instruments and socio-economic incentives which fall within the anticipatory/ preventive, polluter pays principle and economic/efficiency cost effectiveness. It has been recognised by governments and multilateral environmental institutions that financial instruments (aggregate abatements costs, user charges levied on emissions; financing emission charges; tradable permits; and emission charges (environmental taxes) as enforcement instruments) within market economies can be an effective means for high environmental quality control. These have to be seen within the framework of cost-effectiveness, administrative costs, distributional impacts and policy making contexts (Lohman, 1994:58-62; Weizsacker and Jestinghaus,

1992). The distinction between control and command and economic incentive policies implies that different policy instruments (coded instrument choice) can be used in different contexts, and also depends on the social, political and economic contexts (developing or industrialized) within which financial instruments and socio-economic incentives are undertaken, that is, subsidies, affordable and efficient public transport, incentive charges on emissions, demarcated bicycle tracks and walk paths etc (Mohamed Salih, 1997a).

An overarching anticipatory/preventive environmental policy instrument, with a strong link to cost-effectiveness and distributional impacts is Environmental Impact Assessment (EIA) which can attain manifold policy objectives, including: a) the regulation and hence control of impacts of human activities on the environment, b) the maintenance of environmental quality, c) the prevention or regulation of natural and human-made environmental damage, d) to react to unforeseen environmental damage, e) the regaining of the assimilative capacity of the environment, and f) the development of sustainable human interventions in various sectors of the economy (Barlett, 1989; Wood, 1995; Hanley and Spash, 1995; Galpin, 1995).

Since the late 1970s, regional and national environmental laws are increasingly influenced by the global environmental fora, negotiations, conventions, treaties and declarations and the emergence of the concept of the global environmental governance (Commission for the Global Governance, 1995). According to *Our Global Neighbourhood* (1995: 3-7), global governance:

- is viewed primarily as intergovernmental relationships, but it must be understood as also involving non-governmental organizations (NGOs), citizen movements, multinational corporations and the global capital market. Interacting with these are global mass media of dramatically enlarged influence;
- there is no single model or form of global governance, nor is there a single structure or set of structures. However, recognizing the systemic nature of the issues with which it deals, it promotes systemic approaches in dealing with major global concerns; and
- the aims of global governance include reforming and strengthening the existing system of intergovernmental institutions, improving its means of collaboration with private and independent groups. Among other things, global governance will strive to subject the rule of arbitrary power-economic, political,

or military – to the rule of law within global society.

Global environmental governance operates within the confines of and is consistent with the principles laid down in global governance legal instruments, conventions and international environmental policy proclamations. *Our Global Neighbourhood* (1995: 216) indicates that global environmental governance emphasises the need:

- to put to action a global legal, intellectual and institutional groundwork for a concerted effort to achieve sustainable development;
- to establish a Trusteeship of the Global Commons (TGC) and to be made responsible for acting on behalf of all nations, including the administration of environmental treaties; and
- to enact an Earth Charter for world endorsement pending the outcome of the global efforts exerted by the Earth Council and Green Cross to achieve this endeavour.

It has become obvious that one of the most conspicuous outcomes of the Earth Summit is that the stage has been set for global governance aided by diverse trends of eco-globalism to become the dominant paradigm and the most influential actor on the global scale. According to Rich (1994: 273), 'the premise of the earth summit was that global environmental management can work. The project, as it emerged out of Rio, has two principal approaches – a growing body of international environmental treaties among nation-states and increased foreign aid for environmental protection and management as well as poverty alleviation, channelled and managed mainly through the Global Environmental Facility (GEF) and other financing arms of the World Bank'. Such notions do imply not only that the national economy has already been rendered subject to capital, but also that sustainable development has been defined according to the logic of capital and the dominant structures that support it (for more details refer to Mohamed Salih, 1997c).

As has been mentioned earlier, it is almost impossible to implement environmental policies without effective, highly responsive and committed environmental institutions. In most countries, environmental institutions concern themselves with what is called the dart-board theory i.e. overlapping circles on interrelated national, provincial, district and local/village (depending on the administrative structure of the given country) institutions. These institutions also represent a hierarchy of authority and decision making powers.

Politics of Conservation Strategies

Despite its noble endeavour, the notion of sustainable development has been contested by various environmental actors in the South and the North. Some Southern states and intellectuals portray sustainable development as a neo-colonial design devised by the North to deny the South the right to development. They argue that since the North is responsible for most of the environmental damage caused so far, it should be responsible for the cost of the sustainability and maintenance of the global environment (Shiva, 1993; Agarwal and Narian, 1991; Mohamed Salih, 1997c). According to this school, the North has exceeded its own waste assimilative capacity (i.e. the ability of the environmental life support systems, including the global commons to accept waste and render them harmless). The problem is politically prompted by the question as to how to regulate the use of the global commons (seas, oceans and the atmosphere), which have historically existed as a free-access common property for the disposal of wastes unequally produced at the global scale. Global environmental policies are contested on similar basis to those on which national environmental policies are contested, that is, in terms of losers and beneficiaries from a given environmental policy orientation, the nature and consequences of the policy instruments used to implement it. Environmental policy instruments impact differently on different economic, social and political interests and activities.

Sustainable development policies are described by Eckersley (1995) as an extension to free market environmentalism which defends decentralized private choice as a superior economic management system to the command and control approach. Eckersley (1995: 15) also argues that, 'Free market environmentalism shares two common features with the ethos of sustainable development and structural economic policy reforms: (i) Deregulation, free trade and the liberalization of the economy and (ii) Reduction of state intervention and the liberalization of market forces'. Free market environmentalism is consistent with the current drive towards globalization (GATT) and regionalisation, including the emergence of competitive regional trading blocs (EEC, NAFTA, ASEAN etc). It is criticised by environmentalists on the basis that its ethos may encourage excessive use of natural resources for economic growth without due consideration to environmental destruction. This is said to be particularly true for developing countries which depend on primary commodity production for export because that is where their comparative advantage lies. Economic competition, according to devout environmentalists, may create political tensions between poor and rich as well as the dominant trading blocs, if international regulatory environment policies and laws appeared to affect trade competitiveness of certain states or regions (Mohamed Salih, 1997c). In the same vein, Goldin and Winters (1995: 11) lament that, 'Policies for sustainable development raise problems of international coordination in two fashions: first, global or trans-border pollutants such as carbon dioxide or nuclear fallout cause direct negative externalities with all the well-known problems of free-riding in policy decisions. In their most extreme form these are reflected in the tension between OECD countries' desire that developing countries should curb their emissions of green house and ozone-depleting gases, and the latter's understandable concerns not to do anything that might jeopardise their developmental objectives. Second, even if pollutants are wholly local, there are international implications to the extent that sustainability spills over into competitiveness. If exports or direct foreign investment generate rents, non-cooperative environmental policies will be set at non-optimal levels if they have any impact on such flows'.

What is apparent in the above analysis is that two main paradigm shifts are closely associated with resource dynamics and sustainable development. The first is the shift from infinite growth to limits to growth and from limits to growth to limits of earth's absorption capacity. The second is the shift in global ethics from development to sustainable development. These two paradigm shifts have been accompanied by divergent societal responses in the contested South-North perspectives. The first is that in some parts of the South, the shift has been from development to stagnation and from stagnation to crisis. The second is that in the North, the shift has been from simple modernity to the ethos and apprehension of second modernity labelled as a risk society. Risk society notion has been elaborated by Ulrich Beck's (1995) *Ecological Politics in an Age of Risk*. As in second modernity, the ecological politics of risk society is global which diminishes the opportunities to insure against risk. In the past, for most people, hazards had their roots in the local circumstances; or in traditional and far away societies, they could be imputed to cosmic occurrences. But the second modernity brings with it hazards which make the insurance principle less tenable. Global resource dynamics produce global socio-economic and political processes beyond containment within the boundaries of one nation state: global warming, HIV/AIDS, famine, overpopulation, period crisis, world financial crash, nuclear accidents, etc. Global resource allocation, use and management and environment and global risks which follow from that are largely without boundaries.

It is argued elsewhere (Mohamed Salih, 1997a, b) that inter-state and regional economic bloc tensions over the use or evasion of their international obligations to sustainable development as a global ideal are labelled international environmental conflicts, which could result in far-reaching environmental security problems (Thomas, 1992; Kakonen, 1992). Martinz-Alier (1991: 134-135) asserts that, 'the politics of environmental politics is a direct consequence of the shift of environmental costs or benefits to other social groups (national or foreign), or to future generations...as such politics determines not only environmental policy, but also the environmental agenda'. The instruments used to settle these conflicts include legal instruments (international environmental law) and environmental diplomacy (environmental fora and negotiations). In short, like other public policies, environmental policies (local, national and international) have a differential effect on people welfare, employment opportunities, incomes, and national interests and market competitiveness *vis-à-vis* other nations and states.

On the other hand, the social constructionist perspective as advocated by Hannigan (1996) does not prelude the arrival of an ecologically post-modern society in which, according to Spaargaren and Mol (1994: 341), ecological modernisation is possible. The most obvious limits to the realisation of this is the likelihood that it might deal with only the industrial dimension of modernity and neglect dimensions of capitalism and surveillance which narrows the concept of nature to its role in maintaining the sustenance base of human activities.

Conservation Science and the Role of Committed Research

An important aspect of the debate on conservation science institutions in general is the institutionalisation of global conservation science and technology which gave legitimacy to a counter critique of science and technology, not only as instruments capable of enhancing human domination over nature but also as front runners in conservation. For instance, Yearley (1996: 114-115) made the point that in the age of imaging, conservation science has become an instrument of a legitimation crisis in the efforts to deconstruct/ reconstruct nature. Others (e.g. Taylor, 1970; Beck, 1995; Yearley, 1996; Lash, 1996) have questioned the ability of science and technology to cure the environmental problems created by their use or misuse.

It is contended elsewhere (Mohamed Salih, 1997d) that notwithstanding the general critique of science, conservation science is a particular brand of science that has been utilised in the quest for the protection and conservation of nature, including biodiversity. On the one hand, without science and technology, humanity might have not experienced the current environmental crisis. On the other hand, it is difficult to reconstruct what other types of environmental and non-environmental crises humanity would have experienced without the development of science and technology.

Therefore, it is obvious that any attempts to protect and conserve the environment should be mindful of the role of research conducted both by natural and social science. It is impossible to envisage conservation without conservation science, particularly in areas where tradition and indigenous knowledge are ill-equipped to cope. Essentially, the magnitude of the environmental problems which exist today are beyond the ability of one science discipline (social or natural) or tradition (modern or indigenous) to grasp fully all conservation parameters and consequences. Conservation involves a variety of complex issues such as the role of modern technologies for species and genetic conservation, genetic banks, protection of endangered species, marine life, large mammals, watersheds, tropical forests, etc. Conversely and as mentioned earlier, there are also pertinent socio-economic and cultural issues whose consideration is essential for the success of conservation strategies, plans and policies.

Claims to justify the importance of social science research contribution to conservation has been meticulously elaborated by Tolba and EL Kholy (1992) as follows:

- There are major social, economic and political impacts and dimensions to the environmental crisis. Although most environmental problems require some understanding of the natural world, and accordingly demand expertise in the natural sciences, this demand is by no means exclusive. It is now evident that social conditions bring about environmental problems and that social and political changes are necessary for their solutions. Economically, environmental concerns are inseparable from industrial, agricultural and other types of investment and international finance.
- Social sciences have already developed theoretical analysis which can be used to research the environmental problems. For instance, environmental problems in the South could be analysed in terms of structural dependency, underdevelopment, unequal exchange, international division of labour, and the political economy of resource management. There is the critique of the sociology of scientific knowledge and its dependence on a preconceived and pre-defined notion of scientific authority. And social science relevance to the environmental movement as a social movement responding to social and political demands resulting from the environmental crisis.
- Environmental problems have specific social, economic, political and health consequences which can only be studied by social sciences. It is for these and other reasons that the social dimension of environmental studies is increasingly gaining recognition the world-over. The social sciences domain of the environmental studies concentrates on at least three areas:

  - how human activities may alter the environment;
  - how societies and individuals are affected by environmental change; and
  - how human societies may adapt to or intervene to modify its rate of direction.

These areas of research are generally linked with various aspects of human behaviour, political activity, economic structure, social organisation, institutions etc. It is now generally recognised that the challenges that face social scientists are more difficult to overcome than earlier anticipated.

Considering the insistence of some natural sciences on the monopoly of knowledge over conservation strategies, a warning has been made (Mohamed Salih, 1997a and 1997d) against the glorification of the universality of scientific validation in conservation science because science itself is a social construct (and not less ideological than tradition) often influenced by human interests (Beck, 1995; Eder, 1996; Dickens, 1996). In a recent publication, Pepper (1996: 241) has demonstrated with several instances that, 'the findings of science mirror or even depend on the influences from society, . . . and are relative to the society from which these scientists come'. If carried through, this discussion has far-reaching implications than questioning the objectivity of science, to that of the possibility of a subjective interpretation of scientific claims, thus revealing the unspoken linkages between and the contestation of scientific claims on the basis of objective or subjective human interests (Mohamed Salih, 1993). As a new phenomenon, but more differentiated than has been noted by Hannigan (1996: 86):

> Epistemic communities are not only bound together by a common technical expertise but they also share a number of casual and principled beliefs. In the case of environmental issues, these communities of knowledge have been largely comprised of ecologists who share a common belief in the need for a holistic analysis – a view which carries over to the policy advice.

However, it should be noted that the institutionalisation of conservation science has played an important role in advancing the debate and interest in at least two ways: a) the realisation that conservation requires a more involved human agency, and b) the recognition that concerns with conservation has much broader significance for development, environment as sustainability as well as present and future human well-being.

Researchers are increasingly resistant to mono-disciplinary tendencies aiming at separating people's traditions from the science-nature debate. However, it is encouraging that this tendency is increasingly in the decline with the realisation that conservation is concerned with both people's aspirations for development and global responsibility to conserve and protect the ecosystems, habitat, genetic resources and species on which they subsist. In some instances, the epistemic communities which inform the state-centred/globally-driven conservation science are more concerned with the protection of nature than community development and the welfare of the indigenous people who protect their natural habitat for centuries.

Such tendencies might have prompted the critiques that despite its universal claims, conservation science has not (probably not intentionally) given much attention to exploring the particularity of socio-cultural, economic and political environments within which the universal laws of science/nature are applied (Mohamed Salih, 1997d). Naughton-Treves and Sanderson (1995: 1273) comment in a similar situation that even in the domain of natural science:

> Land management and species conservation, the primary numeraries of conservation are increasingly driven by global political economic values, because those values are generated off-site, they are not sensitive to the biophysical prerequisite of conservation over the long term.

Natural and social conservation science, now, and more than ever before, has the responsibility of asserting itself to be more than an instrument for the justification of the global nature of the environmental crisis. It has to be transformed into a liberation science, an instrument for empowering people to conserve their environment. In this respect, social and natural conservation science ought to be subjectively responsive to the variety of human situations, conditions and traditions within which the natural capital exists and where it should duly be protected and conserved. This cannot be done if the social and natural conservation science establishment is not conscious of the subjectivity inherent in other interactions with their natural and built-environment.

Conclusion

The implications of the current paradigm shift for sustainable development is that post-development is not necessarily post-modern and that images of the global environmental crisis began to bear consequences on national and local resource management, allocation management and planning. In today's highly interdependent world, global paradigm shifts in resource dynamics manifest themselves in shifts in resource perceptions at regional, national and local levels. In an excellent paper on Environment and Development in Africa, Okidi (1993: 25–26) advocated the need for a vital management triangle for innovative management of natural resources which links researches, institutions and governments. He calls for African governments to 'initiate and formalize links and working relations between researchers, policy makers and economic producers (including farmers and industries) to ensure information flow and consultation among the three in matters of management of environment and natural resources'.

Reflecting on environmental policy context, principles and practice in Africa reveals that they have produced mixed results even in their proponents' view. In the African context, and due to poverty and underdevelopment, business and international trade operators often misuse economic policy instruments, including market-based instruments such as a) taxes on emissions and tradable emissions permits and b) resource use-permits. However, I'm equally doubtful whether economic policy instruments without legal instruments to enforce them could provide sufficient safeguards for effective environmental quality control. It is up to African countries to initiate action to realise the noble idea proposed by Okidi *(ibid.)*.

References

Adams, W.M. 1990. *Green Development: Environment and Sustainability in the Third World*, Routledge, London and New York.

Agarwal, A. and Narian, S. 1991. *Global Warming in an Unequal World: A Case Study of Environmental Colonialism*, Centre for Science and Environment, Delhi.

Bartlett, R. V. 1989. Impact Assessment as a Policy Strategy, in Bartlett, R. V. (Ed), *Policy Through Impact Assessment: Institutionalized Analysis as a Policy Strategy*, Greenwood Press, New York, Westport, Connecticut and London.

Baumol, W. J. and Oats, E. E. 1975. *The Theory of Environmental Policy*, Prentice-Hall, Engelwood Cliffs, New Jersey.

Beck, U. 1995. *Ecological Politics in an Age of Risk*, Polity Press, Cambridge.

Blaikie, P. M. and Brookfield 1989. Land Degradation and Society, Methuen, London.

Brenton, T. 1994. *The Greening of Machiavelli, The Evolution of International Environmental Policies*, The Royal Institute of International Affairs, London.

Brundtland Report, 1987. *Our Common Future*, World Commission on Environment and Development, Oxford University Press, Oxford.

Capra, F. 1985. *The Turning Point*, Flamingo, London.

Carson, R. 1962. *Silent Spring*, Houghton Mifflin, Boston.

Clark, Gordon L. 1995. 'Global Competition and Environmental Regulation: Is the Race to the Bottom Inevitable?', in R. Eckersley (Ed), *Markets, the State and the Environment*, Macmillan, Basingstoke.

Commission on Global Governance, 1995. *Our Global Neighbourhood, Report of the Commission on Global Governance*, Oxford University Press, Oxford.

Dickens, P. 1996. *Reconstructing Nature*. Routledge, London.

Eckersley, R. 1995. *Markets, States and the Environment: Towards Integration*, Macmillan Education Australia, Basingstoke and London.

Eder, K. 1996. *The Social Reconstruction of Nature*, Sage Publications, London.

Ehrlich, P. and Ehrlich A. 1996. 'Betrayal of Science and Reason: How Anti-Environmental Rhetoric Threatens Our Future', Island Press, A Shearwater book, Washington DC.

Galpin, Alan, 1995. *Environmental Impact Assessment: Cutting Edge for the 21st Century*, Cambridge University Press, Cambridge.

Giddens, A. 1996. Urbanism, Globalisation and Environmental Politics', in D. Goldblatt (Ed), Social Theory and the Environment, Polity Press, Cambridge.

Goldin, Ian and Winters, L. Allan 1995. *The Economics of Sustainable Development*, Cambridge University Press, Cambridge.

Guha, R. and Martinez-Alier, J. 1997. *Varieties of Environmentalism: Essays North and South*, Earthscan, London.

Hanley Nick, and Clive, L. Spash 1994. *Cost-Benefit Analysis and the Environment*, Edward Edgar, Aldershot.

Hanley Nick, Shogren, Janson, F. and White, Ben 1997. *Environmental Economics in Theory and Practice*, Macmillan Press, Basingstoke and London.

Hannigan, J. A. 1995. Environmental Sociology: A Social Constructionst Perspective, Routledge, London and New York, IUCN 1997, *Indigenous Peoples and Sustainability*, International Books, Utrecht.

Kakonen, Y. 1992. (Ed), *Perspectives on Environmental Conflict and International Relations*, Pinter Publishers, London and New York.

Lash, S. Szersznki and Wynne, B. 1996. (Eds), *Risk Environment and Modernity: Towards a New Ecology*. Sage, London.

Lohman, Lex de Savorinin 1994. 'Economic Incentives in Environmental Policy: Why are They White Ravens?' in J. P. Opschoor and R. K. (Eds), *Economic Incentives and Environmental Policies: Principles and Practice*, Kluwer Academic Publishers Dordrecht, Boston and London.

Martinz-Alier, J. 1991. 'Ecological Perception and Distributional Conflict', in F. J. Dietz, F. van der Ploeg and J. van der Straaten (Eds), *Environmental Policy and the Economy*, Northern Holland, Amsterdam, London, New York and Tokyo.

Meadows, D. 1972. *The Limits to Growth: A Global Challenge; a Report to the Club of Rome*, University Books, New York.

Meadows, D.H., Meadows, D.L. and Randers, J. 1992. *Beyond the Limits: Global Collapse or a Sustainable Future*, Earthscan Publications, London.

Mohamed Salih, M. A. 1997a. 'Global Ecologism and Its Critics', in Thomas, C. Peter Wilkin (Eds), Globalisation and the South, St. Martins, New York, Macmillan Press, Basingstoke.

Mohamed Salih, M. A. 1997b. 'Natural Capital', In P. O'Hara (Ed), Encyclopaedia of Political Economy, Routledge, London and New York.

Mohamed Salih, M. A. 1999. 'Introduction', in M. A. Mohamed Salih and S. Tedla (Eds), *Environmental Planning, Polices and Politics in Eastern and Southern Africa*, St. Martins, New York, Macmillan Press, Basingstoke.

Naughton-Treves, Lisa and Sanderson, S. 1995. *Property Politics and Wildlife Conservation*, Cambridge University Press, Cambridge.

North, R. D. 1995. *Life on a Modern Planet: A Manifesto for Progress*, Manchester University Press, Manchester and New York.

Okidi, Charles 1993. *Environment and Development in Africa*. Africa Technology Centre, Nairobi.

Opschoor, J. B. 1999. 'Foreword', in M. A. Mohamed Salih and S. Tedla (Eds), *Environmental Planning, Polices and Politics in Eastern and Southern Africa*, St. Martins, New York, Macmillan Press, Basingstoke.

Opschoor, J. B. and Turner, R. K. 1994. 'Environmental Economics and Environmental Policy Incentives: Introduction and Overview', in J. P. Opschoor and R. K. Turner (Eds), *Economic Incentives and Environmental Policies: Principles and Practice*, Kluwer Academic Publishers Dordrecht, Boston and London.

Pepper, D. 1993. *Eco-socialism: From Deep Ecology to Social Justice*, Routledge, London.

Reid, David 1995. *Sustainable Development: An introductory guide*, London: Earthscan.

Rich, Bruce 1995. *Mortgaging the Earth: The World Bank, Environmental Impoverishment and the Crisis of Development*, Earthscan, and London.

Sachs, W. (Ed) *Global Ecology: A New Arena of Political Conflict*, Zed Books, London and New Jersey, London and New York.

Shiva, V. 1993. 'The Greening of the Global reach; Global Environment or Green Imperialism ?', in W. Sachs, (Ed), *Global Ecology: A New Arena of Political Conflict*, Zed Books, London and New Jersey, London and New York.

Taylor, Annie and Thomas, C. 1999. *Global Trade and Social Issues*, Routledge, London and New York.

The Group of Lisbon 1995. *Limits to Competition*, The MIT Press, Cambridge, Massachusetts, and London.

Thomas, C. 1992. *The Environment in International Relations*, Energy and Environment Programme, The Royal Institute of International Affairs, London.

Tolba M. K. and O. El Kholy 1992. *The World Environment 1972 - 1992: Two Decades of Challenge*. Published by Chapman and Hall for United Nations Environmental Programme, London, Glasgow, New York, Tokyo, Melbourne and Madras.

United Nations, 1990. *Global Outlook 2000: Economic, Social and Environmental*, United Nations Publications, New York.

Utton, Albert, E. and Henning, Daniel, H. (Eds), 1973. *Environmental Policy: Concepts and International Implications*, Praeger Publications, New York, Washington and London.

Weizsacker, E. U. von and Jestinghaus, J. 1992. Ecological Tax Reform: A Policy Proposal for Sustainable Development, Zed Books Ltd., London and New Jersey.

Wood, Christopher 1995. *Environmental Impact Assessment: A Comparative Review*, Longmann Scientific and Technical, Essex.

Yearley, S. 1996. Sociology, Environmentalism, Globalisation: Reinventing the Globe, Sage Publications, London, Thousand Oaks and New Delhi.

Chapter 21

Sustainable Human Development and the North-South Dialogue[1]

J. B. Opschoor

Introduction

Since I lived and worked here, and even since I was here last in 1994, much has changed in Botswana. Levels of economic activity have gone up as visibly as the cities and towns have expanded. Income per capita is now more than twice as high as it was then. And, compared with a few years ago, its competitiveness has gone up substantially. That's fine. On human resources and human capital, I was told today that that has grown too, but less than income per capita. Unfortunately, the country has run into health problems that hardly existed when I was here.

At any rate, macro-economically, the country is doing well. Less so, perhaps, is when it comes to economic diversification or getting rid of income inequality. But what would interest me also is, has the country become more sustainable? Yesterday, I bought the 'Environmental Statistics' of Botswana – an interesting follow up to work that my own successor at NIR, Jaap Arntzen did 15 years ago, the 'Profile of Environment and Development in Botswana'. At a first glance, it gives a mixed picture but more research is required to say anything at all meaningful. But it does indicate that much progress has been made in monitoring environmental change, in environmental policy development, and even the number of environmental NGOs. Flying into Botswana 9 days ago, I learnt that Botswana's first lady heads a tree planting and waste awareness campaign and I was told by informed sources that environmental awareness generally in Botswana is now far more deeply entrenched

[1] This chapter is based on a public lecture delivered by the author at the University of Botswana on 9th February 2001.

than it was when I was here. So, apparently it was a good thing I left Botswana when I did: things have only improved since then!

I would like to address you on Sustainable Development and the North-South debate on it. In doing so I have to inevitably bring in globalisation and what it means for the type and intensity of economic development, and the social and especially environmental repercussions thereof. One last remark before I really start is that I will often derive my examples from the Climate Change discussions. I have been, and still am, deeply involved in these, as a researcher, as a participant in the Dutch delegation to the UN Conference on Environment and Development (UNCED) in Rio in 1992, and more recently, as a review editor of one of the reports, the so-called IPCC (Intergovernmental Panel on Climate Change), which is going to come out soon, on what science can say with respect to climate change and about ways of mitigating it.

Strategies towards fair and sustainable global socio-economic development require environmentally and culturally tenable patterns of production and consumption as well as a fair distribution of income and access to its sources (human, social, natural and produced capital). They also imply an institutional structure that will achieve and maintain such patterns and distributions. The latter may require as much as what has been called a new 'global partnership' (OECD/DAC, 1998) for 'sustainable human development' (UNDP, 1996, 1999), but also, I would argue, an enhanced capability to 'manage globalisation' and hence institutions tackling the driving forces of unsustainability and other forms of human insecurity.

Development as a concept, or, more precisely, underdevelopment, goes back to Truman's inaugural address in 1949. The phrase has subsequently been picked up by many, including economists and economic institutions that typically see development as something that was either equal to economic growth, or something that economic growth would at least be a pre-condition for a proxy of development in a much broader sense.

There have been deviant opinions and theories. I mention two:

- The structuralist school with its theories on centres and peripheries in economic evolution as causing structural inequalities in exchange and identifying power differentials as the sources of continuing underdevelopment. The Institute of Social Studies (ISS) has historically felt related to this perspective, as is clear from the list of its Honorary Fellows.

- The 'distributionalist' school. Within the debate on growth or redistribution I should mention Kuznets, who, in the 1950s developed his theory laid down in the curve named after him, that income inequality would decline automatically as average income would go up. This argument conveniently took away attention from distributional issues, as according to the Kuznets curve, or law, economic growth would automatically reduce it. So, stop worrying too much about it, was the message. Statistical and econometric work since then has shown this to be wishful thinking rather than a law or even a reality. UNDP statistics show that with global economic growth inequalities even become bigger.

So much for these deviant opinions. At the moment, the mainstream views on development, especially by economists, is captured by the Washington Convention and carried forward by WB and IMF to all corners of the world, through Structural Adjustment. Essentially it is microeconomic, monetarist, neo-classical analysis focussing on wrong monetary policies and market imperfections imposed largely by government regulations on the economic process. Liberalisation and privatisation plus reduction of state recurrent expenditures are seen as essential features of development; its agent no longer being the state but rather private investors, domestic or foreign.

To carry on with our set of notions, let's move to human development. As we now see it, human development is more than economic development and the latter is more than economic growth. One speaks of *human* development when: (i) the range of social, economic and political choices of groups and of individuals is expanded, and (ii) a decent standard of living is assured in terms of education, nutrition and health, freedom, democracy and human security, as well as 'sustainability' (UNDP, 1996).

Some Words About 'Sustainable Development'

The phrase goes back to 1980, when it was launched in IUCN/WWF/UNEP's 'World Conservation Strategy'. This strategy aimed at 'achieving sustainable development through the conservation of living resources'. I must say I wasn't aware of this at the time. In 1980 my colleague at NIR, Letsholo Kgathi (now at the Okavango research Station) was very busy studying the way in which people in the villages of Kgatleng - especially the poorer people - were coping with the then prevailing drought. But when I went to the Institute of Environmental Studies in 1982, that document was almost the first one I read.

According to the World Conservation Strategy, apparently, development could be made sustainable only if living, or renewable resources were conserved, that is, managed in such a way that their availability over time would be ensured at least at a given level. An elaborated version of this interpretation would later be called 'strong sustainability': keeping at least constant over time some initial amount of available natural capital. Or, put in other words, it called for development choices that would at least not restrict the environmental space available for future generations.

In 1987, the Brundtland Commission, in its report, *Our Common Future*, launched a significantly different notion of sustainable development. Here this was defined as 'development that meets the needs of the present without compromising the ability of future generations to meet their needs'. This, to an economist, reads as: per capita satisfaction or levels of income and consumption need to be at least maintained over time. And this would entail that the total capital base (the sum of physical capital, human and social capital and natural capital would be constant, on a per capita basis). This is what later was referred to as 'weak sustainability', as the amount of natural capital would no longer and by necessity have to be preserved. That could be run down as long as it could be substituted by produced capital or human capital (a factory for a forest, so to speak) without jeopardising the income or consumption that would be generated. More elaborate definitions in the Brundtland report[2] do mention that, among other things, 'natural resource management' is one of the components of sustainable development, but that is a far cry from stating that they need to be preserved at certain levels.

Almost a decade later, OECD/DAC (1998) produced its views on: 'Shaping the 21st Century: the contribution of development co-operation'. It too speaks of sustainable development. We read there that key factors in sustainable development include: improved capacity for managing economic and social policies; heightened attention for accountability, the rule of law and human rights, etc.,

[2] The World Commission on Environment and Development describes *sustainable* development as '.... a process of change in which the exploitation of resources, the direction of investments, the orientation of technological development, and institutional change are all compatible and enhance both current and future potential to meet human needs and aspirations' (after WCED 1987).

as well as 'appreciation for environmental sustainability'. I'd say that these are all wonderfully nice things, but to subsume all of that under sustainable development is pulling the teeth out of an already tamed concept. As my colleague Wolfgang Sachs once said: the locus of sustainability has shifted subtly from environment to development. With that shift the perceptual paradigm changed: from conservation of natural resources to conservation of development (Sachs, 2000). One may regret this - as many do in Europe - or one may welcome this, as many do in developing countries and in the USA (though for very different reasons).

Economic Development and Environmental Sustainability

This paradigm shift in relation to sustainability mirrored events taking place on the surface of the economic process and its interface with the environment. The UN conference of 1992 has brought the promise of environmental improvement, but often, and in significant respects, the inverse is what happened since 1992.

The review in 1997 by the United Nations' General Assembly of the results of five years of environmental efforts in the wake of the UN Conference on Environment and Development (UNCED, Rio de Janeiro) presents a disconcerting picture: '. . . the overall trends for sustainable development are worse today than they were in 1992' (UNGA, 1997). The state of the environment '. . . has continued to deteriorate' (ibid.), especially in areas such as toxic substances, waste, greenhouse gases; production and consumption patterns; poverty-related use and management of natural resources such as soils and fragile ecosystems, and biodiversity. According to the UN General Assembly, this instability of the overall environmental trends increases potential obstacles to future economic and social development, especially in developing countries.

The 'political' assessment as summarised here, is backed up by numerous detailed scientific studies. For instance, UNDPCSD (1997) observes increasing environmental pressure due to energy use (with still rising levels of carbon emissions) and increased threats of damage to natural resources such as land, water and forest cover, declining per capita grain harvests, increasing water withdrawals as a proportion of water availability; it is expected that these trends may continue well into the 21st Century. RIVM (1997) reports a build-up of environmental pressures and a growing risk of depletion of renewable resources and increasing water scarcity. UNEP (1997) provides an assessment of environmental trends in 7 world regions and for a range of resources or environmental systems such as land, forests, biodiversity, freshwater, marine and coastal zones and the atmosphere. In almost all of these combinations, the environmental situation had not improved and in more than a half, it had in fact deteriorated. And Tol (1997) calculates that the damages due to climate change (if carbon levels are allowed to double) in developing countries could be almost twice as high (in absolute terms) as in the OECD countries.

In Sub-Saharan Africa, the productivity of the savannahs has dropped by 35 per cent (according to my good host, Prof. Darkoh of Environmental Sciences). Over 50 million people in this part of the world live on land so degraded that they are frequently at risk from famine (Darkoh, 1999). These are two of the examples he gives of what he calls the 'dismembering of the subcontinent's natural resource base', as a consequence of globalisation and global environmental change.

So much for the facts on environmental change. Let us dig a bit deeper and look at the driving forces behind these trends. One of these – and I have to restrict myself here to this one only – is economic growth. This section reviews what is known at present about the relationships between economic growth, structural adjustment, trade and foreign investment on the one hand, and the environment on the other.

The Environmental Kuznets Curve

The relationship between environmental pressure and *per capita* income is often assumed to take on an inverted U-shape (see, for instance, World Bank, 1992) referred to as the 'Environmental Kuznets Curve', estimated as a quadratic function with diminishing environmental pressure beyond a threshold level of income per capita ('de-linking'). De-linking can go on as long as the rates of reduction in environmental throughput per unit of income exceed productivity growth rates. However, technological efficiency improvements may show declining revenues and/or increasing costs that will eventually even suppress positive learning curve effects. Assuming economic growth tendencies to remain in force, from some point onward, economic activity and environmental pressure may 're-link' (Opschoor, 1990). Results of empirical studies indicate that several specific types of environmental pressure have gone down with rising average incomes, beyond some threshold, and others have continued to rise (see De Bruyn and Opschoor, 1997; De Bruyn et al., 1997; De Bruyn, 2000). There is no evidence for the expectation that delinking occurs endogenously or automatically. This holds *a fortiori* when environmental change is studied at a more aggregate level (e.g. that of material throughput). Moreover, where absolute de-linking has taken place, it does not always appear stable under conditions of sustained economic growth. Finally, the impacts of de-linking in a global and dynamic setting are unknown (Arrow *et al.*, 1995).

In conclusion, contrary to what would have been a convenient situation, correlations between economic growth and environmental improvement aimed at testing the hypothesis of a parabolic relationship do not lead to unambiguous results. Sustained economic growth is not necessarily environmentally sustainable, nor will it automatically become sustainable.

Structural Adjustment and the Environment

Environmental repercussions of stabilisation programmes are basically indirect, unpredictable in direction and depend on the period during which the programme was carried out – with plenty of examples of adverse environmental repercussions (Opschoor and Jongma, 1996; Reed, 1992, 1996). Up till recently economy-wide policy reforms (such as in stabilisation and structural adjustment) have had little regard to their environmental consequences (Arrow et al., 1995) with adverse consequences attached to them. The World Bank's recent suggestion to develop 'complementary policies' amounts to an approach that can at best be characterised as retrofitting economy-wide reforms where integrated sustainability oriented interventions might have long run efficiency benefits. Moreover, structural adjustment programmes may entail a decline in the institutional capacity to regulate the interaction between economic development and environmental and social capital due to reductions in government expenditure.

Thus, even though macro economic policies may (or may not) be a necessary condition for development, they certainly are not a sufficient condition for development to become sustainable. More is needed, especially in the form of institutional reforms - both nationally and internationally - and of rethinking along qualitative lines of development and adjustment strategies, before economic reforms will become sustainable over a long period of time.

Pollution Havens?

Foreign direct investment (FDI) can bring about significant environmental improvement if it is accompanied by the introduction into the host country of new, environmentally efficient technologies. Negative impacts may also be expected where regimes are lax or tolerant, and property rights are poorly defined. Many environmentalists expected this latter tendency to prevail and industry to move to locations with little environmental protection, especially in developing countries (the 'Pollution Haven Hypothesis').

Do we in fact see dirty sectors expanding (and/or regulation lagging behind) in developing countries? Zarsky (1999) finds evidence that both domestic and foreign firms in developing countries are constantly improving their environmental performance in response to national regulation and/or local community pressure. Pollution haven situations are as transient as low wage havens were, for reasons of economic dynamics: they draw in more production and hence drive up the preference for environmental regulation, and thus are self-defeating and vanishing.

Relatively, pollution and resource-intensive industries such as petrochemicals and mining may be examples of where FDI does give rise to pollution haven-type phenomenon. The World Wildlife Fund has done a study on the mining industries and their environmental regulation in the Pacific Region, e.g. the destruction of forests and rivers in Indonesia, the destruction of 50 per cent of the fish yields in several rivers in the Philippines (WWF 1999). Other resource intensive sectors elsewhere have given rise to similar phenomena (e.g. oil spills in Ecuador and Nigeria, risks of environmental destruction in Chad and deforestation in many countries).

So, the pollution haven debate cannot be conveniently aggregated away. At the sectoral and regional levels one may discern what more macro-oriented studies find hard to see at the national and macro level.

Globalisation

Since the 1980s (or even much earlier, as some argue), the accelerated rise of globalising markets along with the arrival of an information based economy carried by new information and communication technologies, has profoundly changed the international order (Sachs, 2000). Economically, globalisation results in a single integrated economy shaped by corporate networks and their financial relationships. It represents a deepening of the integration of national economies into the global economy. It results in eroding vertical sovereignty (relationships between states and the private sector: 'While globalisation integrates markets, it fragments politics' (Reinicke, 1997). For hundreds of years, the political and economic order was dominated by the nation state and determined by the territoriality that this entailed. As Sachs says: 'like a container' the nation-state was to hold society in all its layers, creating a self-enclosed entity which engaged with other such entities through international diplomacy and international trade. These containers finally burst open with globalisation (Sachs, 2000). Goods, money, information, etc moved across borders, giving rise to a transnational social and economic space. Economies and countries are becoming integrated by forces related to interactions of and transactions between transnational corporations (TNCs), in the setting of eroding regulatory countervailing powers and in the absence of appropriate levels of international (global and regional) governance.

OECD believes that closer linkages between states or economies, notably through trade and investment, will promote mutually beneficial economic growth. In fact, '. . . a dramatic increase in global prosperity would be in prospect' (OECD, 1997d). In a recent Human Development Report (UNDP, 1999), it is recognised that globalisation opens new markets (especially financial ones), and operates with new tools and new actors (e.g. WTO, trans-national NGOs) as well as new rules (including multilateral agreements on trade and intellectual property). Globalisation is also seen as creating threats to human security: financial volatility, job and income insecurity, health and personal insecurity, political and community insecurity, cultural and environmental insecurities. The institutional change inherent in the current mode of globalisation crowds out care, erodes social cohesion and furthers ecologically unsustainability. On the social side, for instance, a split opens up between the globally oriented middle class and marginalised or even excluded people who are not 'on-line':

'While globalization removes barriers between nations, it erects new barriers within nations' (Sachs, 2000).

OECD (1997a-d) has explored several scenarios of plausible global economic futures until 2020 (compared with 1995), including a 'high performance scenario' (HPS) of accelerated growth and a 'Business as Usual' (BaU)-scenario. Essentially HPS, is a scenario of further growth due to accelerated liberalisation of trade regimes and international finance ('Linkage-led development' or LLD). With HPS, the OECD calculations expect world GDP to grow at 227 per cent, almost twice as high as would be the case in a business as usual development. Per capita income in the world would rise by close to 60 per cent in BaU, and with almost 2.5 times as much in the high performance scenario. In the HPS-scenario, trade will increase more than 1.5 times as fast as production, especially in the non-member economies (NMEs), and with a relatively large increase of OECD-NME-trade. Incomes per capita would rise with 100-150 per cent in developing countries, and even with 270-340 per cent in the case of high performance growth.

However, carbon emissions will rise to nearly 2.5 times the 1990 level (of 6 gigatonnes), in 2020. In the BaU-scenario, carbon emissions would grow too, to 10 billion tonnes but the high performance scenario adds almost 5 billion tonnes to that. This is 25 per cent above the level accepted implicitly at Kyoto.

In fact, the environmental impacts of more FDI in a liberalised, globalising world, are likely to be even more problematic. LLD might lead to standstill, rollback or non-enforcement of national environmental legislation. It might also open up national stocks of natural resources to exploitation by foreign capital. This would allow such investors to appropriate rents on these resources and to put pressure on national governments to much more rapidly sell out these resources than is compatible with sustainable development, thus potentially hampering future development prospects. A fundamental approach to curbing market-embedded mechanisms of cost shifting is called for.

I have already quoted Darkoh (1999), who very aptly summarises the situation for developing countries and for Africa in particular:

- a dependency on primary production in exports;
- patterns of production and consumption – the developing world emulating those of the North – that are on a collision course with the natural environment;

- structural adjustment forcing countries to increase exports and cut state budgets;
- FDI replacing ODA, etc.; and
- in his words, 'As globalisation intensifies . . . and as urbanisation and industrialisation gain momentum, in these countries, the developing world faces the worst effects of economic development – widespread pollution and land degradation'.

Strategy Responses

Practitioners of development in the mainstream institutions stay within the context of the Washington Consensus. In line with the World Bank assessment of aid and similar evaluations, a new mantra can be heard: without good governance, aid cannot be effective, or (in a more up-beat version): poverty, poor social conditions and environmental degradation are concentrated in countries with poor policies and weak institutions. *Market failure* is seen as the main cause of social and environmental externalities. OECD/DAC (1998) appears to advocate broadening this mainstream approach in an enlightened perspective, referred to as 'the global partnership'. Development must be integrated in policies to facilitate the 'full participation' by poor countries in the global economy. Fairly concrete targets for social and economic policies are proposed, putting ownership of the development process in the hands of stakeholders.

Others go substantially outside or beyond the mainstream. Emmerij (1997) emphasises that economic globalisation must be matched by a globalisation of social policy. He calls for an up-dated 'growth with redistribution' approach. Elsewhere, Emmerij (1998) advocates for regional and global 'contracts' (and funds, filled by, e.g. Tobin Taxes) and calls for a reform of the Bretton Woods institutions. UNDP (1999) proposes an agenda for action including: reformulated and strengthened social policies, reducing the threats of financial volatility, global action to tackle global threats to security, the reversal of marginalisation of poor, small countries, the creation of a more inclusive system of global governance including WTO reforms (legal aid, ombudsman facilities and an anti-monopoly arrangement), a World Environmental Agency, a World Investment Trust, a reform of the General Assembly to allow for representation of civil society. One can envisage an environmental extension of this approach. Carley and Spapens (1998) have developed a global approach towards sustainable patterns of production and consumption. Their objective is to shift economic activity towards a position (by 2050) where each country is living within its available 'ecological space'.[3] This will require:

- dramatic changes in production and consumption patterns (and consumption habits and values) of the rich -especially in the North;
- a reassessment of North-South-relationship on the basis of an equity principle; and
- a constructive critique of markets and associated global frameworks of regulation (of trade and investments) in view of their insensitivity to key social and ecological concerns.

Managing Globalisation

Markets increasingly predominate in processes of global change. Without countervailing power, this will exacerbate problems of inequality and unsustainability, as well as other forms of human insecurity. These challenges imply that one cannot leave global change processes to a mechanism that is driven by competitiveness and towards efficiency alone, that is inclined to give in to cost-shifting tendencies and hence is prone to directing economic development to create potentially huge problems elsewhere and tomorrow. Key issues emanating from this review are the *sustainability, manageability, coherence and fairness* of processes of social and economic change that the world will experience. I will flag the former two in the remainder of this chapter.

Globalisation, the State and Development Strategy

There is nothing 'natural' about the tendencies of liberalisation and deregulation that characterise globalisation. Globalisation is as much a project of social construction as any alternatives would (have) be(en).

---

[3] This space is defined as' the total amount of energy, non-renewable resources, land, water, wood and other resources which can be used globally without lasting environmental damage and without impinging on the rights of future generations, within the context of equal rights to resource consumption and concern for the quality of life of all peoples of the world.

In view of the risks associated with the downside of globalisation, the tendency towards declining state power is one that must be addressed. The state still is to be the cornerstone of development strategies, albeit a state that has political support and allies with a strong civil society. The reality of globalisation is that it is weakening the state.

In fact, there is a need for *revitalised states* with improved capabilities and capacities in the institutional, political, technical and administrative domains, and that are in dialogue with an intellectually as well as institutionally empowered civil society. States are to provide the legal and regulatory frameworks in which ideally, market structures and market processes will be embedded and subjected to societal control of some sort. States are to guarantee the rule of law, of security of contract and property as well as access to resources (including natural resources) to users today as well as tomorrow. States are to be able to take responsibility for their economic policies as well as for their social and environmental actions. Only then does it make sense to consider partnerships with private sector enterprises, especially with trans-national corporations, to be potentially instrumental in achieving sustainable human development.

Globalisation will have to be 'managed' (Ricupero, 1999), which means:

- the realisation of smooth, gradual and balanced patterns of structural change in domestic economies favouring sustainability and equity;
- a gradual and controlled incorporation of vulnerable national economies in the process of economic globalisation; and
- institutional change or reform towards enhancing the capacities to implement health, educational, labour and environmental policies conducive to achieving sustainable human development.

Associated with a revitalised state is the need for a strong, *well-mobilised civil society*. In fact, one may discern an emerging pressure to check market forces, a growing 'countervailing power'. The challenge is to identify and introduce institutions in the domain of formal and informal social and economic interactions that can change globalisation to become conducive to sustainable and more human development. Economic growth and the patterns of production and consumption should be made to become less unsustainable and this cannot be done without political and civil power countervailing and curbing economic forces. Mechanisms and institutions are needed to direct globalisation and to curb it, where necessary (Opschoor, 1996b; Dwivedi, 1998). This may entail a trajectory with less spectacular overall growth figures but with a more human and sustainable pattern of development. Such new institutional arrangements would have to ensure fair trade with prices internalising environmental costs; they would also have to allow political control over capital movements so as to curb international cost shifting or rent-grabbing, and the intertemporal deterioration of relative comparative advantages.

Sustainable Development and the North-South Debate

Sustainable development was put firmly on the international agenda by the Brundtland Commission in its report to the UN General Assembly in 1987. It was a core element of discussion at the subsequent UN Conference on Environment and Development at Rio in 1992. As we have seen, it had already become a compromised concept, somewhere between the old notion of development through economic growth on the one hand, and its original connotation of natural resource preservation. This was largely due to the fact that the notion of sustainability came from environmental organisations, with a support base largely in the North. The South had already, since the early 1970s, indicated that it was worried about ecological notions and proposals because they might prove to become stumbling blocks on the road to economic development and growth. 'Sustainable Development' was what came out in 1987 and 1992 (at least in the legally non-binding 'Declaration of Rio', particularly in some of its principles, for example:

- a right to development, as well as to a healthy environment;
- non-interference in other nations' environments; and
- common but differentiated responsibilities of various categories of nations to ensure the sustainability of development.

We have seen how this has been watered down in the OECD/

DAC perspective of development. We have also seen that in a slightly more idealistic mode, it has remained clearly visible in the notion of 'human development' as forwarded by the UNDP. Meanwhile, UNCED had taken place and had also, in addition to the Rio Declaration, given birth to a formidable Agenda 21 (with 700 pages of actions and proposals and framework).

Conventions on Climate Change and Biodiversity

The notions inherent in the original concept of sustainability had already been under fire from developing countries' governments because they feared that these ecological considerations would draw away not only attention but also financial resources in the ODA-domain, from what they considered to be more in their interest: the conventional type of projects and programmes of economic and infrastructural development. In other words, they would reduce resources available for poverty-alleviation and economic growth. Environmental issues per se were given lower priority than these other problems.

In the preparations for UNCED after 1987, some additional huge differences in opinion surfaced already at the level of the agenda setting:

- Climate change and Biodiversity were not the prime concerns of developing countries (why not put things like water and land degradation on the agenda?).
- The issue of the urgency of these particular environmental issues in absolute terms (i.e. how bad is climate change, if it is there at all? How important is biodiversity? etc).
- The issue of the responsibility for these environmental issues once they were accepted as items on the agenda: the polluter should pay, hence the industrialised countries should pick up the bill.

Generally, the response of OECD countries was:

- To already widen the agenda in Rio and to promise an even broader agenda for international environmental negotiations afterwards.
- Within this wider agenda, to give explicit attention to issues such as financing sustainability and technology transfer, and to make

promises on both scores.

- To agree to the need to talk about environmental issues in relation to poverty. In fact, about half of the measures proposed in Agenda 21 (in terms of the budgets attached to the proposals) related more to poverty-alleviation and development in general than directly to environmental measures. And much was made of the so-called win-win aspects in environmental measures: they would, it was said, be not only environmentally desirable, but also socio-economically effective.

Meanwhile, it has become clear that, on the whole, the progress since then has been very slow, if there was movement at all:

- On the first issue, there have been negotiations on a number of broad issues of concern to developing countries, and environmental agreements are in the making on issues other than climate change (CC hereafter) and biodiversity.
- On the second, no substantial new and additional means have come forward. The GEF has not been big enough in relation to what Agenda 21 had made developing countries to expect, and what is worse is that since 1992, ODA has been reduced to a fraction of what it was. So the opposite of the new and additional funds actually flowed.
- Much of the proposals in Agenda 21 remained where they were: in Agenda 21. Hence, one should not be surprised with the findings of UNGA 5 years after Rio: no positive overall environmental trends.

It is to be hoped that the up-coming Rio+10 evaluation next year in South Africa is going to reveal more progress after 1997 than before, but I would not really know why that should be the case, in view of the economic growth in the setting of globalisation, and the direct and indirect environmental repercussions that these might have given rise to. Even though there are signs that in several places the environmental efficiency, that is, the environmental pressure per unit of consumption or production (e.g. energy content, pollution load, etc) has improved, yet, through its sheer dominance in volume terms, the growth in production and consumption may even more likely have overridden these efficiency gains. In my own country, for instance, we are nowhere near realising the reductions in energy use and carbon emissions that we would have to achieve if we are to come close to the targets proposed in the context of the Kyoto-protocol in follow up to the FCCC.

And that brings me to the latter issue: climate change (CC). I will not go into the technicalities of this more than is absolutely necessary, but would like to use it as a case study on sustainability-oriented debates between the North and South.

CC is a concern that has to do with human emissions of greenhouse gases (GHGs) (including CO2) – related disturbances in temperatures and weather regimes in the century ahead of us. Some parts of the world are likely to seriously suffer from increased droughts, cyclones, and other changes in climate and weather. Some other scientists doubt these predictions. It is the long-term aspects of CC as well as the need to start soon to have impacts 20 or 30 years from now, that necessitate CC-mitigative and adaptive strategies to be implemented soon. And as the phenomenon of CC as well as the causing of it is world-wide, the CC problem needs to be addressed at the international level. CC is recognised as one of the Global Environmental Problems. That is why many countries wanted it tabled at Rio (FCCC with 181 signatories, but not yet enough ratifications).

In order to facilitate international decision-making two processes were set in motion:

- an on-going assessment of the state of the art in CC-science and economics: the Intergovernmental Panel on CC; and
- a series of negotiations at the international level, by the FCCC-CoPs, fed by the results of IPCC.

(I happen to be involved in the latter process, not any more – after Rio – in the former).

The FCCC had 181 signatories at the UNCED but is not yet ratified though. Further elaboration by FCCC-CoP is in by now 5½ sessions, the last one held but suspended in The Hague, in November 2000 (and to go on somewhere later this year) under Minister Pronk of the Netherlands. As the South African Minister for Environment, Rejoice Mabudafhasi said: it was a disappointment (*Sowetan* 17 January 2001). It failed to reach an agreement to consolidate and ratify the reductions proposed earlier, in Kyoto. There, proposals had been made to have the industrialised countries (OECD) reduce in 2008-2012 with 5 per cent GHG-emissions compared to what they were in 1990; other countries were not bound to reduction targets, according to Rio principles (especially the one on 'differentiated responsibilities'). And no further progress could be made related to the mechanisms with which these reductions were to be pursued (the so-called 'flexible instruments' of Clean Development Mechanism, joint implementation, and emissions trading. And as the South African Minister said: not only the climate is going to suffer as a result, also the poor people of the world will be among the losers (*Sowetan*, January 17, 2001).

Let us look for a while at these emissions and reductions. Scientists hold that the earth can absorb 13 tonnes of CO2 – one of the main GHGs – annually. Reductions like those proposed at Kyoto would most probably fall far short of what would be needed to curb GHG-related changes in atmospheric temperature and weather regimes. That would certainly be the case if developing countries would continue to grow economically without leap-frogging to cleaner and leaner technologies and without stopping to, as my colleague Darkoh has called it: 'emulating the consumption and production patterns of the North. In fact, Sachs (2000) calculates that if all countries followed the OECD ones in emitting, per capita, 11.4 tonnes CO2 per year, then the emissions of the current 6 billion people would be almost 70 billion tonnes, or 5 times the amount that scientists call safe. And if in a few decades from now all global citizens would emit at the slightly reduced level of about 10 tonnes, they would still generate almost 8 times as much as what is considered safe now.

Sachs' conclusion (supported by Darkoh, 1999) is that the production and consumption patterns of the North are NOT sustainable and should NOT be emulated by the South. Sachs concludes: 'The North cannot be held up as a model any longer; the trail blazers (along the economic development path, JBO) are without a compass. The recognition of the finiteness of the earth has been a fatal blow to the idea of development as envisaged by Truman'. Not everyone agrees, though – in fact, many would argue that here some environmentalists are really putting brakes on rising living standards in the South. That is not the case, but why are these environmental arguments not convincing?

- The damage calculations show that there is great damage in store, related to CC, especially in developing countries but the assumptions underlying it are still being challenged, despite efforts by the IPCC to take away as much uncertainty as it can. This is

NOT done from the South alone, it is even more forcefully done from the North. So, the stringency of the Carbon-constraint on the ecological space available to humankind is challenged. May be the earth can handle much more than these 13 billion tonnes per annum. Why this challenge? Because great interests are at stake in the energy and transportation industries, and threats to current lifestyles in the North are at least as painful as threats affecting future lifestyles in the South. Non-oil producing countries in the South still can free-ride on these academic doubts, and obviously the oil producing countries - mainly in the south - have even more interests in challenging academic prognoses and damage calculations based thereon.

- Even so, these figures are impressively large; why does not every government in the South see CC as a prime risk? To begin with, not all countries suffer to an equal degree. Southern Africa does as do hurricane/tycoon/cyclone-prone Asian countries or island states such as the Maldives. But others would not, or much less. And even where the damage is likely to occur, it will occur, spread over time, with most of the damage in future. Not much is necessarily visible now, or attributable to CC. The Red Cross may believe (as it does) that already now we have 3 times as many bad weather incidents attributable to CC than in the 1960s, with 9 times as much damage, but not everyone knows this and even fewer are convinced. Future damages discounted to a present value today will evaporate as the discount rates go up, and it may be assumed that discount rates or the underlying time preference, are related to income levels. So here too, interests and perceptions differ, also within the South, depending in this case on where one lives and how much future generations and future events count on one's perceptions today.
- Then there is a rights-based approach. OECD citizens and enterprises have been emitting GHGs all along and have become wealthy on that basis. Developing countries should be granted the same right to develop and hence they should be allowed to continue environmental pollution. In the case of GHGs and CC, the principle of differentiated responsibilities has been translated into categories of countries with and without duties to reduce emissions. So that seems OK, but the US Congress refuses to ratify the Kyoto protocol because it wants countries like China and India to also accept targets.
- Another type of discussion is going on about the question: who

is to pick up the bills for CC-preventing measures? It is clear that most of the emissions originate in OECD countries. And if one looks at accumulated emissions over time, most certainly the industrialised countries are responsible for more than 75 per cent of all emissions. OECD then should pay for the costs of CC or measures to curb it. That is the position taken largely by developing countries and interesting schemes to achieve that have been proposed. But OECD countries point at the fact that economic growth very quickly changes the map when it comes to showing the contributors to climate change. Very soon, Russia will have surpassed the US and China, India, Brazil and other countries are emerging not only as economies but also as Climate changers! So, the fight then is one over cost-sharing mechanisms.

Conclusions

The following are the conclusions of this chapter:

- There are marked differences in positions vis-a-vis what constitutes a relevant issue. Ecological constrains or the notion of a finite environmental space are thus regarded with suspicion.
- There are marked differences with respect to the relative urgency even within a set of accepted environmental problems.
- There are strong ethical cases made with respect to the right to development and hence the right to use more environmental space: the North has always done it.
- Similar arguments can be put forward to underpin positions about cost-sharing ratios etc. After all: the North is largely to blame, **and**: the North has the means (the South **not**).
- There is divergence of opinion on environmental matters and their relevance in the North and between countries or regions in the North, as much as there is in the South.
- Given the fact that many in the North are concerned about these global environmental problems and that co-operation with developing countries is essential, countries in the South are actually sitting on a diamond mine. They can cash in on this Northern concern, whether that concern is justified or not. They thus have a strong set of cards to play the game with. Some would conclude that they are already using these cards very smartly.

References

Arrow, K., B. Bolin, R. Costanza, P. Dasgupta, C. Folke, C.S. Hotelling, B-O. Jansson, S. Levin, K-G. Maler, Ch. Perrings and D. Pimentel, 1995. 'Economic Growth, Carrying Capacity and the Environment'. Science, 268, 520-21.

Bruyn, S.M. de, J.B. Opschoor and J.M. van den Bergh, 1997. 'Developments in the Throughput-Income Relationship: theoretical and empirical observations', *Ecological Economics,* 20 3, 255-269.

Bruyn S.M. de and J.B. Opschoor, 1997. 'Economic Growth and Emissions: Reconsidering the Empirical Basis of Environmental Kuznets Curves', *Ecol.Econ.* 15, 2, 161-177.

Bruyn S.M. de, 2000. *Economic Growth and the Environment: An Empirical Analysis,* Kluwer Academy, Press, Dordrecht, Boston, London.

Carley M. and Ph. Spapens, 1998. *Sharing the World: Sustainable Living and Global Equity in the 21st Century,* Earthscan, London.

Cleveland H., H. Henderson and I. Kaul,1996. *The United Nations: Policy and Financing Alternatives,* Elsevier Science and Apex Press (UN), New York.

Daly H.E. 1996. *Free Trade, Capital Mobility and Growth versus Environment and Community,* ISS Public Lectures, 26 September 1996, Institute of Social Studies, The Hague.

Darkoh M.B.K. 1999. 'Role of Business in Sustainable Development'. *Desertification Control Bulletin,* 35, 2-6.

Darkoh M.B.K. 1997. 'Trade, Environment and Sustainable Development'. *Malaysian Journal of Tropical Geography,* 28, 2, 67-82.

Dwivedi O.P. 1998. 'Governmental Response to Environmental Challenge: A North-South Perspective', in Jabbra J.G. and O.P. Dwivedi, (Eds), 1998. pp. 1-15.

Emmerij L. (Ed) 1997. *Economic and Social Development into the XXI Century,* Inter-American Development Bank and John Hopkins University Press, Washington, DC.

Emmerij L. 1997. 'Development Thinking and Practice', in Emmerij L. (Ed), 1997. pp. 3-40.

Emmerij L. 1998. *The World Economic Challenges at the Threshold of the XXI Century,* Keynote Address, Internat, Scientific Conference Institute for World Economics, Budapest, Feb 27-28 1998.

Grunberg I. 1998. 'Double Jeopardy: Globalization, Liberalization and the Fiscal Squeeze', *World Development,* 26, 4, 591-605.

Jabbra J.G. and O.P. Dwivedi, (Eds), 1998. *Governmental Response to Environmental Challenges in Global Perspective,* International Association of Schools and Institution of Administration, IOS Press, Amsterdam.

Kaul., I. 1993. 'Development Trends and Challenges: The Implications for Development Research', In Zentrum fur Entwicklungs-forschung, Wissenschaft im Interkulturellen Dialog. Documentation des Symposiums, 13-14 December 1993, Bonn University.

Luttik J. 1998. 'Accounting for the Global Economy: Measuring World Trade and Investment Linkages', Basingstoke: MacMillan, in association with OECD 1997a, *Economic Globalisation and the Environment,* OECD Paris.

OECD, 1997a. *Economic Globalisation and the Environment,* OECD, Paris.

OECD, 1997b. *Globalisation and the Environment: Preliminary Perspectives.* Proceedings of a workshop, Vienna, January 1997, OECD, Paris.

OECD, 1997c. *Towards a New Global Age: Challenges and Opportunities,* OECD, Paris.

OECD 1997d. *Globalisation and Linkages to 2020: Challenges and Opportunities for OECD Countries,* (Linkages II Study) OECD, Paris.

OECD/DAC 1998). *1998 Development Co-operation Report: Staying the Course Toward Development Results Through Partnership,* DAC, OECD, Paris.

Opschoor, J.B., 1990. 'Ecologische duurzame economische ontwikkeling: Een theoretisch idee en een weerbarstige praktijk', in P. Nijkamp and H. Verbruggen (Eds), *Het Nederlands Milieu in de Europese Ruimte,* Stenfert Kroese, Leiden, pp. 77-126.

Opschoor J.B. 1996a. *Sustainability, Economic Restructuring and Social Change,* Inaugural Address 23 May 1996, ISS, The Hague.

Opschoor J.B. 1996b. 'Institutional Change and Development Towards Sustainability', in R. Costanza, O. Segura, J. Martinez-Alier: *Getting Down to Earth: Practical Applications of Ecological Economics.* Island Press, Washington, DC, 1996, pp. 327-351.

Opschoor J.B. and S.M. Jongma 1996. 'Bretton Woods Intervention Programmes and Sustainable Development', *Environment and Development Economics.* 1, 2, 183-203.

Reed D. ed 1992. *Structural Adjustment and the Environment*, Earthscan Publishers, London.

Reed D. (Ed) 1996. *Structural Adjustment, the Environment and Sustainable Development*, Earthscan Publishers, London.

Reinicke W.H. 1997. 'Global Public Policy', *Foreign Affairs*, 76, 6, 127-138.

Ricupero R. 1999. 'Beyond the Unification of Markets. A Global Community of Cooperation and Shared Knowledge for Security and Development: Some Personal Reflections', Report of the Secretary-General of UNCTAD to UNCTAD X, UNCTAD, Geneva.

RIVM (State Institute for Public Health and Environment) 1997. *The Future of the Global Environment: A Model-based Analysis Supporting UNEP's First Global Environment Outlook.* UNEP/DEIA/TR.97-1, UNEP, Nairobi 1997.

Sachs, W. (2000). 'Development: the Rise and Fall of an Ideal'. Article for the Encyclopedia of Global Environmental Change, Wuppertal Papers No 108, August 2000, Wuppertal Institute for Climate, Environment and Energy, p. 29.

Scholtens L.J.R. 1998. 'Stokt de Kapitaalstroom?' Economisch-Statistische Berichten, 12 June 1998.

Singer H.W. 1997. 'Is Development Still Relevant?' Emmerij 1997: 507-511.

Streeten P. 1997. 'Globalization and Competetiveness: Implications for Development Thinking and Practice', Emmerij (1997): pp. 107-147.

UNDP, 1996. *Human Development Report 1996*. Oxford University Press, New York, Oxford.

UNDP, 1999. *Human Development Report 1999*. Oxford University Press, New York, Oxford.

UNDPCSD (UN Department for Policy Coordination and Sustainable Development) 1997. *Critical Trends: Global Change and Sustainable Development*, UN S/EA/225, New York.

UNEP (United Nations Environment Programme), 1997. *Global Environmental Outlook*, Oxford University Press, New York, Oxford.

UNGA (United Nations General Assembly), 1997. *Programme for the Further Implementation of Agenda 21*, adopted by the Special Session of the General Assembly 23-27 June 1997 (Advanced text, 1 July 1997), UN, New York.

WCED, 1987. *Our Common Future*, Oxford University Press, Oxford.

World Bank, 1992. *The World Development Report*, Oxford.

World Bank 1998. *Assessing Aid*, Oxford University Press, New York.

World Bank, 1999. *The World Development Report*, Oxford.

Williamson, J. 1997. 'The Washington Consensus Revisited', Emmerij 1997: 48-61.

WWF International, 1999: 'Foreign Investment in the Asia-Pacific Mining Sector', (Draft).

Zarsky, L. 1999. 'Havens, Halos and Spaghetti: Untangling the Evidence About Foreign Direct Investment and the Environment', Conference Paper for Conference on Foreign Direct Investment and the Environment, OECD, The Hague, Netherlands, 28-29 January 1999.

Index